RECONSTRUCTIONS

RECREATING SCIENCE AND TECHNOLOGY
OF THE PAST

EDITED BY
KLAUS STAUBERMANN

National Museums Scotland

Published in 2011 by
NMS Enterprises Limited – Publishing
a division of NMS Enterprises Limited
National Museums Scotland
Chambers Street
Edinburgh EH1 1JF

Publication format:
© National Museums Scotland 2011

Individual chapters: © Michael Bailey, Elizabeth Cavicchi, Jan Frercks, Peter Heering, Heather Hopkins, Fraser Hunter, Søren Neilsen, Mark Osterman, Philip Steadman, Klaus Staubermann, Doron Swade and Michael T. Wright 2011.

Websites cited checked March 2011.

Images: © 2011 (as individually credited); see pages 272–74.

No part of this publication may be reproduced, stored in a retrieval system or transmitted in any form or by any means, electronic, mechanical, photocopying, recording or otherwise, without the prior written permission of the publisher.

The rights of Michael Bailey, Elizabeth Cavicchi, Jan Frercks, Peter Heering, Heather Hopkins, Fraser Hunter, Søren Neilsen, Mark Osterman, Philip Steadman, Klaus Staubermann, Doron Swade and Michael T. Wright to be identified as the authors of this book have been asserted by them in accordance with the Copyright, Designs and Patents Act 1988.

British Library Cataloguing in
Publication Data
A catalogue record of this book
is available from the British Library.

ISBN: 978 1 905267 48 4

Publication layout and design by
  NMS Enterprises Limited – Publishing.
Cover design by Mark Blackadder.
Background image: Elevation of output apparatus with inking rollers top right (c.1849) (© Science Museum/SSPL)
Foreground images: The *Planet* locomotive replica (© courtesy, Museum of Science & Industry in Manchester); The completed reconstruction of the Deskford Carnyx (© National Museum Scotland); A fixed and washed negative is dried over an alcohol lamp (Courtesy Scully & Osterman); Difference Engine No. 1 (1832) in Meccano (2006) (Doron Swade); *The Sea Stallion from Glendalough* in rough seas (Dougie Petrie).

Printed and bound in Great Britain by Bell and Bain Ltd, Glasgow.

Published by National Museums Scotland as one of a series of titles based on NMS scholarship and partnership.

For a full listing of titles and related merchandise, please contact:

www.nms.ac.uk/books

Contents

Contributors IV

Introduction VI

Foreword XII

    The Antikythera Mechanism:
    Reconstruction as a medium for research and publication
    MICHAEL T. WRIGHT 1

    Using experimental archaeology to answer the 'unanswerable':
    A case study using Roman dyeing
    HEATHER HOPKINS 21

    The sound of Iron Age music: Reconstructing the Deskford Carnyx
    FRASER HUNTER 50

    The *Sea Stallion of Glendalough*: Reconstructing a Viking-age longship
    SØREN NIELSEN 59

    Learning through replications: The *Planet* locomotive project
    MICHAEL R. BAILEY 83

    Reconstructions as experimental history: Historic computing machines
    DORON D. SWADE 103

    The spiral conductor of Charles Grafton Page: Reconstructing
    experience with the body, more options, and ambiguity
    ELIZABETH CAVICCHI 127

    Experience and self-reflection:
    An electrical-historiographic-didactic experiment
    JAN FRERCKS 171

    Materialised skills: Instrumental development and practical experiences
    PETER HEERING 194

    Aim at the stars, reach the Nicol: The Zöllner photometer
    KLAUS STAUBERMANN 209

    Reconstructing the painting studio of Johannes Vermeer
    PHILIP STEADMAN 234

    Mid-nineteenth-century photographic studio technique:
    Why recreate nineteenth-century photographic technology?
    MARK OSTERMAN 251

Image credits 272

Associated keywords 275

Contributors

Michael R. Bailey, MBE, DPhil, MA, is a consultant, author and lecturer on early railway technology. He is a Newcomen Fellow and Past-President of the International Society for the History of Engineering and Technology, and a Vice-President of the Stephenson Locomotive Society. He is also a Trustee of the Museum of Science and Industry in Manchester.

Elizabeth Cavicchi teaches at MIT's Edgerton Center through exploring natural phenomena and recreating historical experiments. Postdoctoral research at the Dibner Institute followed her EdD from Harvard University; masters' degrees at Harvard, Boston University, and MIT; undergraduate studies at MIT. A visual artist, she responds to creativity in history and learning.

Jan Frercks is a private-docent at the Department of Humanities and Educational Sciences at the University of Braunschweig and a grammar school teacher in Flensburg. His research interests include the history of physics and chemistry in the eighteenth and nineteenth centuries, the history, philosophy, and didactics of scientific experiments, and the relation between research and teaching.

Peter Heering is Professor of Physics and Physics Didactics at the University of Flensburg. His research focuses on the analysis of experiments from the history of sciences with the replication method, the relation between research and teaching experiments, and the implementation of the history of science in science education.

Fraser Hunter is the Principal Curator of Iron Age and Roman collections at National Museums Scotland. His main research interests are in Celtic Art, Iron Age metalwork, and the interaction of the Roman Empire with the groups around its borders.

Heather Hopkins graduated with a BSc in archaeology in 2002 and a PhD in engineering in 2007, both from the University of Bradford. Heather now presents and publishes her research widely. She was among the first to realise the archaeological potential of the virtual replication of physical properties and used it to reconstruct an entire industry.

Søren Nielsen is a trained boat builder. He has built and sailed reconstructions and Nordic traditional crafts since the mid-1980s. Since 1993 he has been head of the Viking Ship Museum boatyard and since 2008 head of the department of maritime crafts and reconstruction at the Viking Ship Museum in Roskilde, Denmark. His department reconstructs and builds boats for other institutions and individuals, and acts as consultant for reconstruction projects in and out of Denmark.

Mark Osterman is Process Historian at George Eastman House, International Museum of Photography in Rochester, NY. Osterman is recognised internationally for his research in the technical evolution of photography from the silhouette to making gelatine emulsions. His own images made with arcane processes are represented in many institutional and private collections.

Klaus Staubermann is Principal Curator of Technology at National Museums Scotland. He trained in philosophy and enjoys doing things with his hands. He has researched and published on historic reconstructions and employs historic replicas as part of the Museum's public engagement events.

Philip Steadman is Professor of Urban and Built Form Studies at University College London. He has also taught at Cambridge and the Open University. He has written a number of books on mathematics in architecture and art, in particular the geometry of perspective. His book *Vermeer's Camera* was published by Oxford University Press in 2001.

Doron Swade, MBE, PhD, MSc, CEng, CITP, FBCS, is an engineer, historian and museum professional. He masterminded the construction of a mechanical calculating engine designed by Charles Babbage in the nineteenth century. He was Curator of Computing and later Head of Collections at the Science Museum. He has published widely on Babbage, curatorship, and the history of computing.

Michael T. Wright, MA, MSc, FSA, was educated at Shrewsbury School, Oriel College (Oxford) and Imperial College (London), gaining degrees in Physics and in History of Technology. He worked at The Science Museum from 1971 until 2004, latterly as Curator of Mechanical Engineering. His main interests lie in mechanism and in tools and workshop technique, and in the history of both these broad topics, combining practical work with reading, writing and lecturing.

Introduction

HISTORIC RECONSTRUCTIONS OVER the past decades have become a decisive tool in the history of science and technology. This collective volume, based on a work-shop held at National Museums Scotland in 2009, brings together key studies of recently completed reconstruction projects. These cover subject areas as diverse as physics, computing, photography, communication, transport and military. Contributors present their reconstruction projects, as well as explore how their work has provided them with a new or deeper understanding of the skills, practices and tacit knowledge involved in the making and use of material culture. They analyse competing traditions in the manufacture and use of artefacts and reflect on the changing social roles of inventors and practitioners. The book covers questions such as: What are the motives for doing reconstruction projects? What can we learn from placing these projects in a broader cultural context? And what are the uses of 'failed' reconstructions? Contributors also discuss the relevance of their projects for the public understanding of science and technology.

Employing reconstructions is not a recent invention. Experimental archaeology is perhaps the discipline with the longest track record in re-working historic artefacts and practices. From the nineteenth century onwards archaeologists attempted to recreate the techniques and crafts of the past and produced the first literature raising methodological questions. During recent decades, by linking archaeology with anthropology, reconstruction projects have adopted an increasingly social nature.[1]

Until the second half of the twentieth century, maritime history was the most spectacular area in the area of reconstructions. Boats of all sizes and periods, from canoes to submarines, were reconstructed. Moreover, impressive journeys on these vessels have been undertaken in order to understand historic craftsmanship, navigation and exploration. Although this work is to a large extent the outcome of experimental archaeology, it is now regarded as a discipline in its own right. Part of this book will look at the various techniques developed, as well as experiences gained in this fascinating field.[2]

INTRODUCTION

Reconstructions in history of science have attracted immense attention during the past three decades. From experiences with replicating historic instruments for teaching purposes (the Oldenburg School), and reflecting about the general replicability of scientific experiments, reconstructions in the history of science have become a cornerstone of this discipline.[3]

Reconstruction projects can have significance for contemporary science. For example, historic Indian American drugs are under investigation in order to develop new and more effective painkillers. Further, analysis of the contents of ancient cadaver stomachs provides evidence of the diets of historical peoples; these inferred diets explain other effects in the cadaver data, such as malnutrition and its impacts on the body. This information can also be used to recreate historic diets such as early Egyptian beer.

Moreover, virtual reconstructions are under development using computer modelling resources. Intriguingly, these virtual reconstructions disclose considerable overlap between the virtual model and evidences of the real event or activity. Virtual realities today play a significant role in the reconstruction of historical practices. They help to visualise complex phenomena or objects as well as spaces and environments. Virtual reconstructions have become especially important in the understanding of historic weather and climate changes where they have produced a lot of decisive information. One of the shortcomings of virtual reconstructions continues to be the limited possibility of interaction. Already the purely visual and audial experiences available through early virtual reconstructions are superseded by more sensual and tactile representations. Future technology will surely enhance the authenticity of virtual representations of historical events and works.[4]

Understanding of historic artefacts is impossible without understanding the practice attached to, or embodied in them. Practice here can mean both the manufacturing of the artefact and its uses. As philosophers of science have pointed out, the question of if and how historic skills can be understood by interaction with the historic object is one of the most challenging in the history of science and related disciplines. The case studies presented in this volume will help to shed some light on the relation of reconstructed objects and the historic practices imbedded in them.

Of course, there is a wider hermeneutic dimension to historic reconstructions. Many questions arise: How do we set material reconstructions in a broader context? How do we use materials today to reconstruct such immaterial practices as the thinking of the peoples who used similar materials in the past? Motives and choices of materials play an equally important role for the philosopher, as they do for the practising historian. Debates of the past years have looked at the role of tacit and local knowledge in the making and uses of artefacts in the history of science and technology. More recently, studies have focussed on skill and knowledge transfer. All of this will be reflected in the case studies presented in this book.

Although the case studies presented in this book differ in period, location and type of apparatus, the approach taken for the reconstructions reflect a similar method. This method includes the examination of existing artefacts and records, the construction of the replica, its use, and the interpretation of the experience gained. All authors carefully examine the design and materials of the original artefacts and employ reconstructions in order to learn about the functioning, manufacturing and uses of the historic device, machine or instrument. Maintenance is another crucial aspect addressed by many authors. It is this use and maintenance of the replicas that helps us to understand their role and significance in the past. Thereby, reconstructions become probes in historic skills, performance and culture.

Michael Wright in his reconstruction of the Antikythera Mechanism presents his project as a tool for research and a medium for publication. He begins by noting and analysing details of the original mechanism, aiming to bring them together subsequently into a new reconstruction. The act of replicating components stimulated ideas as to why they were formed and what purpose they served. As a by-product, it prompted Wright to consider the tools and techniques of the original maker. The working of the resulting physical model provides Wright with incontrovertible proof that the reconstruction is, at least, practicable.

Heather Hopkins in her study of Roman Dyeing sets out with a fierce debate about the scale of dye manufacture in Roman Pompeii. Her study is the first to reconstruct the relevant parts of the industry to determine its parameters and gauge its capacity. Her use of replica dying gave rise to further investigation and an ergonomic assessment of the original apparatus in Pompeii was undertaken. The replica was then amended to explore the effects of changes in design. Her analysis answers specific questions about the size of the dyeing industry, but illustrates also the application of her approach and techniques to answer 'unanswerable' questions in experimental archaeology.

Fraser Hunter has reconstructed a Carnyx, an Iron Age animal-headed trumpet from Deskford and one of the treasures of National Museums Scotland. His paper looks at the evidence behind this recent reconstruction project, and sets it in the wider context of other discoveries and of other work done in music-archaeology. His striking work recreates the sounds projected by this ancient trumpet.

Søren Nielsen discusses the *Sea Stallion from Glendalough*, a reconstruction of Skuldelev 2, a 30-metre long warship from the Viking Age. The ship is the tangible result of a reconstruction process, where the considerations and reflections concerning practical solutions, workmanship, properties of the needed material and consumption of resources are equally important results. The reconstruction included sailing *Sea Stallion* from Denmark to Dublin and back in rough weather. Building and sailing the ship provided Nielsen not only with detailed knowledge of the construction and handling of the long ships; it also gave him new insights into local shipping communities.

Michael Bailey describes the replication of one of the earliest class of main-line steam locomotives, the *Planet*. This replica locomotive, constructed complete with carriages for passengers, now operates on a track at a restoration of the original Liverpool Road Station in Manchester, England. This site was the terminus of the Liverpool and Manchester Railway, the world's first main-line railway which opened in 1830. The replica was designed to imitate the original class as much as possible. The project of reconstructing the locomotive produced a lore of learning about the design and materials of the original which could not be inferred in any other way. By actually operating the restored locomotive, project volunteers found themselves relearning driving and firing skills. Engineers in the project used present-day test equipment to conduct performance trials on the restored locomotive and determine the power output and efficiency of the locomotive's design.

Doron Swade puts his project in context by summarising recent major reconstructions of historic computing machines: the Second World War Colossus; the 'Bombe' at Bletchley Park; the Manchester 'Baby'; Thomas Fowler's nineteenth-century mechanical calculator; Konrad Zuse's Z3 and Charles Babbage's Difference Engine No. 2. Babbage's machine, the subject of Swade's own reconstruction, was designed in the 1840s but never built until Swade commenced his project. These new/old machines pose intriguing questions. What is the historical and social utility of such constructions? What, if anything, is learned that could not otherwise be learned? What category of object are they – fictitious antiques or new primary sources? What kind of history do such constructions represent? Swade's paper describes the Babbage project and discusses issues triggered by it and related reconstruction projects.

Elizabeth Cavicchi studies bodily experiences based on the replication of a home-made spiralled conductor built by the Harvard medical student Charles Grafton Page in 1836. Surprisingly Page felt shocks everywhere, even where no direct battery current passed. Bodily contact across greater spiral spans yielded greater shocks. Where Page reported increased shock, Cavicchi encountered variable signals. She traces Page's subsequent developments of the spiral and its applications further to illustrate the productivity of these efforts. Diverse experimental options, together with ambiguity in what is observed, provide her with fertile ground for exploring these complex phenomena.

Jan Frercks examines a laboratory course in the history of science he taught at the University of Jena. Students rebuilt and used electrical and galvanic instruments of around 1800. The lab-course is used as an experimental test of the uses of reconstructions. The core question is whether replicating past experiments allows better learning about electricity or about past scientific practice. The students were only loosely guided during their experimental work, but they were encouraged to reflect on their own behaviour and experiences. During the participants' retrospective assessment of the lab-course, documented in a written report and in a comment at the end of this article, it became clear that they differed significantly from the Frerck's interpretation.

Peter Heering writes about his experiences with historic solar microscopes. Solar microscopes are demonstration devices that were particularly popular during the second half of the eighteenth century. Through practising with replica instruments, it became possible to develop an understanding of the broader visual culture that can be related to the solar microscope. Moreover, it became possible to understand the technical developments of the instrument. Here, aspects such as user-friendliness seemed to have played an important role and newly-acquired skills of the demonstrator were incorporated into the design of the device.

Klaus Staubermann has replicated a nineteenth-century astro-photometer designed by Karl Friedrich Zöllner, Germany's first professor of astrophysics. This photometer was based on a Nicol polariser, an optical device invented by William Nicol in Scotland in the early nineteenth century which became one of the key components of nineteenth-century optical experimentation. Staubermann's work demonstrates how historic observers interacted with new designs and how different practitioners using new types of instruments could reach visual agreement. He discusses the reconstruction of the photometer and explores how doing this has provided him with a new or deeper understanding of the skills, practices and tacit knowledge involved in the making of nineteenth-century astronomical photometry.

Philip Steadman describes the reconstruction of a room in a house in the Dutch city of Delft, used as a painting studio by Johannes Vermeer in the 1660s and '70s. He presents two possible routes for a reconstruction. The first is from contemporary drawings, maps and an inventory of the house made by bailiffs after Vermeer died. The second is through Vermeer's own paintings, since he showed the room in as many as twelve of his pictures of interiors. The accuracy of perspective representations in Vermeer's paintings is such as to make it possible to determine the sizes of windows, floor tiles, the height of the ceiling, and so on. Steadman shows that the reconstructions via the two routes agree closely. The interest for art history is the light thrown on Vermeer's working methods, specifically his use of the camera obscura. In effect the room itself was an instrument through which the painter explored the effects of light and shadow, for whose rendering he is so much admired.

INTRODUCTION

Mark Osterman examines mid-nineteenth-century photographic studio techniques in the final chapter of this book. His paper is based on primary research and hands-on manipulation of photographic techniques used in commercial photographic studios throughout the industrialised world from the early 1850s to the late 1870s. The photographic method in common use during this era was the pairing of hand-coated wet-plate collodion negatives on glass and the gold-toned albumen print. Osterman's reconstructions are influential among many professionals today, including historians, curators of photographic collections, photograph conservators and fine arts photographers. For example, fine arts photographers employ Osterman's reconstructive practices as an alternative and complement to contemporary digital imaging techniques.

The aim of this book is to encourage students, teachers, curators and scholars to engage in historic reconstructions. The case studies presented here give an overview of the width and breadth of reconstruction projects carried out during the past two decades. They are intended as both stimulation and guidance for future reconstruction projects. The authors hope that their contributions will help to develop this exciting and instructive method further.

This book would not have been possible without the help and support of many individuals: Jane Carmichael, Alexander Hayward and Lesley Taylor for lending their generous support to both the workshop and this volume; Katarina Grant and Maureen Kerr for their help with organising and facilitating the workshop; and Lynne Reilly for her help and expertise with copy-editing this volume; and all contributors of this book for making it possible.

The workshop that led to this book unfortunately was overshadowed by my grandfather's death. We had said our goodbyes earlier, but I would like to dedicate this book to his memory; much of what I know about doing things with my hands, I owe to him.

Notes

1. For a recent account, see P. J. Cunningham, J. Heeb and R. Paardekoeper (eds) (2008): *Experiencing Archaeology by Experiment* (Oxford: Oxbow Books).
2. A good introduction and overview is given in J. Bennet (ed.) (2009): *Sailing into the Past* (Barnsley: Seaforth Publishing).
3. Note, for example, P. Heering, F. Rieß and C. Sichau (eds) (2000): *Im Labor der Wissenschaftsgeschichte* (Oldenburg: Universität Oldenburg Verlag).
4. A good example of how to combine the real and the digital is Martin Kemp's work on modelling Leonardo. M. Kemp (2006): *Leonardo da Vinci* (London: V&A Publications).

Foreword

JANE CARMICHAEL
Director of Collections
NATIONAL MUSEUMS SCOTLAND

THE MULTI-DISCIPLINARY NATURE of the collections of National Museums Scotland means that we are accustomed to working across very different subject areas. However, the papers presented at the workshop 'Reconstructions' held in 2009 and now brought together for publication, were particularly wideranging. They instigated some fascinating adventures in reconstructing historical items and an astonishing range of insights across many subject areas. In many cases, they bridged the traditional divide between the sciences and the humanities to the benefit of both. These 'reconstructions' push the boundaries of engagement with museum objects and bring them to life, not just for the sake of it, but in pursuit of enquiry. It is a rich seam to explore and we hope that this publication will illustrate and inspire others with its potential. Klaus Staubermann, Principal Curator of Technology, is to be congratulated on his successful realisation of this new strand of collections research at National Museums Scotland and the resultant network of national and international contacts who have contributed to this book.

MICHAEL T. WRIGHT

The Antikythera Mechanism
RECONSTRUCTION AS A MEDIUM FOR RESEARCH AND PUBLICATION

Introduction

THE ANTIKYTHERA MECHANISM, a small but elaborate Hellenistic instrument that displayed indications of astronomical events as functions of date, embodies by far the earliest known application of toothed gearing.[1] Due to its fragmentary, ruined state it has proved hard to understand, and it remains a difficult artefact to interpret and to appreciate without the use of supporting material. For these reasons it was an apt subject for reconstruction. In developing the reconstruction that forms the subject of this chapter, the author has contributed significantly to the understanding of the instrument, and at the time of writing his models remain the best illustrations of the original in all its detail.[2,3]

The instrument had two dials, front and back, interconnected by toothed gearing driven by hand. The back dial bore two displays. The upper exhibited an astronomically-based calendrical cycle; and the lower showed another temporal cycle offering a means of predicting the possibility of lunar and solar eclipses. The full extent and exact nature of the display of the front dial is less clear. The outer of its two concentric rings showed the days and months of the year, so that a pointer might indicate the date. For the

FIG. 1.1 The Antikythera Mechanism, reconstruction by M. T. Wright (front view). The inner dial ring represents the Zodiac, divided into 360 degrees and with the twelve signs named. The outer ring is the annual 'Egyptian' calendar of 365 days, used by Hellenistic astronomers, with the months named. Pointers show the positions of Moon, Sun, Mercury, Venus, Mars, Jupiter and Saturn, and the day. In the boss of the Moon hand, a small rotating ball shows the phase of the Moon, and the scale around it shows the day of the lunar month. (Michael Wright)

inner ring, representing the Zodiac, there were undoubtedly pointers indicating the positions of Sun and Moon in a geocentric scheme, and the motion of the Moon-pointer was cleverly modified to model Hellenistic single-anomaly lunar theory. Near the centre, the rotation of a small half-white, half-black ball displayed the phase of the Moon. That much is certain, but we see clear evidence of the loss of other mechanical parts that drove further pointers on this dial, indicating, we argue, the places in the Zodiac of the five planets then known. We suggest also that the motion of the Sun-pointer was modified to model the single-anomaly solar theory. Our initial purpose in realising our reconstruction as a full-sized model was to satisfy ourselves and to demonstrate to others that this reconstruction, more elaborate than any other and accounting more fully for the detail found in the original, was fully practicable.

We do not attempt here either to reason the case for our interpretation or to explain the detailed functioning of the instrument.[4,5] Our present purpose is to discuss particular points that arose in the course of our work which are of a more general interest and relevance.

Discovery, and earlier attempts at reconstruction

We begin with a brief account of the discovery of the Antikythera Mechanism, of the context in which it was found, and of previous attempts to reconstruct it.[6]

Antikythera is a small island off the South coast of mainland Greece, in the middle of the strait connecting the Aegean Sea to the Western Mediterranean Sea. There, in 1900, sponge divers found statues in deep water just off shore, the largest items of a cargo of mixed luxury goods in an ancient shipwreck. The wreck itself is now dated to the second quarter of the first century BC, most securely on the basis of coins found at the site. Much of its cargo appears to be contemporary with the wreck itself or just a little earlier, but a few items of large-scale bronze statuary are thought to be older still. Inconclusive evidence and historical considerations suggest various dates for the Mechanism, ranging from the second century BC to nearly the time of the wreck.

The Greek government hired the divers to recover what they could, and a great mass of varied material, the 'Antikythera Treasure', was taken to the National Museum in Athens, where much of it is on display. The fragmentary bronze Mechanism was noticed amongst this material only later, when engraved lettering (in Greek) and traces of a few wheels were seen on its surfaces. It became clear that the instrument had an astronomical purpose, and some mistaken commentators described it as an astrolabe. Over the ensuing decades there were several attempts to offer reconstructions of it, but all – based only on literally superficial examination – were sketchy and speculative. In 1959 Derek de Solla Price was the first to suggest an approach to the correct general layout,[7] but only in 1974, after radiography had revealed much hidden detail, could he build on his own earlier description by offering an account of the internal mechanism.[8] Price's reconstruction is now known to be wrong in many respects, but it was the first to reflect at all closely the actual degree of intricate internal detail. His paper was widely read, causing a ripple of sensation through much of the scholarly world.

Our research

Many scholars found it difficult to adjust their perception of Hellenistic mechanical achievement, and there was a degree of incredulity that the instrument described by Price could have been made so early, using only simple tools. The present author was struck by the idea that one could, through model-making, demonstrate that this difficulty was illusory. At the same time, however, he was dismayed to find that parts of the argument through which Price arrived at his reconstruction were, at best, unintelligible. The doubts thus cast on its correctness seemed confirmed a few years later, when a model made under Price's supervision, borrowed for temporary exhibition at the Science Museum, proved both unconvincing and unworkable. Thereupon the author became determined to examine the original for himself.

This examination of the original fragments was carried out in collaboration with the late Allan Bromley of the University of Sydney. We began with a minute direct examination using the naked eye and various types of magnifier. Direct measurements were made of the positions of many surface features, using a simple measuring jig. Straightforward and stereoscopic photography was extensively employed. The fine radiographs which Price had used, prepared by the Greek physicist Charalambos Karakalos, were not available for our study. Therefore we made our own.

The fragments of the Mechanism are essentially flat, its gears and other components lying in a number of closely-spaced layers. Conventional radiographs of good resolution could be made only with the rays passing roughly at right-angles to the layers. They showed most features clearly, but could not show the depth of each feature within the mass. Karakalos had attempted stereoscopic radiography but had found it of limited use. We therefore devised and made apparatus for use with the conventional X-ray apparatus available at the Museum, allowing us to exploit linear tomography. Using this technique, one makes radiographs in which the features in a single chosen plane yield sharp images, while features out of the plane are more or less blurred. By taking sets of exposures, with a small adjustment of the apparatus between each so as to shift the chosen plane by a fraction of a millimetre, we prepared sequences of images in which the depth of each of the several features in the specimen might be found. With application, these sequences allowed us to build up a mental picture of the arrangement in depth.[9]

Having accumulated a great deal of material, we then faced the problem of its analysis. Bromley took the material to Sydney for this purpose, but his attempt proved abortive, due largely to his advancing terminal illness. Some years later the author retrieved as much of this material as could still be found, and began again.

Reconstruction as a research tool

It was clear that Price had been led astray through having formed too firm an idea as to how the instrument must have functioned. Once that idea had taken root, he had then laid emphasis on evidence that supported it, and – perhaps unknowingly – had distorted or suppressed evidence that led in any other direction. He offered arguments, now seen to be specious, in support of this manipulation; and he presented further ideas which distracted the reader's attention, and perhaps his own, from its extent.

It was important to avoid the same error of self-deception, and the author's prior experience offered a different way forward. As a museum curator, he was often obliged to deduce the purpose of an unidentified artefact from a study of its detail. The same approach was applied in this case; everything that had been written about the instrument was ignored, and attention was directed wholly and simply to a study of the fragments themselves.

The detail of each part, once grasped, was then modelled, and the several small assemblies thus created were brought together. The original is incomplete, and so this process could take us only so far.

FIG. 1.2

The Antikythera Mechanism, reconstruction by M. T. Wright (back view). The spiral scales show long sequences of months. The 235 of the upper display represent a civil calendrical cycle of 19 years, indicating for each month whether it has 29 or 30 days, and for each year whether it has 12 or 13 months. The subsidiary dials record (right) a 4-year cycle of pan-Hellenic games and (left) the 76-year Calippic cycle. The 223 months of the lower spiral represent an eclipse cycle after which the pattern of eclipses repeats. Eclipse-predictions marked in the relevant months state whether it is to be a solar or lunar eclipse, and a time in hours. The subsidiary dial offers a correction, since the times of all eclipses shift about 8 hours for each successive cycle.

(Michael Wright)

Thereafter, as a second phase, lost features had to be restored in order to make sense of what had been copied. This phase of reconstruction would have been very much harder had it not been for the stimulus of mechanical necessity to fit parts to the model, and the resulting direct confirmation of the appropriateness of each step taken. Indeed, had the work been carried out merely as a 'paper exercise' or in any other way, it would still have been necessary to build the instrument before the reconstruction could be published with any confidence.

Nothing but actually making the instrument and trying it out in practice could have shown, so directly and so definitively, that the restored parts fitted and that the whole was workable. Computer modelling, for example, as usually carried out, generates no more than a moving diagram. In such a virtual model the parts have no mass and no inertia. There are no loads and there is no friction. There is no deformation and no breakage. There is no wear. The computer-generated virtual model has great value, but for our purpose it is not an adequate substitute for reality. It might be added, that for us it represents a far more difficult procedure. In any case, it is a fact that in this instance physical reconstruction was a central feature of the heuristic method.

Most of the detailed discussion that follows relates to the author's first model which was, as we have described, built during the process of reconstruction and as a means of developing that reconstruction. As a pragmatic choice, it was built of scrap materials that lay to hand.

A second model has been more tidily made, of new material. Initially it differed only slightly from the first. A few details of the original had been better understood, and could be modelled more accurately. Better design solutions had been found for some restored parts, which either were more serviceable or conformed more closely to precedents found in the original fragments; or, interestingly, they offered both advantages. The model was exhibited in this state.[10,11] Meanwhile, further study offered a function for a previously neglected detail, which suggested in turn that, while our restoration of mechanism to drive pointers for the positions of the superior planets was generally correct, the original incorporated an unsuspected subtlety embodying a more advanced form of planetary theory. The second model has consequently been disassembled and, at the time of writing, is being modified to illustrate this further feature. A preliminary announcement has been made,[12] but a paper intended to describe this development more fully remains in preparation.

Dimensions

It is desirable to build any reconstruction to the same size as the original where practicable, so as to be free of any effects due to scaling; and in this case, in which the original is of moderate size, it would have been simply perverse to have chosen any other scale. Obviously the ideal is to make the parts of the reconstruction to precisely the same proportions and sizes as those of the original; but for the Antikythera Mechanism that is probably an unattainable ideal.

The precise measurement of what remained was a non-trivial problem. Only some features, and some of the layout, could be measured directly. For the rest, we relied on radiography. Radiographic images, being shadow-pictures, are always larger than the features that they show; but with a little extra information it is possible to scale from them to a reasonable approximation. One can scale from some other feature of known dimension lying nearly enough in the same plane as the one to be dimensioned or, knowing the geometry of the radiographic set-up (distance of the target in the X-ray tube from the feature in question and from the plate), one can invoke the geometry of similar triangles in correcting the size of the image to the size of the original feature. Both techniques were of some use, although it happened that by the time the analysis was made some radiographs, and (more crucially) information on the radiographic geometry, had been lost.

Although it remains in principle possible to derive a full set of dimensions from our data, this course has not been followed; to do so would be both tedious and expensive. A more recent study made by others, using modern CAT apparatus, should make it very much easier to find dimensions, but we have had no access to the results of this work.

Even if we might determine the precise present dimensions of the individual components, it happens that almost none of them remains undamaged. Many are nearly enough complete for the intended geometry to be clear, but whether that intended geometry was accurately carried out remains open to doubt. The original shape and size of other parts is much less certain. Throughout, the thickness of components is a particular problem. There are rather few points at which thickness can be measured directly, and even then the state of the remains is such that it is not easy to estimate whether the measured thickness is the same as the original thickness; it may have been reduced by corrosion and other damage, or increased by the accumulation of marine deposits or corrosion products. In most cases the free edges, at which it was possible to make some direct estimate of thickness, are in fact very obviously wasted.

There is the further important question of fits and clearances which, though often overlooked, are essential points in the construction of mechanism. No doubt the parts were adjusted to one another with as much care as the maker's skill and experience allowed, and it would be most valuable to be able to establish this aspect of his working practice; but unfortunately the state of the original precludes hope of ever being able to do so.

It is possible that some basic dimensions, at least, were related to a length-standard current when and where the instrument was made. Such a discovery would be interesting, perhaps leading to insight into aspects of workshop practice or adding to our knowledge of the place and time of manufacture. This study would, however, demand the expenditure of much time and effort. Therefore, because it holds no relevance to our primary interest in the function of the instrument, it has not been undertaken.

While we should prefer to replicate the dimensions of the original exactly, it is worth reflecting on the value of investing time in attempting any particular degree of precision. Here, our first priority was to check that ideas for the several parts 'made sense'; that the individual parts were compatible; and that the whole was practicable. The great pitfall to avoid was the publication of an unworkable reconstruction. Fortunately, for this purpose alone the precise dimensions of the model were relatively unimportant. It was desirable though to maintain a close relationship between model and original, so that later we might hope to use the model as a medium for communicating a true impression of the original to the public without having to rely on their understanding of the insignificance of any observable compromise.

Overall, however, there was little point in going through the very time-consuming process of establishing some dimensions closely, when many others would have to be estimated. The pragmatic solution was to model the instrument on the basis of close approximations to the dimensions and proportions of the original, and it proved possible to do so on the basis of very few absolute measurements. Beyond these we worked from photographs and radiographs, and largely by eye. In fact, no working drawings were prepared; beyond those few check-measurements, the parts were made to 'look right'.

Following replication of the parts that remain, we addressed the task of devising restorations of the many that are lost. Here we note only that the dimensions and proportions of the extant parts were taken as guides to the design of all further parts. A separate section is devoted to the more general problems of restoration.

Materials

The problem of replicating the material specifications of the original is yet more difficult than that of copying its dimensions, and seems likely to remain so. Arguably it is at least as important. The choice of materials for an artefact can have a profound effect both on the ease of manufacture and on its satisfactory function. Our reconstruction might be made to the right design and to the right dimensions, using techniques agreed to have been available to, and likely to have been used by, the workman of that time; but if it were built of inappropriate materials, then doubt might remain as to whether it reflected

FIG. 1.3

The Antikythera Mechanism, **fragment A** (front view). This is the earliest photograph of the largest fragment. Comparison with **Fig. 1.4** shows how cleaning has revealed further features, while considerable material has been lost. In particular, note the traces of the wooden case, to lower left.

(Michael Wright)

truly either the construction or the function of the original. We consider, however, that the pragmatic compromise that we adopted is acceptable.

It is often stated that the instrument was made of bronze. That is what we would expect for this sort of light metal construction from antiquity. The surfaces of the fragments certainly have the appearance of the corrosion products of bronze, and the radio-opacity of the material is right for bronze. Analyses of some small detached fragments, reported by Price, suggested a bronze with a fairly low proportion of tin (and little else), but unfortunately the tin-content could not be determined with precision.[13] The range within which it was thought to lie is, however, appropriate both for workability and for mechanical strength. So although, strictly speaking, we have no guarantee that these fragments came from the instrument at all, we suppose – provisionally – that they are indicative of the material used for the working parts of the Mechanism.

It appears, however, that some parts were made of a different alloy. Non-destructive tests have recently been carried out on some fragments small enough to be placed in the chamber of an electron microscope.[14] The pieces selected are all ones that bear inscriptions: parts of components called 'door plates' or 'cover plates' which seem to have formed external covers, lying over the dials. A preliminary announcement indicates that these sheets were made of an alloy with little copper and a very high proportion of tin: a pewter, which is certainly very different in its mechanical properties from the bronze indicated by the previous analyses. So soft a material would have been useless for most of the mechanical parts of the instrument, but a good choice for sheets subjected to little stress which were to be extensively engraved with lettering. The sheets might have been backed by wooden boards to give the covers sufficient mechanical strength; but this remains no more than a guess because no trace of any means of fastening has been noticed on these pieces.

The new analysis, revealing the presence of a material of very different properties from the bronze previously found, raises the possibility that, quite apart from accidental variations such as those due to the reuse of scrap or the preparation of separate batches of alloy, different materials may have been chosen for individual working parts in the attempt to exploit their individual properties. We do not know, for instance, whether it was then recognised that friction and wear are reduced if dissimilar metals are used for parts that rub on one another.

Some observations may be made on the way in which the parts were formed. Just a few parts – spacers and bosses – are of substantial thickness, up to about 10mm. These may have been prepared as castings, but the odd profiles of some suggest that they were in fact pieces of scrap material, cut down to fit. Arbors may have been cast as rods before being dressed to the required forms and sizes, or they too may have been cut from larger preexisting pieces.

Most components were, however, of a fairly uniform thickness, about 1.5mm, and these appear all to have been cut from sheet, bent up where necessary. We do not know whether they were all cut from a single sheet or from several sheets, perhaps made from separate melts. Such sheets would have been cast as slabs which would then have been hammered down to the required thickness, perhaps finished by being scraped or scoured clean.

Aside from their composition, we remain uncertain of the metallurgical state of the components. We would wish to know whether the metal was left as cast, or underwent hammering or other cold-working; and if so whether it was subsequently re-annealed. The stiffness (influencing the strength of components in bending) and the hardness (dictating the resistance to wear) can vary enormously according to such treatment.

For pieces made of sheet, the slab must have required annealing at intervals during the process of beating out; but we cannot lay down any rule as to whether the finished sheet would then have been annealed or left in a work-hardened state. Some non-planar parts were clearly bent up from sheet, which could have been done successfully only after annealing. Typically, however, we find that they have subsequently cracked and broken along the line of the bends. This suggests strongly that they were not annealed after bending, leaving the metal more highly stressed in the region of the bend and liable to the propagation of cracks due to stress-corrosion. In any case, the fact that the material of these parts was beaten out into sheet and then bent up, shows that it was not a particularly high-tin bronze, for which this manipulation would have been difficult or, with a very high tin-content, impossible. None of this, however, tells us anything about composition of the arbors or of any other parts that may not have been cut from the same sheet.

It may be that further analyses will be carried out in due course, but difficulties will remain. The instrument has spent nearly two thousand years in seawater and considerable chemical changes, such as leaching, may have taken place. On examining radiographs, one sees a bright line around the margin of most components, which seems to indicate the migration of tin (of higher atomic number than copper, and so of greater radio-opacity) from the mass to the outside of each part. It follows that any analysis which samples only the surface layer is likely to suggest a deceptively high tin-content.

Faced with the possibility that the parts of the original were made from a range of bronze alloys, and with continuing uncertainty as to the exact composition of any one part, we needed a pragmatic solution to the problem of choice of materials for reconstruction. The properties of bronzes vary greatly, depending on the tin-content and the presence of other constituents. As we have indicated above, it was important both to avoid enhancing the operation of the instrument by choosing metals of a higher performance than the original, and to show that we did not evade supposed problems of construction by choosing metals that were significantly easier to work. However, the mechanical behaviour of a low-tin bronze, as indicated by the early analyses, differs rather little from that of modern commercial brass[15]; consequently, for the parts made from sheet-metal, readily-available sheet brass was used. Not only is it far cheaper than bronze, but enough brass of a suitable thickness was already to hand in the workshop; most of the parts were cut from a kicking-plate found on a discarded door, while the main frame plate, a rather thicker piece, was cut from an old office name-plate. Some of the spindles were made of drawn brass, and a few were of soft mild steel. Other thick parts were made of brass scrap.

All the brass used was of rather inferior quality. Compared to the bronze supposed to have been used in the original, it is less stiff and less hard. The same may be said for the few pieces of soft steel used. The parts of the model are therefore less strong, and will wear faster, than they would if made of bronze. Brass sheet is more easily bent to shape, but this difference is not significant; bronze calls for more frequent annealing, but otherwise it is merely a matter of how hard it must be struck. Brass is easier to cut out (using hammer and chisel), but in the accurate formation by filing of gear-teeth and other delicate profiles, it is actually easier to work with the firmer bronze, provided one's files are sharp. The most important point in our choice of brass was that, far from being enhanced, the performance of the instrument was likely to fall short of what it might have been using bronze. If a reconstruction in brass worked, one could be sure that a similar instrument made of bronze would work better, and last longer.

The design of lost parts

Since the Antikythera Mechanism is highly fragmentary, in preparing a reconstruction we were obliged to design replacements for lost parts; and in doing so we introduced some degree of conjecture. We may distinguish three different levels of restoration, carrying with them different degrees of certainty.

The first concerns that part of the internal mechanism which is largely preserved, comprising gearing that drives dial displays for which we have direct evidence: on the front dial, pointers for the places of the Sun and Moon in the Zodiac and the date, and the display of the phases of the Moon; and on the back dial, pointers for the calendrical and eclipse cycles and for the preserved subsidiary dials. While many surviving components are damaged and incomplete, the restoration of most of their forms called for little comment and has provoked no significant controversy. This is not to say that all would-be reconstructors agree; but to date we believe that in all cases of disagreement our observations have proved at least as satisfactory as any others. Thereafter it remained necessary to devise replacements for further components and details, now absent from the original, needed to allow this part of the instrument to function as intended. Here there may be more scope for continuing discussion over the detailed design, even though it is remarkable how far precedents may be cited from among the parts that remain; and in any case the small scope for changes means that they can have only a minor effect on form and function.

Our second level of restoration concerns interconnected questions about the instrument's overall structure and casework. Here the evidence is harder to identify, and there consequently may be less agreement. Small traces of woodwork indicate that the metal mechanism was fitted within a wooden case. Early photographs show more than now remains,[16] but there is enough evidence for the reconstruction of a rather complicated, stepped design in place of the simple rectangular design favoured by others.

Nearly all the surviving mechanism, comprising most of the largest fragment of the instrument (fragment A, Figs 1.3 and 1.4), was mounted on a rectangular frame plate which lay between the two dials. Fragment A also includes a piece of the back dial plate. Down one side, between the frame and dial plates, lies part of a wooden batten separating the two. To the outside, covering the edge of the frame (but not of the dial, which was evidently wider), lies part of a broader piece of wood which formed the side of the case. The early photographs show a third piece of wood, against the end of the frame plate and meeting the second piece at right angles. It is now lost, but it has left a print in the mass of corrosion products surrounding the frame plate, showing that the two pieces came together in a mitred joint. This detail can be understood only as an external corner, showing that the case embraced the frame plate closely all round. We have no indication of its depth, but it certainly did not extend to the rear beyond the inner face of the back dial plate which is seen to be both wider and longer than the frame plate. Details seen in the recently-discovered fragment F – a corner of the back dial plate with traces of wood surrounding it – confirm our conjecture that the woodwork was stepped out to accommodate the large back dial.

Rotating arbors extended from the internal mechanism through the dial plates to carry pointers, but there is no evidence that either dial plate was fastened to the frame plate by metallic components. The simplest assumption – always to be preferred where there is no contraindication – is that each of these three plates was fitted separately to the wooden case. In our reconstruction they rest in rebates in the wooden case; the square front dial drops in to the front of the case, while the taller rectangular frame plate and still larger back dial plate are both inserted from behind.

We know that the front dial was inserted from the front because on the back of the remaining corner of the front dial (preserved in fragment C) is a small bolt which was worked from the front by a thumb-button. We have fitted four such bolts, one to each corner of the dial; but equally there might only have been two at adjacent corners, the opposite edge having a joint or being held in place by a ledge. We adopted the simplest possible shape for the front of the case. Its rectangular section, dictated by the form of the frame plate, continues to the front where it is rebated to accept the square dial and two further pieces that fill the openings above and below the dial. In the first model the filling pieces are metal plates, but in the second they are wooden panels.

The frame plate is held into its rebate from behind by two battens, modelled on the fragmentary one observed in the original, and the back dial lies directly over them, in its larger rebate. Too little of the back dial remains for us to have evidence of the means of holding it in place. Conjecturally, we have fitted it with four bolts modelled on the one from the front dial mentioned above. Possibly this dial was not so readily removable as the front dial, but for our purposes, in demonstrating the construction and function of the internal mechanism, it is convenient to be able to take it out without using tools.

The contrate wheel of the original, through which motion was transmitted to all the internal gearing, is found in fragment A stuck to the frame plate by the cohesion of corrosion products, but there is no fastening to hold it there. We suppose that the 'cradle' in which it seems to lie is the remaining part of a circular socket which was fixed to the side of the wooden case; and this is how we have modelled it. The inner part of the socket holds the contrate wheel in place, while the narrower outer part forms a bearing for the stem of a wooden driving knob. The knob is adopted in place of the winch handle suggested by others, since the available evidence suggests that the winch handle was a mediaeval invention.

Our third and final level of restoration refers to further features, beyond those obviously required by the functions listed above. There is evidence for such parts, but it is rather far from obvious and other researchers have tended to suppress it; nevertheless its presence cannot be either denied or ignored. We refer mainly to additional mechanism, driving further pointers on the front dial to indicate the places in the Zodiac of the five planets known in antiquity. Our reasoning for these inclusions has been given in outline elsewhere, and we intend to publish in greater detail on the relation between the design of our restoration and the development of Hellenistic astronomy. Here we confine ourselves to a short discussion of practical considerations.

The restoration of epicyclic mechanism to the largest wheel of the original – the Mean Sun Wheel, seen on the front of fragment A – suitably proportioned (with the necessary large epicycle) to drive an indicator for the planet Venus – makes sense of several features that otherwise remain unexplained: not least, the size of the wheel itself. The addition of two further epicycles driven by the same epicyclic gear train, for Mercury and for the solar anomaly, satisfies ideas of consistency and completeness while remaining compatible with the physical evidence. These three epicycles are carried round on the Mean Sun Wheel, as the astronomy requires, at the rate of the Sun (one revolution representing one year); and they are rotated, each at its own speed, by gearing that derives its motion from engagement with a fixed central wheel. We make these points to illustrate the origin of our design for further mechanism for the planets Mars, Jupiter and Saturn.

For these three planets ('superior planets', as opposed to the 'inferior planets' Mercury and Venus) the astronomical model is inverted; the epicycle

FIG. 1.4

The Antikythera Mechanism, **fragment A** (front view). This recent photograph of the largest fragment (about 205mm largest dimension) may be compared with **Fig. 1.3**. Fragment A contains most of the surviving gears; several may be seen lying on the frame plate, below the spectacular large wheel with four arms. Many more wheels lie, largely hidden, behind the frame plate. One revolution of the large wheel, which lay behind the front dial, represented one year. It drove the pointers for the Sun's position and the date, and it, together with the traces of further mechanism that was mounted on it, is the foundation of Wright's reconstruction of planetary indications. (Michael Wright) is carried round at its own speed, but it rotates on its own axis at the speed of the Sun. For each planet, this entailed repeating the ensemble restored to the Mean Sun Wheel, with the function of the gearing that drove it inverted, and with the size of the epicycle and the gear ratio chosen to suit the particular planet. Thus three epicyclic assemblies were added, but no wholly new mechanical ensemble was introduced because every feature could be related to a precedent already found in the instrument.

The mechanism for Mercury and Venus occupied space which would otherwise have been void, but the inclusion of the extra stages, for Mars, Jupiter and Saturn, called for a small increase in the depth of the case beyond what would otherwise have been necessary. However, that dimension was not defined by what remains of the instrument, so no difficulty arose. Moreover, current work by others, who are reading more of the fragmentary inscriptions on the outside of the instrument, seems increasingly to indicate that the instrument did indeed display planetary motion or events in some way. We maintain that our reconstruction fits the evidence well, and better than any other that has yet been offered.

The simple arrangements described in this section have proved wholly satisfactory in practice, as the model has been taken on extensive tours on which it has repeatedly been demonstrated and disassembled in front of audiences.

Workshop tools and technique

Reconstruction proceeded without any need to prepare drawings. After taking thought, and in a few cases making a rough sketch, we found that each new part or assembly simply 'designed itself' and fell into place. It is an essentially modern idea that mechanical work must or should be drawn out before it is made, and we ought not to be surprised that it proved possible to dispense with drawing when reconstructing an instrument that was probably not drawn out in the first place. To have achieved this without any drawing, seems to us to lend weight to the validity of the outcome.

This validity would however have been called into question if its construction appeared to depend on the use of tools or techniques not available to the maker of the original. Rather than attempting a strict re-enactment of Hellenistic workshop practice, we thought it appropriate to use our own familiar tools, but to remain aware of the probable differences between ours and the ancient ones, and to select tools and techniques that would keep these differences small. Most of the work was in fact done using a fairly small selection of hand tools, of which sufficiently similar ancient specimens are known or attested to by the existence of characteristic toolmarks on artefacts.

Many historical examples, drawn from the Roman Imperial period, are a little more recent than the Mechanism; but the demonstrable heavy dependence of Roman material culture on Hellenistic technique, the stability of well-established craft practices, and the inertia due to the conservatism of craftsmen, all allow us some historical latitude. To answer a challenge often made, tools of steel were widely available, even though the underlying metallurgy was not understood until modern times. The main difference between the tools of the Hellenistic mechanic and our own is, we suggest, one of efficiency and perhaps of durability, not one of capability. Perhaps the greatest single advantage that the modern mechanic has over his Hellenistic counterpart is in the use of a firm screw-vice to hold his work.

There were many holes to be drilled, including a ring of 365 very small ones in the front dial plate (hidden behind the calendar ring). All these holes could have been made using simple drills, such as the traditional watchmakers' bow drill. We, however, used a small mechanical drill press for making most of the holes.

There is little doubt that some parts of the original were turned in a lathe, and we certainly found it convenient, if not absolutely necessary, to use one too. The arbors, bushes and similar parts were turned in a foot-lathe built in about 1820, employing hand-held cutting tools and techniques like those habitually used in making clockwork. This machine is far more handy than any Hellenistic lathe, but the two would have been not too distantly related, and we emphasise that in any case none of the parts really had to be turned. In this context the lathe can be regarded simply as a labour-saving device.

Some commentators have seen the making of the gear wheels as a matter of difficulty or at least of wonderment, supposing that such gear teeth could have been made with sufficient precision only by the use of some sort of appliance or machine. The wheels of the original are, though, not particularly precise; we judge that they are just good enough to run adequately. The form of the teeth may have been reasonably consistent, but the angular separation between adjacent teeth, which can have been very little affected by either wear or damage, is quite variable. Appearances are consistent with the teeth having been cut by hand using a file or slip of abrasive stone, after being marked out in a separate operation. Whether this were done by copying from some sort of pattern such as a division plate, or originally (that is, not by copying) for each wheel, is debatable. We have presented an argument for believing that a division plate may have been employed, but even so the teeth would still have been cut by hand.[17] In any case, we have demonstrated on many occasions that wheels of this size may be divided into any chosen number of parts, with sufficient accuracy, using no more than a pair of compasses; and the whole operation can be performed quite fast. When the writer was younger (and so

RECONSTRUCTIONS

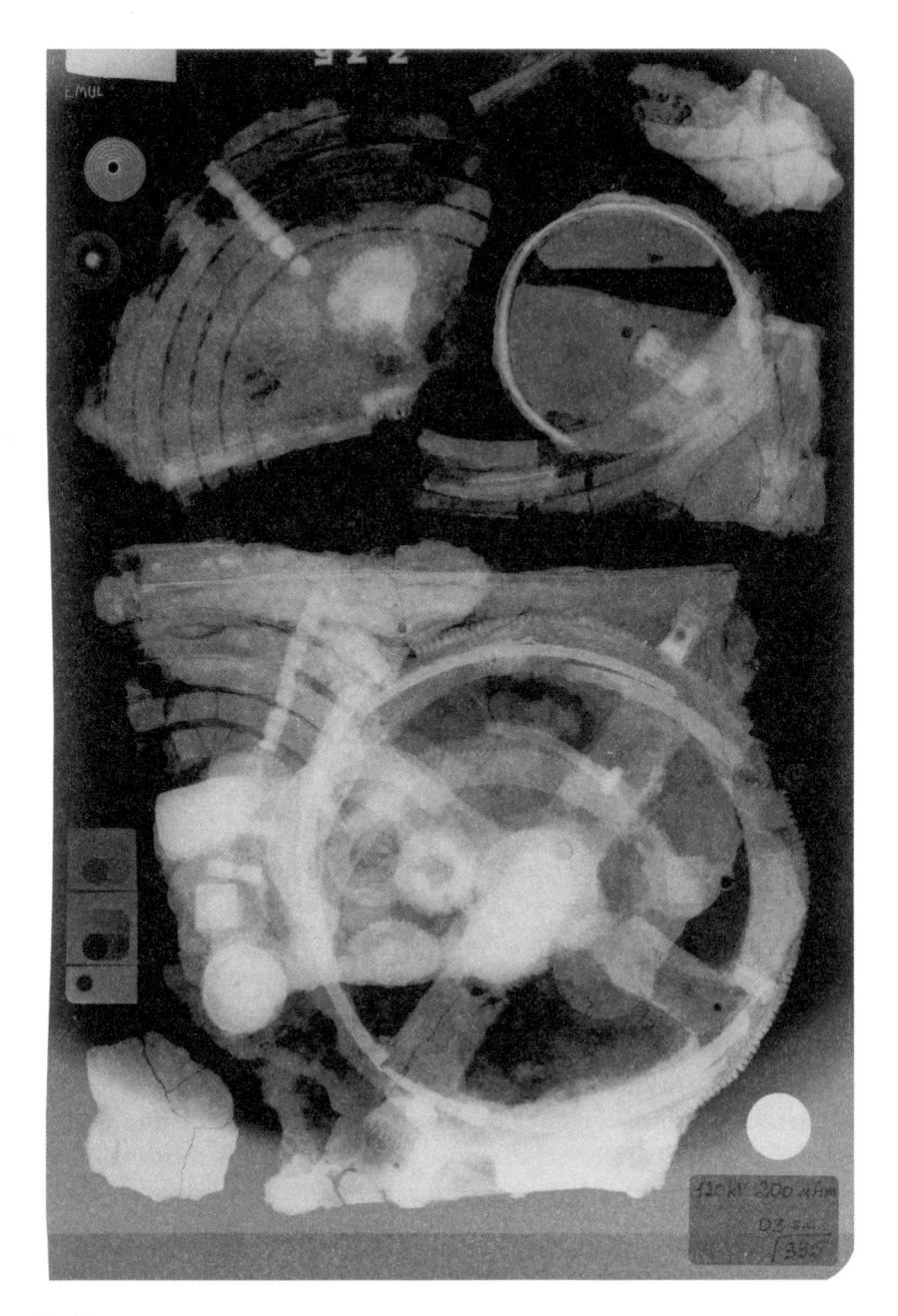

FIG. 1.5

The Antikythera Mechanism, radiographic plate prepared by A. G. Bromley and M. T. Wright. The following **fragments** are seen: **A** (below), **B** (above left), **C** (above right), **E** (top right), and a small piece now fragment called **19** (bottom left). The latter formed part of a sheet that lay over the back dial, and it bears an inscription alluding to calendrical cycles displayed on the dial. The plate also includes images of prepared radiographic control pieces.

(Allan Bromley and Michael Wright)

did not need reading glasses) he found that, taking both operations together, the work could be done at the rate of about 30 seconds for each tooth.[18] The point did not have to be proved again, and so for this reconstruction nearly all the teeth were cut by simple machinery.

Otherwise, we used a machine-tool only for cutting the long S-shaped 'spiral' slot in the back dial plate. The geometry is actually a chain of semi-circular arcs struck from different centres. Probably the maker of the original scribed the centre-line of the slot and filed it out, perhaps first removing the bulk of the material by 'chain-drilling' a series of closely-spaced holes. For expediency, we cut the slot using a slot-drill in a light vertical milling machine, the plate being rotated under the drill about the desired centre for each semicircle.

Joints in the original can be seen to be secured using cotters, rivets formed on the ends of arbors, rivet-pins and nails. Possibly some were also soldered. The details of the original fastenings were replicated where they could be identified, and similar details were adopted elsewhere, with this exception: screws were substituted for some rivets so as to facilitate disassembly. Careful thought was taken in introducing these unauthentic – and anachronistic – fastenings, to ensure that they should not enhance the performance of the altered assemblies in any significant way.

Experience of wear and breakdown

The first model was built in stages, as our reconstruction progressed. The front dial, with most of the gearing driving its display, was first assembled in the winter of 2001–2002, and it has been widely demonstrated ever since, a period of nine years at the time of writing. Not the least interesting thing about it has been its history of breakdowns, almost always of the same components in the gear-train leading from the one-turn-a-year motion of the Sun and date hands to the one-turn-a-month motion of the Moon hand: an increase in velocity of over twelve times. It is not at all surprising that this happens, in view of the almost unbelievably crude form of the gear teeth, and the far greater difficulty in obtaining a satisfactory action when (as in this train) the driven wheels run faster than the drivers. Some of these wheels have now been replaced five times. Reconstruction has allowed us to experience and appreciate at first hand shortcomings in the design and execution of the original.

Conclusion

In our study of the Antikythera Mechanism, physical modelling was an important feature of the process of reconstruction, and reconstruction was in turn essential to the development of our understanding of the instrument. The experience of making the mechanism led us to significant insights as to why and how parts were fashioned as they were. Without this, and the stimulus of being obliged to restore lost parts to create a coherent and functional whole, our present grasp of the nature of the original, of its good features and its shortcomings, would have been hard to acquire. The experience of working the model and demonstrating it to others has developed a heightened awareness of the way in which the original might have been viewed and used in its own context.

These ends were best achieved by modelling the instrument as faithfully as was practicable. In doing so, we accorded other desiderata a lower priority. For example, in viewing this model, with its gears in closely-spaced layers and enclosed in a wooden case as in the original, the onlooker is denied any immediate appreciation of the internal arrangement and function. That end could better have been served by fitting up the wheel-work in more widely-spaced layers, in a skeletonised frame or one made of transparent material, as others have done; but to do that would have been to compromise seriously the basic validity of the reconstruction.

Computer animation has advantages in communication with the public.[19] The virtual model is portable, clean, easy to store, easy to stop and start whenever one chooses and to watch for as long as one wants. The internal mechanism is seen to move, without the inconvenience of the box getting in the way. However, the virtual model has no friction and suffers no wear. There are no questions of mass, inertia, weight, strength, stiffness or working loads. It is no more than a moving diagram. Without the validity conferred on it by the working of the physical model, it would be of very little value, and it lacks much of the appeal of 'the real thing'; which, we find, the physical model does most definitely possess. It is, however, an immensely valuable expository device, to supplement the physical reconstruction.

Words in print seem likely to remain paramount as the formal records of our thoughts, and we have much more to write about the Antikythera Mechanism. Before that, though, we believe it has been valuable to show other people that the reconstruction works, and to show how it does so. The fact that it works does not guarantee that the reconstruction is right, but it is a sine qua non; more generally, the sight of the reconstructed instrument in motion is the most convincing demonstration possible that it is at least plausible. If a picture is worth a thousand words, an artefact is worth many pictures and a working instrument is worth many more. To put the point differently, there is a case for accepting physical reconstruction as a medium for publication.

Notes

1. Athens, National Archaeological Museum, inventory number X.15087.
2. The author's website www.mtwright.co.uk, under development, is intended to provide further information and a full list of his publications.
3. The author's first model has been demonstrated on a number of documentary programmes broadcast on television. A short amateur videorecording, made by Jo Marchant, has been posted on YouTube. It may be found on Dr Marchant's website www.decodingtheheavens.com under 'The Device'.
4. All the author's significant papers to 2006 are listed with full bibliographical details in a review paper: Wright, M. T. (2007): 'The Antikythera Mechanism reconsidered' in *Interdisciplinary Science Reviews*, vol. 32, no. 1, pp. 27–43. Further papers are in preparation.
5. Important contributions to the present understanding of the instrument have been made by members of the Antikythera Mechanism Research Group, who have published two papers to date: Freeth, T., et al. (2006): 'Decoding the ancient Greek astronomical calculator known as the Antikythera Mechanism', in *Nature*, 444 (30 November 2006), pp. 587–91; Freeth, T., et al.(2008): 'Calendars with Olympiad display and eclipse prediction on the Antikythera Mechanism', in *Nature*, 454 (31 July 2008), pp. 614–17. See also the Group's website: www.antikythera-mechanism.gr
6. An accessible account may be found in: Marchant, J. (2008): *Decoding the Heavens* (London: William Heinemann).
7. Price, D. J. de S. (1959): 'An Ancient Greek Computer', in *Scientific American*, vol. 200, no. 6, June 1959, pp. 60–67.
8. Price, D. J. de S. (1974): 'Gears from the Greeks', in *Transactions of the American Philosophical Society*, vol. 64, no. 7, 1974; reprinted as an independent monograph, Science History Publications, New York 1975.
9. Wright, Bromley and Magou (1995).
10. Galluzzi, P. (ed) (2009): *Galileo: Images of the Universe from Antiquity to the Telescope*, (Florence: Firenze, Giunti). Catalogue entry II.4.3, p. 135. Exhibition held at Palazzo Strozzi, Florence, March–September 2009.
11. Fansa, M. (ed) (2009): *Ex Oriente Lux? Wege zur neuzeitlichen Wissenschaft* (Mainz am Rhein: Philipp von Zabern). Catalogue entry VI.6, p. 382. Exhibition held at Landesmuseum für Natur und Mensch, Oldenburg, October 2009–January 2010.
12. Wright, M. T. 'A practical approach to studying the Antikythera Mechanism', XXIII International Congress of History of Science and Technology, Budapest, July–August 2009. No proceedings were issued, but an independent paper is in preparation.
13. Price (1974), Appendix I (pp. 63–66).
14. Zafeiropoulou, M. and P. Mitropoulos, 'The Antikythera Shipwreck, the Treasure and the Fragments of the Mechanism', XXIII International Congress of History of Science and Technology, Budapest, July–August 2009. No proceedings were issued.
15. A useful overview of the properties of a wide range of brasses and bronzes may be found in Holtzapffel, C. (1842), *Turning and Mechanical Manipulation*, vol. I (London), p. 266 ff.
16. The earliest photographs are those published in Σβορώνος, I. N. (1903): *National Museum in Athens* (Athens), subsequently published in German as Svoronos, J. N. (1908): *Das Athener Nationalmuseum* (Athens: Beck & Barth). Some are reproduced, poorly, in Price (1974).
17. Wright (see note 12).
18. Wright, M. T. (1990): 'Rational and Irrational Reconstruction: the London Sundial-Calendar and the Early History of Geared Mechanisms', in *History of Technology*, 12, pp. 65–102.
19. Several computer animations, illustrating different reconstructions of the Antikythera Mechanism, can be found on the web. Particular attention is drawn to one prepared by Mogi Massimo Vicentini, based on data provided by M. T. Wright, which may currently be accessed at http://brunelleschi.imss.fi.it/galileopalazostrozzi/multimedia/TheAntikytheraMechanism.html

Bibliography

Field, J. V. (1990): 'Some Roman and Byzantine Portable Sundials and the London Sundial-Calendar', in *History of Technology*, 12, pp. 103–35.

Field, J. V. and M.T. Wright (1985): 'Gears from the Byzantines: a Portable Sundial with Calendrical

Gearing', in *Annals of Science*, 42, pp. 87–138.

Hill, D. R. (1985): 'Al-Bīrūnī's Mechanical Calendar', in *Annals of Science*, 42, pp.139–63.

The two items above are reprinted in Field, J. V., D. R. Hill and M. T. Wright (1985): *Byzantine and Arabic Mathematical Gearing* (London: The Science Museum).

Field, J. V. and M.T. Wright (1985): *Early Gearing* [exhibition catalogue] (London: The Science Museum).

Wright, M. T. (1990): 'Rational and Irrational Reconstruction: the London Sundial-Calendar and the Early History of Geared Mechanisms', in *History of Technology*, 12, pp. 65–102.

Wright, M. T. (2000): 'Greek and Roman Portable Sundials: an Ancient Essay in Approximation', in *Archive of History of Exact Sciences*, 55, pp. 177–87.

Wright, M. T. (2002): 'A Planetarium Display for the Antikythera Mechanism', in *Horological Journal*, vol. 144, no. 5, pp. 169–73, and vol. 144, no. 6, p. 193.

Wright, M. T. (2003): 'Towards a New Reconstruction of the Antikythera Mechanism', in Proceedings of Conference: *Extraordinary Machines and Structures in Antiquity*, Ancient Olympia, August 2001 (S. A. Paipetis, ed.), (Patras: Peri Technon), pp. 81–94.

Wright, M. T. (2002): 'In the Steps of the Master Mechanic', Proceedings of Conference: *Ancient Greece and the Modern World*, Ancient Olympia, July 2002, (University of Patras), pp. 86–97.

Wright, M. T. (2003): 'Epicyclic Gearing and the Antikythera Mechanism, part 1', in *Antiquarian Horology*, vol. 27, no. 3 (March 2003), pp. 270–79.

Wright, M. T. (2004): 'The Scholar, the Mechanic and the Antikythera Mechanism' [Printed digest of the S.I.S. annual Invitation Lecture, delivered November 2003], in *Bulletin of the Scientific Instrument Society*, no. 80, pp. 4–11.

Wright, M. T. (2005): 'The Antikythera Mechanism: a New Gearing Scheme', in *Bulletin of the Scientific Instrument Society*, no. 85, pp. 2–7.

Wright, M. T. (2005): 'The Antikythera Mechanism, in *Archaiologia & Technes*, 95 (June 2005), pp. 54–60.

Wright, M. T. (2005): 'The Antikythera Mechanism: evidence for an ancient tradition of the making of geared instruments', in *Eureka! Il genio degli antichi* (E. Lo Sardo, ed.) [exhibition catalogue, National Archaeological Museum, Naples, July 2005–January 2006] (Napoli: Electa), pp. 241–44.

Wright, M. T. (2005): 'Epicyclic Gearing and the Antikythera Mechanism, part 2', in *Antiquarian Horology*, vol. 29, no. 1 (September 2005), pp. 51–63.

Wright, M. T. (2005): 'Counting Months and Years: the Upper Back Dial of the Antikythera Mechanism', in *Bulletin of the Scientific Instrument Society*, no. 87, pp. 8–13.

Wright, M. T. (2006): 'The Antikythera Mechanism and the early history of the Moon Phase Display', in *Antiquarian Horology*, vol. 29, no. 3, pp. 319–29.

Wright, M. T. (2006): 'Understanding the Antikythera Mechanism', Proceedings of Conference: *Ancient Greek Technology*, Athens, October 2005 (Athens: TEE), pp. 49–60.

Wright, M. T. (2007): 'The Antikythera Mechanism reconsidered', in *Interdisciplinary Science Reviews*, vol. 32, no. 1, pp. 27–43. This paper is available at the following website: www3.imperial.ac.uk/portal/page?_pageid=73,7692654&_dad=portallive&_schema=PORTALLIVE#MrMichaelWright

Wright, M. T., A. G. Bromley and H. Magou (1995): 'Simple X-ray Tomography and the Antikythera Mechanism', *PACT (Révue du groupe européen d'études pour les techniques physiques, chimiques, biologiques et mathématiques appliquées à l'archéologie*, or *Journal of the European Study Group on Physical, Chemical, Biological and Mathematical Techniques Applied to Archaeology)*, vol. 45, pp. 531–43. (The issue, entitled *Archaeometry in South-Eastern Europe*, was devoted to the proceedings of the 2nd Conference under that title, held in Delphi, 19–21 April, 1991.)

Wright, M. T. and A. G. Bromley (1997): 'Current Work on the Antikythera Mechanism', Proceedings of Conference: *Ancient Greek Technology*, Thessaloniki, 4–7 September 1997, pp. 19–25.

HEATHER HOPKINS

Using experimental archaeology to answer the 'unanswerable'

A CASE STUDY USING ROMAN DYEING

The temptation, when a problem arose, was always to reach for the volume of Vitruvius, conveniently to hand, rather than go to the site, often inconveniently distant, to look at the actual remains, much less, to cross the quad to that, psychologically at least, even more distant ultima Thule, the Faculty of Engineering. (Trevor Hodge, 1992)

TO UNDERSTAND THE scale of manufacturing of Pompeii would allow an understanding of its economic significance and its place in the wider Roman world. To understand the scale of manufacture as a whole it is necessary to study the processes and outcomes within one specific industry. As the dyeing equipment used in Pompeii is unambiguously identifiable, the dyeing industry can be used to indicate the scale of manufacture as a whole. Textile manufacture is a long process with many stages (Fig. 2.1). Fulleries had multiple uses: they cleaned clothes as well as fulled cloth, and other stages of manufacture may be completed in non-commercial premises and so become archaeologically invisible. The dyeing equipment in Pompeii, on the other hand, is identifiable and has been preserved in the city that it supplied.

Pompeii is a city 'frozen in time'; one of two cities destroyed through the pyroclastic eruption of Mount Vesuvius in AD 79. Herculaneum, the smaller of the two cities, was encased in lava. Pompeii was subject to a rain of pumice that buried the city to first floor level, then subsequent cloud column collapse removed what remained of Pompeii above this level and encased the city in lava. The usual organic taphonomic processes were interrupted and artefact and architectural elements that would ordinarily have been lost have been preserved (Allison, 1992). Both cities remained undisturbed for nearly 1800 years and became a source for artistic plunder. During the nineteenth to twentieth centuries archaeological investigation commenced and is still ongoing.

It is widely known that Pompeii was destroyed and its remains encased through a volcanic eruption. What is not so widely known is that 17 years prior to this event Pompeii had been partially destroyed by an earthquake. This caused great physical, social, economic and legal upheaval in the city. Many commercial and residential buildings were destroyed. The aqueduct that supplied the city ceased to function. Properties became abandoned and derelict. Where a complete line of inheritance had been lost, property became forfeit to the city.

The water supply was restored first to the residential areas of the city. Not only was it advantageous for a dyeworks to relocate to a residential area, but due to the destruction and confusion it was possible for this to happen. Today there are six dyeworks (not seven, as originally thought, *see below*) located in Pompeii. Two are in workshops, two are in the backrooms of villas, and two have obviously 'taken over' the areas formerly used for residential purposes.

The scale of manufacturing of the dyeing industry of Pompeii has been a contentious issue. Moeller undertook a survey of the dyeing workshops and stated that the industry had been so large that Pompeii had exported (Moeller, 1976). Moeller's findings were called into question at the time, most notably by Wild who stated, 'he goes further than the evidence can support him' (Wild, 1977), but as the survey was the most comprehensive seen to date, his theory has persisted. Jongman, using Moeller's data, re-examined his theory and concluded that the industry had been so small that Pompeii would have relied on imports (Jongman, 1988). Later studies re-examined Moeller's data and agreed with Jongman's conclusion. However, what was most noticeable about each of these was that at no point was Moeller's original data questioned or the remains re-surveyed. Furthermore, all of these studies, including Moeller's, were theoretical; modern recipes were applied to a superficial understanding of the apparatus. There was no practical experimental work to understand how the apparatus may actually have reacted to these recipes or the assumed procedure, or to see how much it was actually possible to produce from an apparatus in a given time. The most pertinent assumption had not even been acknowledged, as each of the apparatus had been subject to a volcanic event, buried, then either robbed or conserved on discovery: no complete untouched apparatuses had survived. The assumed design of a dyeing apparatus was an averaged composite of surviving remains. At no point had this design been tested to see if it was even viable.

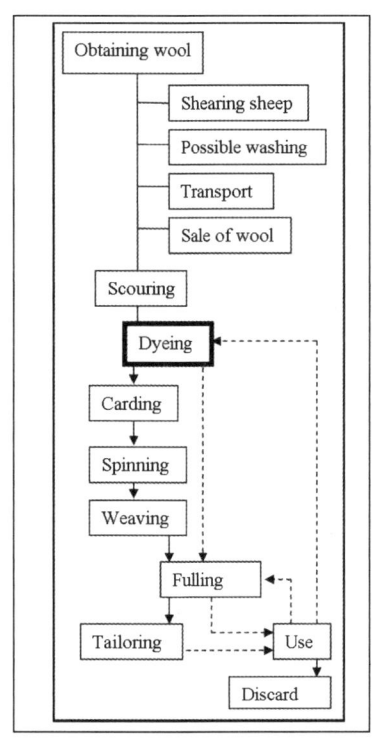

FIG. 2.1 Stages of the process of textile manufacture. (Heather Hopkins)

It was realised that the only way in which it was possible to understand the scale of manufacturing was physically to construct the relevant parts of the industry and use them to see how much it was possible to manufacture. Measurements and photographs had been taken of dyeing apparatuses in a preliminary study in 1994 and these were combined to allow the construction of a replica apparatus (Janaway and Robinson, unpublished). A literary review of contemporary and historic writing, depictions, archaeological artefacts and the analysis from archaeological finds was undertaken and it was concluded that the most often used material for textile manufacture at this time in the Roman world was wool, the most used dye was madder, and the most common mordant was alum. Admittedly the use of these ingredients was advantageous: madder and alum are both allowed as modern food additives and so are seen as relatively safe for experiment.

The wool used and its acquisition were also fairly straightforward. Analysis has shown that the Iron Age British sheep was the equivalent of the modern Soay and that the different shades of its wool were used as an under-colour in dyeing to produce different colours from the same dye (Ryder and Stephenson, 1968; Ryder, 1990). The Roman method was different and the sheep was equivalent to the modern Shetland, a larger sheep with a white fleece that could be dyed any colour (Ryder, 1990). As the University of Bradford is located on the same road as the National Wool Board's storage depot for the North of England, it was a simple matter to go and ask for a fleece.

The choice of recipe proved more problematic. Historically, recipes have been a closely guarded secret and so even contemporary writings such as Pliny's *Natural History* are unreliable; he may have been given an incomplete or incorrect recipe on the assumption that he would not try to use it, or to prevent his success if he did. Few recipes remain, and those that do may not be widely available. For example, it has been reported that there are no contemporary Greek recipes (Monaghan, 2001) and yet several are reported by Lagercrantz (1913) in German and the original Greek. Modern recipes may be just as problematic. At least six recipes for madder dyeing with alum that required pre-industrial equipment were examined and these differed from 'simmer for an hour' to 'boil for six hours and leave overnight' for the same step (Rosetti, 1548; Brunello, 1973; Storey, 1978; Dalby, 1985; Grierson, 1986; Grieve, 1992; Edmonds, 1999, pers. comm.). These recipes were studied and the chemical steps that each was attempting to achieve were discerned and combined to form a single recipe to be used with the experimental apparatus. The quality of water, in particular whether it was hard or soft water and the presence of any impurities, had to be determined. This is a contentious issue, but the present study concluded that despite Pompeii being located on a volcanic outcrop which would ordinarily have given rise to soft water (Arthur, 1986; Peacock,

1989), the water had in fact been hard as it was supplied from the plains via an aqueduct (Hodge, 1981, 1992; Laurence 1994; DeHaan 1997), and that despite the possibility of lead poisoning there was in fact no lead in the water delivered to the dye-workshops in Pompeii (Hopkins, 2007).

While planning the experiment it was asked how much fuel would be required. If a fleece is placed in water, heated and then cooled suddenly, it mats, thereby ruining it. Therefore there has to be sufficient fuel to allow a dye-cycle – that is heating, dyeing and cooling – to be completed without a break. To determine how much fuel would be necessary involves a heat transfer calculation: examining how the heat from that fuel would move through the apparatus, how the apparatus would heat and how the heat would eventually be lost (for full calculation, *see* Watling, 2004). This all depends on what materials would be used to construct the apparatus. The aim of the experiment was to see how a dyeing apparatus would operate and from this how much dyed material could be produced. The quantity that could be produced depended on the length of time needed to heat and cool the apparatus. If different materials were used to the original, they would heat, retain heat and cool in a different way to the original. This would provide an inaccurate result, both in the time taken for a dye-cycle and in the fuel required. If a vast amount of fuel were required for dyeing, fuel may have been a limiting factor. This limitation would be hidden or inaccurately concluded if different materials were used.

Reynolds stated that when constructing a replica of an apparatus, it is necessary to use the original materials that were used in the original artefact that is being replicated (Reynolds, 1999). This is because materials react differently to external physical changes, and that to not use the original materials prevents the replica from being an experiment, but turns it instead into an experience of using something similar. It should be noted that both are important in the understanding of an artefact and its use. When using a replica to undertake an experiment, not only should the experiment have an aim, a null hypothesis and the results be used to answer a defined question, but the replica should be an accurate copy of the original artefact. In this study, use of the original materials presented a problem: the lead and corrosive lime mortar used in the apparatus of Pompeii was vetoed under modern health and safety considerations. It was necessary to substitute in equivalent materials. The dyeing apparatus had been constructed from a rubble or brick surround, held together with lime mortar with a lead kettle. The kettle rested on brick supports (Fig. 2.2). To substitute these materials it was necessary to find equivalents that possessed the same physical properties: that not only had the strength and durability of the original, but also the same thermal conductivity so that they would react the same way to heat. The replica had a brick surround and supports because the value for brick was a mid-value on the

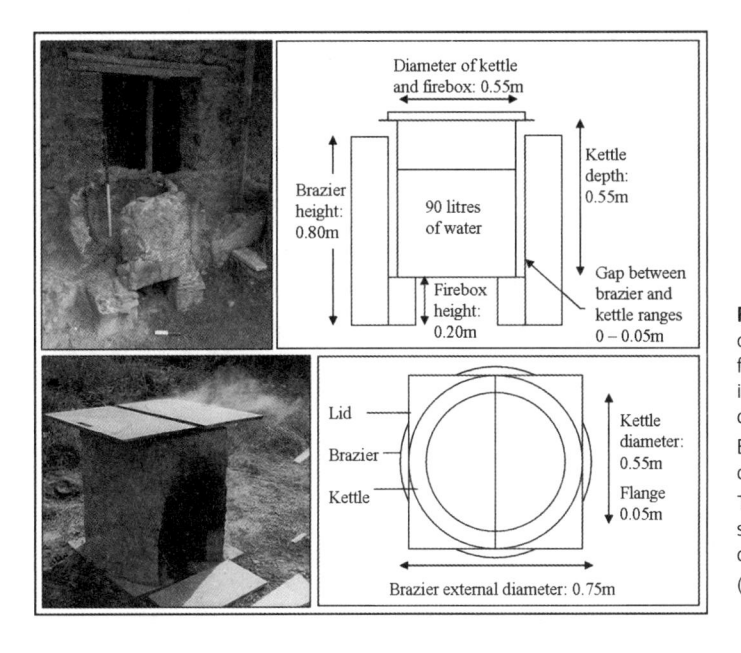

FIG. 2.2 Top left: original apparatus five from property I viii 19 in Pompeii. Scale: 1 m and 0.2 m. Bottom left: replica apparatus. Top and bottom right: schematic of apparatus. (Heather Hopkins)

range for the different stone and brick types used in Pompeii (Hopkins et al., 2005). Stainless steel was used as a substitute for the lead as lead has a thermal conductivity of 35 W/mK while stainless steel has a value of 27 W/mK (Callister, 2000). Stainless steel has a specific heat capacity of 0.460 KJ/(kg.K) W/(M.K) and lead is 0.128 KJ/(kg.K) W/(M.K). When it is remembered that aluminium has a thermal conductivity of 247 W/mK (Callister, 2000), it may be seen how close a match stainless steel and lead really are. The materials and design of the apparatus were the elements important to replicate as the method of using the vat and the physical principles behind its use was what was to be determined. This meant that only the design had to be replicated in its entirety (Mathieu, 2002). It was not necessary to replicate the method of construction, but due to the materials involved and the design it was replicated anyway.

Replica Romans

One criticism levelled at the study was that even though it was possible to construct a replica of a Roman industrial apparatus, it was not possible to produce a Roman who could be seen to use it. It was believed that as modern western Europeans are taller and with a greater reach than the Romans of Pompeii, any figures regarding the operation of the dyeing apparatus would be inaccurate as those taking part in the study were larger. It was noted that it was not possible to ask for or observe the recipes and method used by an original dyer from Pompeii. Because of this it was decided that dyes should not be used in the replications. However, it was still possible to produce a physical replica of a Roman (or slave) from Pompeii.

Although the final stages of the eruption of Pompeii, such as the collapse of the cloud column and the Plinian event (named after Pliny the Younger who described it), occurred devastatingly fast, the volcanic events took over 24 hours. The seismic events, the preceding earthquakes, took months. This meant that at the time of the final entombment of the city, much of the population had already left. However, some of those that could not travel or who sought shelter within the city were discovered in the modern era. As their skeletons remain, it is possible to measure their heights and limb lengths. Although the heights have been published, not all of the limb data has (Capasso, 2001).

Ergonomics is the study of the apparatus–operator interface. It is used in the design of equipment and tools to test before manufacture whether the article will be usable. An item that is designed badly will take longer to use or result in a greater number of mistakes in its use. This would cause damage, cost and slow production. Therefore it would be in the interest of the dyer to ensure that the dyeing apparatus were, if not perfect, at least usable to the dyers.

It is not possible to take a dry skeletal population and apply ergonomic judgements to their measurements. An ergonomic dataset of the physical limits of a population must come from a living, moving population as the additional support and limitation provided by muscles and other tissue will influence the data. Therefore to produce a 'living, breathing Roman', it is necessary to take the Roman dataset and match it to a modern dataset. By a quirk of history the skeletal population of Pompeii and Herculaneum, the population discovered in the streets, buildings and boathouses of these two cities, matches the population of the United States between 1900 and 1970 (Murrell, 1957; Dreyfuss, 1966). Furthermore, a number of social and economic changes occurred in the United States at this time which resulted in a large dataset of the population being recorded. The population as a whole was growing in height. There were an increasing number of women drivers and women in the workforce, such as those that joined the army, and it was realised that modern equipment such as cars needed urgent modification to make them usable to an increasingly divergent population. It is now possible to study both the Roman and American datasets and compare. The figures can then be used to help find a representative 'Roman' to undertake dyeing with the replica apparatus. The average height of a Herculaneum male was 163.8cm, while in the rest of Campagnia it was 164.0cm.

The average height of a Herculaneum woman was 151.7cm, while in the rest of Campagnia it was 152.6cm. The height of the author is 163.83cm – within 0.03cm of the average Herculaneum male. The height of the co-builder is 153.67cm – within 1.07cm of the average Roman female. As these two individuals not only used but also constructed the apparatus, it is possible to state that the replica apparatus could have been used by an original Roman. The limb and reach measurements matched the American data too. The measurements were validated by applying modern forensic techniques used to determine the living height from skeletal remains. Soft tissue adds to the height of an individual, so a range must be given. At the most accurate this range is +/- 4.59cm (Trotter and Gleser, 1958, cited in Bass, 1995:32). This means that the skeletal measurements of the Roman skeletons and those involved in the reconstruction are comparable within the boundaries allowed in modern forensics.

Experiment

Following construction of the apparatus, the first experiment was to determine whether a lid had been used. As a lid has neither been discovered nor depicted, it was presumed that one had not been used. However, this defied logic: the potential for water loss meant that the fleece was likely to be ruined without a lid and the apparatus would be so energy inefficient that it would have been economically beneficial to use a lid just to save on time and fuel. The recipe stated that for a 2 kilogram fleece 90 litres of water should be used (Hopkins 2002, 2007). The kettle had a capacity of 130 litres. It was discovered that after undertaking a single dye-cycle without a lid, 14 litres of water had been lost. This was a sufficient depth to cause the ruination of the fleece. It was calculated that with a lid of only 20mm thickness it was possible to halve the heat lost through the top of the apparatus (Watling, 2004; Hopkins et al., 2005; Hopkins 2007). It was concluded that a lid had been used, but that it was likely to have been constructed from organic material, and that as it began to disintegrate through the effects of the heat and steam it would have been disposed of (Fig. 2.3).

Following the addition of a lid, three dye-cycles were undertaken to determine the time taken to heat and cool the apparatus, the quantity and type of

FIG. 2.3
Left: replica apparatus in use with a lid. Volunteer (Mr Thomas Halliwell) is 1.78m tall.

Right: replica apparatus in use without a lid.

(Heather Hopkins)

fuel used, how an apparatus could be used and maintained, and from this the quantity of dyed fleece that it would have been possible to produce. The fleece was scoured (cleaned) in the apparatus prior to the dye-cycle. The apparatus took 1.5 hours to heat; the recipe stated that the fleece should be simmered at 95°C for 1.5 hours and then allowed to cool naturally. Cooling took over eight hours. It must be presumed that a larger kettle would have taken longer and it should be noted that some kettles in Pompeii had a capacity of 900 litres. It was feasible to siphon some of the water to aid cooling. To clean the apparatus it was possible to lean into the kettle and clean it through the application of hot water, detergent and pressure.

It was discovered that despite charcoal appearing to be the logical choice for heating the apparatus, wood was the fuel used. There was insufficient airflow through the apparatus to allow the complete combustion of charcoal. Wood requires half the oxygen that charcoal requires to burn and so burnt fiercely. There may also have been an economic reason not to use charcoal. Charcoal was required in metalworking as it can be brought to higher temperatures, and it was desired in the home as it is a smokeless fuel, so wood may have been cheaper to buy or obtain: even if twice the quantity of wood were required, it could still have been cheaper to buy wood than charcoal. The fuel used in this experiment was pine off-cuts. This is not to say that the Romans used pine. The aim was to establish the fuel quantity and calorific value required. This dye-cycle took an average of 8kg of pine which has a value of 126195kJ. This early experiment was re-run to give an average and it is these values that have formed the basis for the rest of the study.

Survey

This early experiment informed the practicalities of how an apparatus could be operated and the time and consumables involved in operating it. However, Moeller's early survey had been undertaken without reference to a working replica and so his understanding of the remains that he was surveying was purely theoretical. As no later study had questioned his data, his conclusions had remained untested. There was a need, before the study proceeded any further, to return to Pompeii, and with the aid of a new understanding gained from the experimental dyeing apparatus, to re-survey all of the original dyeing apparatus and their workshops *in situ* in Pompeii. There was a need to test the very foundation that Moeller – and all subsequent authors – had built their presumptions upon.

When viewing a dyeing apparatus, the observer identifies it by applying a set of 'polythetic entitites' (after Clarke, 1978). There is a series of criteria that the viewer has defined, whether they realise it or not, that allows them to compare the object that they are looking at with what they feel a dyeing apparatus should look like. Following use of a working experimental apparatus, that was found to have a viable design only after the remaining fragmentary apparatus had been compared and the designs combined, this set of minimum criteria could be established. A dyeing apparatus had to have:

- Masonry structure
- Kettle or kettle space
- Space for fleece and dye liquor
- Heat source
- Reachable base
- Adequate ventilation

It should be noted that not all dyes require heating. It was possible that a dyeing apparatus may not have held a full fleece, or that it could have been used for cold dyeing. This was considered during the survey. However, the combined requirements for a dyeing apparatus meant that the majority of structures were likely to consist of a masonry outer structure, as this was the most likely structure that had the strength to hold the weight of the kettle and withstand the heating, with a ventilated space below for a fire. Examples have survived in North Africa where instead of a metal kettle the brazier itself has held the liquid, but examples of this type have not yet been discovered in Pompeii (Wilson, unpublished, cited in Hopkins, 2007).

If a dyeing apparatus has been identified, the next step is to assess its location to determine if it was possible to use it as a dyeing apparatus. Following the earthquake of AD62, each 'dyeing workshop' could be located in very different surroundings: a former villa, a workshop on a thoroughfare, a stable. Where the workshop was located would not have affected how much it could physically produce, as long as the criteria required to allow the physical production was in place:

- Apparatus
- Apparatus context
- Property context
- Water supply
- Drainage
- Ventilation

Intangible 'endemic entities' (after Clarke, 1978), the social or economic factors or constraints that are only known within the society that they affect, must be allowed for but cannot be specifically factored in, as by their nature they are unknown. However, it is possible to state whether or not it was possible to use a property that contains apparatus that could have functioned as dyeing apparatus. Dyeing requires a regular water supply and consumables such as fuel. It also requires the disposal of the water once it is contaminated and the complete ventilation of the apparatus. If a property is sufficiently enclosed that this is not possible, then there are doubts as to whether it could have operated as a commercial dyeing workshop.

The location of the dyeing apparatus within the workshop was also a consideration. The workshops were set out in a distinctive U-shaped arrangement design, regardless of the original use of the property and any constraints that this may have imposed. Through the application of the identifying criteria for a dyeing apparatus, for a workshop and observing the layout within workshops, it was possible to dismiss one of Moeller's dyeing workshops (Fig. 2.4). The workshop contained a single apparatus at the centre of the room, when other workshops contained multiple apparatuses in the U-shape. Further study revealed that although the apparatus was of a sufficient diameter to hold a fleece or other materials, it could not have held a kettle that was more than 0.15m deep. The geometry of the vessel that could have been held would not have allowed successful commercial dyeing, and if it had originally been constructed as such it would have been amended in antiquity. The design was instead reminiscent of a take-away kitchen, and given its proximity to the nearby 'public house' it was concluded that this apparatus was more likely to have formed part of the catering industry rather than having been textile related.

Discounting one workshop and apparatus proves the need to return to the original remains, to collect data and test assumptions. The apparatus located in property IX iii 2 demonstrated a further reason for doing this. These apparatuses had been correctly identified as dyeing apparatuses, but their dimensions had been mis-recorded. In their current state, they are too tall to use for dyeing: their geometry is also not of a practical design and ergonomically they would have been very difficult to use. However, when examined closer it was possible to see that new mortar had been applied at some time in the last 200 years. At their original height they would have been well-proportioned and met all of the criteria. Without the use of the replica apparatus, it would not have been possible to determine that the height had been incorrect and the esti-mated scale of manufacture would have been wrong accordingly.

See Hopkins (2007) for a full gazetteer of the dyeing apparatus and workshops in Pompeii. The apparatuses were each photographed, measured, described and ergonomically assessed.

FIG. 2.4 Artefact identified as a dyeing apparatus by Moeller, but discounted during this survey. (Heather Hopkins)

Departing from Archaeology

When recording the apparatus in Pompeii, it was noted that although there was a range of sizes, the geometry (and therefore internal stress) was relatively similar (Hopkins, 2002). The main difference noted was that there appeared to be two different designs: flued apparatus and unflued apparatus. It was noted that the flued apparatus tended to be located within enclosed workshops, whereas the unflued apparatus was located in 'open' workshops. It was reasoned that as the flues were integral parts of the structures of the apparatus, the extra effort and difficulty in their construction, and the presumed difficulty in adding one after the apparatus was completed, pointed to their being necessary to aid ventilation in enclosed workshops by drawing the air through the apparatus. It was decided that to explore what difference would be made by the addition of a flue to an apparatus, the replica should have a flue added. This proved more problematic than first envisaged.

When an experiment is undertaken, it must be set up according to scientific principles and meet a number of criteria. It must be repeatable. It must have a clear aim or null hypothesis (Crumlin-Pedersen, 1999; Mathieu, 2002; Coles, 1973). Only one change to an apparatus or surroundings may be undertaken at any one time (Mathieu, 2002). An experiment that purports to involve a replica must indeed involve a true replica, not an approximate copy with an inaccurate design or materials. As previously discussed, this replica was a true replica as the design was accurate and the materials had equivalent physical properties. However, to replicate the flue a decision had to be made: whether the experiment should depart from scientific replication or from archaeological replication. This is not a decision that would necessarily have to be made in all experimental archaeology reconstructions, but due to the nature of the artefacts involved it was necessary in this case.

The flue on a dyeing apparatus is placed level with the top of the firebox and the base of the kettle to allow the exhaust gases to be drawn away from the fuel and to draw fresh air into the firebox. The logical place to put a flue on the replica apparatus would be at this same place, so as to replicate the design of a flued dyeing apparatus. However, the flued dyeing apparatus to be copied had a taller firebox than the unflued (and replica) apparatus (Fig. 2.5). To place the flue at the base of the kettle would have caused it to be placed at the wrong height. To place the flue at the correct height would mean that it would no longer function as a flue as it would be blocked. But to place it at the correct location for the design to be functional would mean breaking away from archaeological replication as no dyeing apparatus of these dimensions and exact design, that is flued with the shorter firebox, has yet been discovered.

To replicate the full design it would be necessary both to add a flue and to

FIG. 2.5
Left: unflued dyeing apparatus.
Middle: dyeing apparatus with functional flue.
Right: dyeing apparatus with non-functional flue.
(Heather Hopkins)

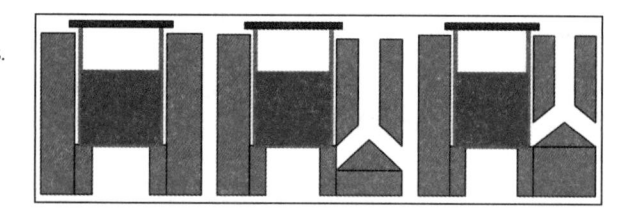

enlarge the firebox. To do this would cause two changes to the replica, not one, and so it would not be possible to argue that there would have been only one variable changed in the experiment.

It was reasoned that as the aim of the experiment was to determine how an apparatus operated, it would be nonsensical to construct an apparatus that could not function. Also, just because a dyeing apparatus has not yet been discovered with these dimensions, it does not mean that one will not come to light in future. The variety already noted in the 40 apparatuses in Pompeii shows that further variations in design may yet be discovered. It was therefore decided to place the flue at the top of the firebox and base of the kettle, to allow the experiment to continue on a strictly scientific basis. This meant that only one variable had been changed: a flue had been added to the replica. Other than the difference in firebox height, the design and dimensions matched dyeing apparatus Five in property VII xiv 17 (Fig. 2.6). The third experiment was a replication of dyeing with a flued apparatus. It was discovered that the fuel used and the time taken were not altered by the inclusion of a flue. However, the exhaust gases could be seen to be drawn up through the flue and it was noted that the flame was easier to control. It was concluded that flues were added to dyeing apparatus to aid ventilation and combustion by drawing air through the firebox.

Maximum output

Following the survey in Pompeii and replica dyeing cycles with flued and unflued apparatus, it was possible to calculate a theoretical maximum for the output of the dyeing industry in a year. As the Romans did not use precisely 90 litres of water for every 2 kg fleece (metric units being a relatively modern invention), but did use a comparable quantity of water otherwise the dyeing would not have worked, it is possible to calculate not just a theoretical maximum, but to place this within a margin of error (Hopkins, 2007). Following use of the replica it was concluded that the dyeing apparatus was used once a day and probably left to cool overnight. The figures below are the maximum that can be produced in a dyeing workshop in Pompeii in a day if each dyeing apparatus is used once at full capacity (*see* Table below).

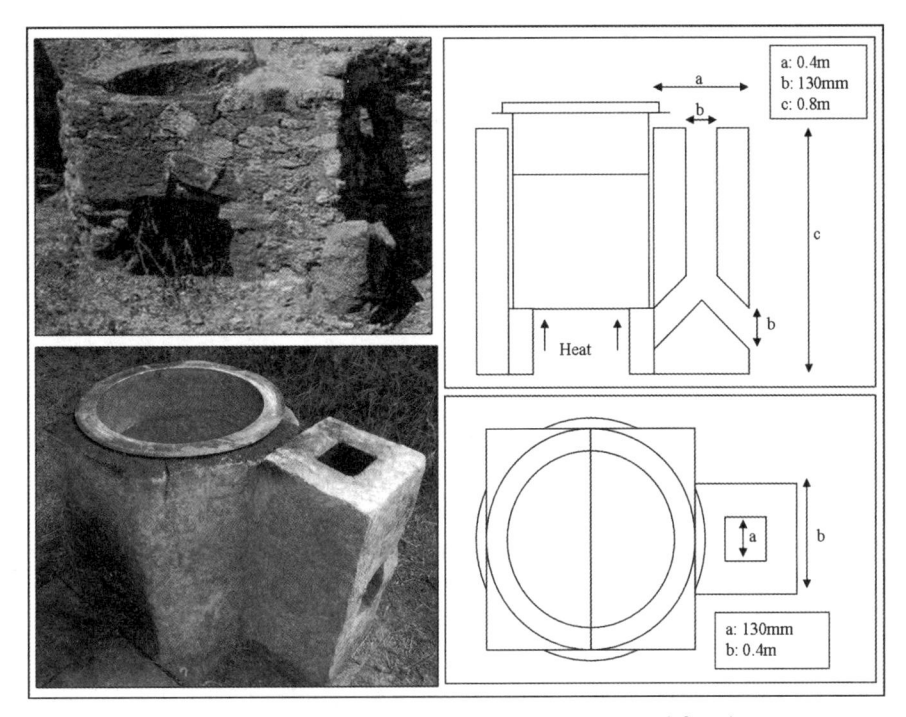

FIG. 2.6
Top left: flued apparatus five from property VII xiv 17. Scale: 1m and 0.2m.
Bottom left: replica apparatus with flue. Right top and bottom: schematic of flued apparatus.
(Heather Hopkins)

If there were 318 working days in Pompeii a year (Grose-Hodge, 1944), allowing for markets, festivals and other days when production would not have taken place:

154 x 318 = 48,972   167 x 318 = 53,106

then between 48,972 and 53,106 fleeces could be dyed annually in Pompeii. By itself, this figure looks large, but the scale of output of an industry can only

TABLE 1

Property	Maximum fleeces	Minimum fleeces
I viii 19	17	15
VI 4	32	28
VI 5	16	13
VII ii 11	41	40
VII xiv 17	42	39
IX iii 2	19	19
Total	167	154

be appreciated if the population that the industry is to supply is also considered.

Figures between 8000 and 20,000 have been suggested for Pompeii, but the figure of 12,000 has been consistently argued (Storey, 1997).

$$\frac{48,972}{12,000} = 4.1 \ (4.081) \qquad \frac{53,106}{12,000} = 4.4 \ (4.4255)$$

If a Roman fleece weighted 2kg:

$$4.1 \times 2 = 8.2\text{kg} \qquad 4.4 \times 2 = 8.8\text{kg}$$

The maximum output of the dyeing industry of Pompeii was between 8.2 and 8.8kg of fleece per person. This may seem a significant quantity, but it is the equivalent in weight to a (large, possibly laundrette) washing machine load. Although not every textile item was dyed and there was a trend for plain clothing with patterns (Bender Jørgensen and Mannering, 2001; Pritchard and Verhecken-Lammens, 2001; Cardon, 2001), soft furnishings tended to be dyed wholly in one colour. When the quantity required to make a single pair of curtains is considered, and it is remembered that the quantity per person would also have to stretch to commercial, government and religious premises, ultimately the quantity of 8.8kg per person appears to be very small. It is possible that further dyeing workshops may be discovered in the unexcavated parts of Pompeii, but as the majority of the city has been excavated and the six that have been discovered are scattered throughout, it is unlikely that so many would be discovered that this calculation would be greatly affected.

Moving into engineering

The figure of 8.8kg per person also relies on the assumption that each dyeing apparatus was fully functional and does not allow for breakages. The material used for the manufacture of the kettle was lead (Monteix and Pernot, unpublished). This was noted as an archaeological novelty, but it was only when it was considered later during calculations for heat transfer that it was realised that the use of lead could have been the limitation on the system of production. Lead is weak, malleable, dense and therefore heavy. During the survey it had been noted that the remaining kettles had concentric circles at their bases (Fig. 2.7). This was dismissed as an archaeological curiosity, but it was believed that it could have been something to do with their being heated and cooled repeatedly (Fig. 2.8). In fact these concentric circles were the proof that the kettles had been subject to a process known as creep.

Lead creeps at room temperature. Creep is time-dependent strain and the rate at which a material creeps increases with temperature (Greenfield, 1972). Lead was an unusual choice for the construction of a kettle that had to hold a minimum of 90 litres of water while being heated. Economically it was a good choice, as lead is a waste by-product of silver production and so would be cheaply available. It is also one of the few, possibly the only, metal that would not 'sadden' a dyed material: lead acts as a mordant, brightening a dye. But even with the economic and chemical considerations, the physical properties of lead should have been a deterrent to its use for a dyeing kettle.

This evidence of creep in lead also highlighted that Reynolds had been more far-reaching than previously thought in his assertion that only the original materials could be used in a replica (Reynolds, 1999). He stated that when building a replica there would be unknown attributes within a material that must be allowed for, and that it is only by using the same materials as the original that this can be done. This replica had matched all of the physical properties of the original except one: lead and stainless steel do not creep in the same way. The only way to study the creep of lead would be to reconstruct the apparatus from lead, something that was banned on health and safety grounds.

FIG. 2.7 Concentric circles in the base of kettle of apparatus five from property VII xiv 17. Scale: 0.2m (Heather Hopkins)

FIG. 2.8 Schematics showing the thermal and weight load the kettle is subject to during a dyeing cycle and the strain this would cause. (Heather Hopkins)

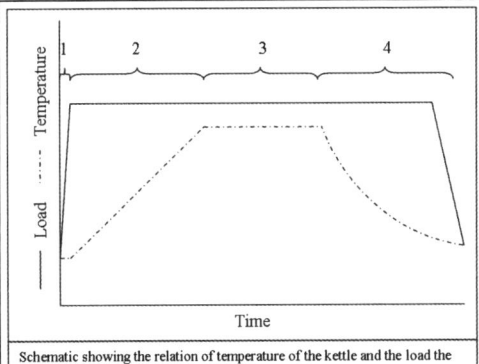

Schematic showing the relation of temperature of the kettle and the load the kettle is subject to during one dye run. The load and temperature are mutually exclusive but change over time as the dye run progresses. The combination of these two factors increases the rate at which the kettle fails.
1: The kettle is loaded with water, then the fire is lit
2: The load is constant, the temperature rises
3: The temperature is held as the dye liquor simmers, the load is constant
4: The fire is extinguished and the vat left to cool. As it approaches ambient temperature the fleece may be removed and the kettle emptied.

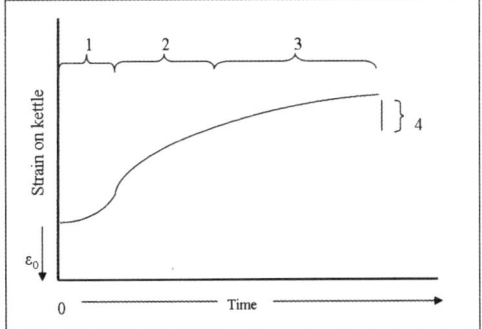

Schematic of strain on lead kettle during one thermal cycle
0: The instantaneous elastic strain caused by loading the kettle
1: The strain caused by heating the water in the kettle, and thereby the kettle
2: Steady strain caused by holding the water, and thereby the kettle, to temperature. This stage is assumed to be 'secondary creep'.
3: Although the effect of temperature reduces as the kettle cools, the strain still increases as the kettle is still under stress.
4: Instantaneous reduction of strain as the water is removed from the kettle. However, as the lead kettle is inelastic the effect of prior strain has not reduced.

While it was believed that the dyeing apparatus would have been heated to a temperature in excess of the melting point of lead, this could not be confirmed without re-running the experiment with thermocouples positioned within and around the apparatus. This was experiment four. It was discovered that the fire at the base of the kettle reached 600°C. Lead has a melting temperature of 327°C. Although the water within the kettle would act as a heat-sink, the sheer weight of the lead and water should have been detrimental to the structure of the kettle. This was puzzling, as the remaining kettles in Pompeii appeared to be intact.

While one door shut, another door opened. Although it was not possible to physically reconstruct a replica apparatus from the original materials, it was possible to construct a virtual replica. Since 1982, Finite Element Analysis has been used to model and test items before their manufacture (Fagan, 1992). This method has been consistently proven dependable in even the most safety-critical of applications, including the testing of aircraft wings. The material to be tested is mathematically divided into discrete parts, 'finite elements', and each is represented by an equation. When an external force or temperature gradient is applied within the computer simulation, the equations calculate its effect on each discrete element throughout the virtual material. It was possible to construct a virtual replica of the dyeing apparatus and then to apply different temperatures and loads over time.

The model was written in the programme ABAQUS. Although previous studies of lead creep had been published, the data presented was unsuitable for use in this model. Empirical data from an earlier study (Sahota and Riddington, 2000) was extended and validated through a virtual simulation. The design, dimensions and materials for the complete virtual apparatus were taken from the dyeing apparatus in Pompeii. The virtual model was constructed (Fig. 2.9). There were three ways in which the dyeing apparatus was subject to loading:

- **Self-weight:** the weight of each component of the apparatus itself;
- **Hydrostatic weight:** the weight of water and the increased pressure of this water by depth;
- **Thermal loading:** the repeated heating and cooling of the apparatus.

The times, temperatures and loads were taken from the replica apparatus. The hydrostatic pressure of 90 litres of water was inputted. Averages of time, temperature and load had to be defined (Fig. 2.10). The damage caused by repeated strain would accumulate, ultimately causing the failure of the kettle. Additional advantages of using a computer simulation of a material include that the external environment can be controlled, the dimensions and materials can be substituted and that it is possible to slow down and speed up time. This

FIG. 2.9
Deconstruction of the apparatus to allow virtual modelling. Line represents cross-section.
(Heather Hopkins)

Metal side, mesh increasing in size from 1mm x 1mm at base to 1mm x 10mm at top

Hydrostatic pressure representing water

Resting interface

Resting interface, tied at points

Base and side coded as one piece

Metal base

Rigid body representing Brick support

Rigid body representing the floor

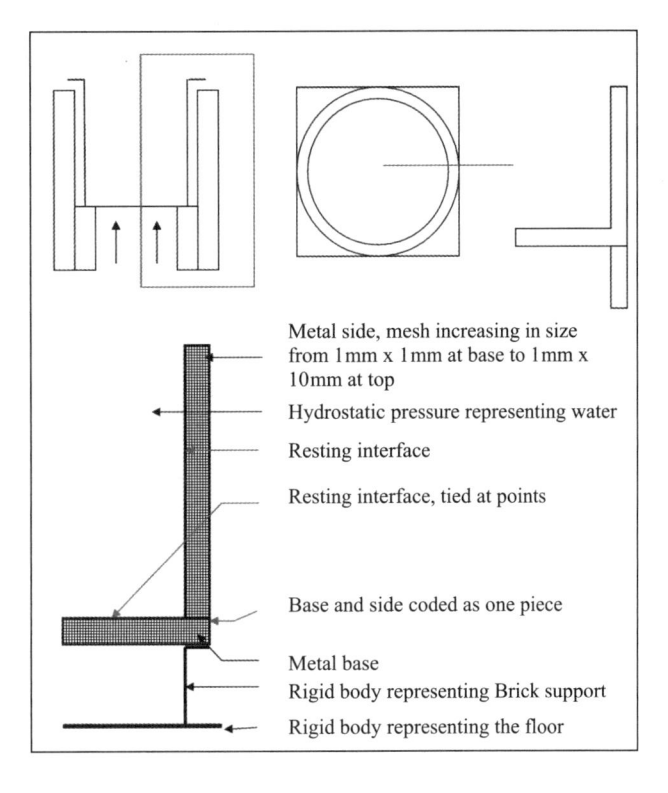

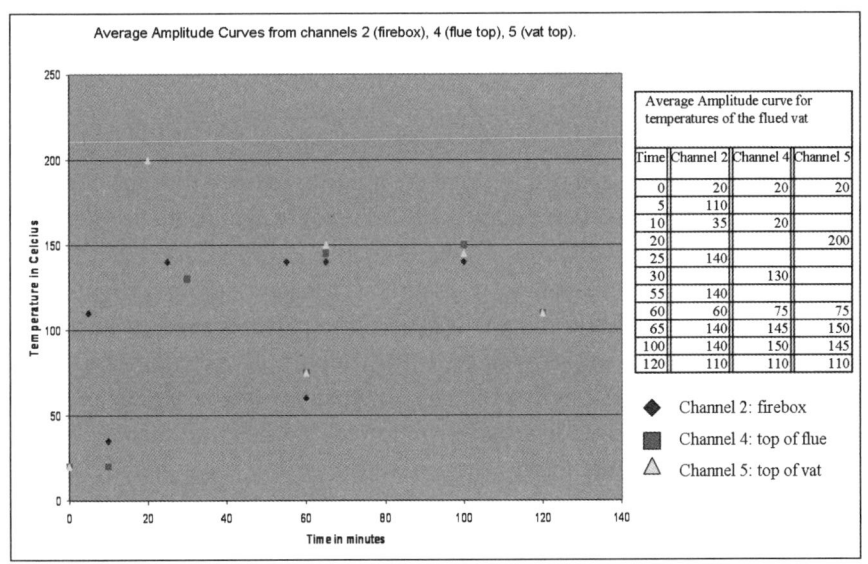

Average Amplitude Curves from channels 2 (firebox), 4 (flue top), 5 (vat top).

Average Amplitude curve for temperatures of the flued vat

Time	Channel 2	Channel 4	Channel 5
0	20	20	20
5	110		
10	35	20	
20			200
25	140		
30		130	
55	140		
60	60	75	75
65	140	145	150
100	140	150	145
120	110	110	110

♦ Channel 2: firebox
■ Channel 4: top of flue
△ Channel 5: top of vat

FIG. 2.10 Defining a temperature average to allow virtual modelling. Temperature variations of the firebox, flue top and vat top throughout the dyeing cycle. Note: The fluctuations in temperature were caused by the use of a real fire and a firebox of a limited size. At the start of the dyeing cycle the fuel was ignited. The peak temperature at 20 minutes followed the fuel igniting completely and the heat transferring through the apparatus. After 60 minutes this fuel had to be replenished. When fuel was added, there was a temporary drop in temperature before the new fuel ignited completely.
(Heather Hopkins)

means that it is possible to observe years of use of an apparatus in seconds, then to slow down and study the point at which it breaks. Through this it is possible to observe the mechanism of failure as a material fails, instead of watching a physical replica fail instantaneously and then trying to reconstruct the process from the pieces. This also has a health and safety advantage: it is possible to see in advance in what time a material is likely to fail and prevent an apparatus from reaching this point.

Results

The apparatus showed no signs of strain when under self-weight (*see* Fig. 2, art section). When the weight of the water was added to the model, the kettle showed signs of strain. Following heating to 95°C, the strain could be seen to have greatly increased. At the end of the first dyeing cycle the strain had increased further and the dyeing kettle had visibly warped. It should be noted that although the lead had 'stretched' – it had been the subject of creep – it had not actually cracked or broken. Also the lead had not warped to such an extent that it had collapsed and it had not yet reached the floor of the firebox. Although the fireboxes were of differing heights, the model replicated a low-ceilinged firebox and so the gap below the lead is significant. It has been suggested that it would have been possible at this point to amend the shape of the lead kettle to allow further dyeing. This would probably have been necessary, as it would have been difficult to build up a fire and to achieve complete combustion with the lead blocking the firebox (Hopkins, 2007). If the lead was left unaltered, the kettle would have continued to be subject to creep and would have reached the base of the firebox by the second dyeing cycle. The intriguing and unexpected result is that the model suggested that although the lead was subject to great stress and showed great strain, it did not actually crack or break. Even when the dyeing cycle was repeated for 160 thermal cycles, the equivalent of six months use, the kettle base did not progress any further down than it had after the second cycle: instead, the kettle thinned slightly at the edge in contact with the kettle support. This demonstrated that the lead was not as fragile as first thought and that it could have been used as a practical kettle. However, it probably required maintenance throughout its working life. When the kettles in Pompeii were reviewed, it was noted that one was the shape expected from this pattern of creep (Fig. 2.11). Other kettles had supports below the base and their creep may have been arrested. It was concluded that the use of lead to manufacture dyeing kettles would not have impacted the output of the dyeing workshops to an extent that it would have limited the overall production of the dyeing industry. The maximum output calculated was still valid.

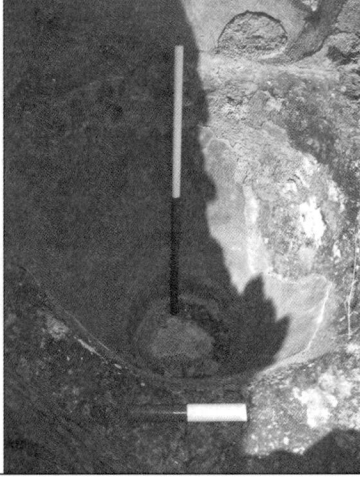

FIG. 2.11
Apparatus six in property VII xiv 17 showing creep.
Scale: 1m and 0.2m.
(Heather Hopkins)

In context

The results alone do not place this work in context, even when the population size of Pompeii is considered. The dyeing industry was part of a manufacturing system, operating within the system of a city (Fig. 2.12). It was influenced by economic, physical and social considerations. When undertaking this study it has been assumed that the industry was operating at maximum capacity. The assumption has been that the inputs and consumables used were inexhaustible, that there was not a shortage, and that these were not a limiting factor. This was a necessary assumption to allow a calculation of the maximum output, but the reality could have been very different: it is possible that the inputs or consumables were limiting factors, or that an intangible economic, social or political factor imposed limitations on the dyeing industry. It must be remembered that in archaeological reconstruction it is possible that elements may be missing and irrecoverable. It is not known at what rate textile was re-dyed rather than dyed as new. It is not known who owned the textile at each stage of production; this would influence the quality control and the possibility of re-dyeing. Intangible elements such as fashion may never be fully understood. It should also be remembered that one-third of Pompeii remains unexcavated and parts of this are situated below the modern town of Pompei. It is possible that further excavation may help to answer these questions and provide a greater insight into the industry, but also that this excavation may never take place.

The Romans had a complex economic system and the dyeing industry would have been a part of this. It has been noted that although there were no Roman words for many of the modern economic terms, they had developed an empirical understanding of the concepts (Edmundson, 1989). The Roman

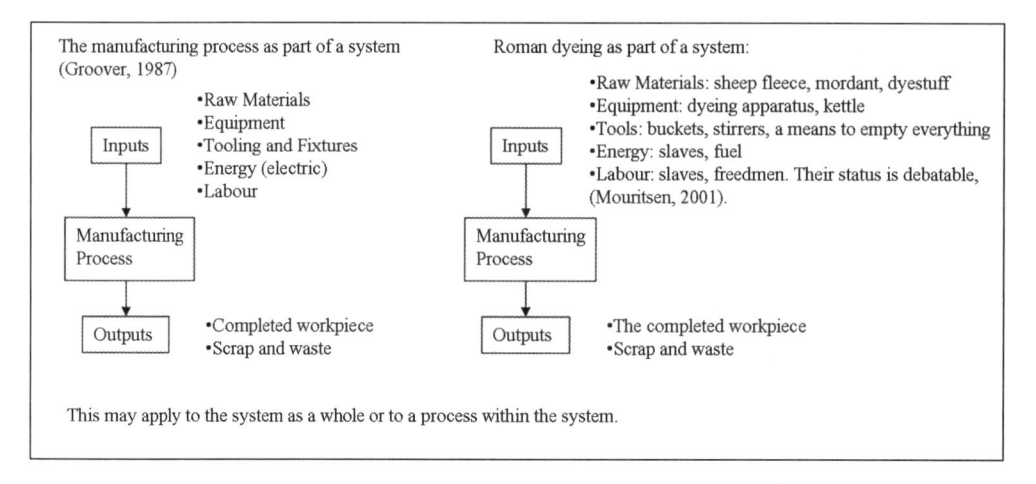

> **The manufacturing process as part of a system** (Groover, 1987)
>
> Inputs:
> - Raw Materials
> - Equipment
> - Tooling and Fixtures
> - Energy (electric)
> - Labour
>
> Inputs → Manufacturing Process → Outputs
>
> Outputs:
> - Completed workpiece
> - Scrap and waste
>
> **Roman dyeing as part of a system:**
>
> Inputs:
> - Raw Materials: sheep fleece, mordant, dyestuff
> - Equipment: dyeing apparatus, kettle
> - Tools: buckets, stirrers, a means to empty everything
> - Energy: slaves, fuel
> - Labour: slaves, freedmen. Their status is debatable, (Mouritsen, 2001).
>
> Inputs → Manufacturing Process → Outputs
>
> Outputs:
> - The completed workpiece
> - Scrap and waste
>
> This may apply to the system as a whole or to a process within the system.

FIG. 2.12 Roman manufacturing as part of a system.
(Heather Hopkins)

Empire was the most bureaucratic seen before the modern day, and there are records showing the development of a complex economic and legal system (Temin, 2001). It must be noted that the Romans operated a 'cottage industry': their manufacture was pre-industrial and before the advent of factories, but that does not mean that it was inefficient. Larger scale workshops of the cottage industry hosted both the greatest volume of manufacturing seen up until that point and the greatest amount that was believed to be possible. The Romans were effective, efficient engineers (Hodge, 1992): if a dyer believed that he could have cost-effectively increased the scale of manufacture to outcompete his rivals, he would have done so.

This study has shown that there had also been an empirical development of some of the theories of modern manufacture: a development of modern industrial ideas in a pre-industrial world. The U-shape formed by the placing of dyeing apparatus in the majority of the workshops in Pompeii was separately developed by Toyota in 1985 (Japan Management Association, 1985; Monden, 1993). This was emulated by other manufacturers due to the advantages of this layout. Although it can take two dyers to start a dyeing cycle, during the cycle the apparatus requires minimal attendance. The layout allows all of the apparatus to be simultaneously monitored by a single person who can be undertaking a second task. The layout means that no apparatus or task is isolated. Toyota took this process further and joined up the U-shapes to allow the efficient flow between manufacturing areas and to allow a single overseer to view different tasks simultaneously. In Pompeii there was not the luxury of space that would have allowed a series of workshops to be combined like this, as the dyeing workshops were constructed in what remained of the city following an earthquake. Constraints on location could have been intangible, such as legal or economic reasons, or tangible such as the limits of space and storage or water supply. The Roman dyers would have had to fit the location and layout of their dye works within the pre-existing physical and socio-economic fabric of the city (Hoffman, 1967:185).

The process of textile manufacture is linear (*see* Fig. 2.1). This process was horizontally divided into task-specific workshops. This may have been due to ownership and specialisation. It is also possible that zoning constraints may have existed. A positive zoning constraint would be the construction and use of all dyeing apparatus together in dyeing workshops, to allow the pooling of resources for the use of such specialist equipment. A negative zoning constraint could be the isolation of dyeing so as not to cause ruination to any finished fabrics. Within the dye workshop itself, processing was also linear. Each fleece had to undergo a number of steps toward dyeing, otherwise each subsequent step could not have been undertaken. These are called 'precedence constraints'. For example, a fleece had to be pre-mordanted before dyeing, otherwise the dye would not bind to the fleece and the colour would not be fast (Grierson, 1986; Grieve, 1992). These steps were closely related – the pre-mordanted fleece had to be still damp with mordant when it was placed within the dyeing apparatus and each step of the dyeing recipe had to be carefully adhered to. It was therefore necessary to plan the processing of each fleece in advance to ensure that the entire procedure could be undertaken without interruption in one step. At present it is not possible to determine whether scouring (cleaning the fleece) and pre-mordanting took place within the dye works: such information is an intangible endemic entity, but it could be suggested that it did.

It is important to distinguish that within Roman manufacturing there was the system and production line within a workshop, and the system and production line of the manufacture as a whole. It may be argued that the Romans horizontally divided the manufacturing process, but across a city, not a factory. The situating of specific apparatus together allowed the Romans empirically to develop both the horizontal division of process and line balancing, which meant that the dyeing industry was possibly the most efficient that it could ever be.

In a linear process, a product is subject to precedent constraints. If each part of the process of manufacture takes a differing amount of time, it is possible that a delay could result in the later part of the process. Line balancing is the arrangement of workstations and processes to prevent or lessen these delays. If a section is not balanced, the slowest part of production would dictate the production rate overall. Each workstation works independently of other parts within the system, and only depends on the completion of preceding items. If a preceding item can be completed within half the time of a subsequent process, it is possible to complete twice as many and then split them between two parallel subsequent workstations (Groover, 1987; Muhlemann et. al., 1992; Hopp and Spearman, 2001). In the Roman method of production, both parallel processing and storage buffers were used in line balancing. This meant simultaneous processing using independent apparatus and the storage of excess produced until it could be used. This would mean that the only limiting factors would be the number of previously completed items and a delay on the provision of consumables. This could be of significance in an industry such as textile processing, where consumables such as fleece were seasonal. It is also possible to stagger the processing of the items, so that while one apparatus is reaching the end of a dyeing cycle another may be just beginning to heat. This meant that theoretically it would have been possible to keep a set of dyeing apparatus operational while only needing a single person to attend them: if ten apparatuses each required ten minutes of attention at the start and finish of the dyeing process, they could be kept in cycle so that there was never a break in use.

The Romans used an (automated) flow-line in production and used 'asynchronous transfer' to move the items from one workstation to another as each process was completed (Groover, 1987). These terms do not mean that machinery was involved: in this context it means that the items were moved sequentially by hand through the linear process. Cycle rates are usually slower in systems that use asynchronous transfer than in other systems (Groover, 1987; Hopp and Spearman, 2001). This is not important in the context of the dyeing industry in the pre-industrial world, as the other more modern systems were unavailable. The design of the workshops allowed for division of processing according to tasks and the limiting of distance and time between stages. It was also possible to integrate a specialism without disrupting the flow of the system.

The method of manufacture was either Job Shop production, where the products are manufactured individually or in small quantities by highly skilled craftsman, or Batch production, where a number of identical products are made and replenished as they are depleted through customer demand (after Groover, 1987; Muhlemann et al., 1992). Batch production is the more cost-effective. The method of production used in Pompeii would have been determined by the rate of demand and by the rate at which customers could place a specific order for products that the dyeing workshop had not previously manufactured. The ownership of a product would influence the inspection stage and the associated feedback loop at each stage of manufacture. The final inspection would have been either before sale or return to the client. If a client owned an item that needed to be re-dyed and the dyeing did not work, it may be possible that the dyers had another attempt. If the item was due for sale, it may be possible that a mis-matched colour was deemed to be natural variation and that the item was put on sale anyway. A failure in a preceding stage may have been removed from the system, amended or allowed to continue, again with the argument of natural variation. Unless records are subsequently unearthed, these are intangible demands. The demand for textiles is unknown, so even if it were possible to determine how much could be produced, the consumption rate would still be indeterminable.

Task division allows specialisation. 'By giving each worker a limited set of tasks to do repeatedly, the worker becomes a specialist in those tasks and is able to perform them more quickly and more consistently' (Groover, 1987). In the textile manufacturing industry this would mean that dyers became extremely specialised. It can be argued that this was necessary as dyeing requires precise chemistry, and these dyers had to dye without the aid of thermometers or pH meters. It is possible to argue that this would have meant dyers were accorded a higher social position despite a great number starting as slaves. It may be supposed that there must have been some form of variation of task, otherwise the records of lead poisoning (written by Vitruvius) would have mentioned the dyers as being at high risk. The social and legal status of dyers is a complex issue. For a fuller examination please refer to Hopkins (2007).

The horizontal division of the manufacturing process could be argued as evidence that the 'industry' was pre-industrial, as the tasks were not physically linked in a flowing process of a type used in assembly lines of the modern era. However, the physical evidence of the location of the apparatus shows that there was already an elimination of duplication, a specialism of task and the likelihood of line balancing which argues in favour of the manufacturing process showing the beginning of industrialisation as recognised today. It may be argued that the industry as a whole had vertical integration between the horizontal divisions, as all of the processes involved in the conversion of raw materials into finished products were present in the city of Pompeii. It is not possible to state the economic or social relationships between each part of the industry as a whole, or the ownership at each stage of production, as these are intangible endemic entities (Clarke, 1978). It is possible to define the physical places, the workshops, in which the processes took place and thus note that the industry had been physically divided by task.

Downtime due to breakages would not have been hugely problematic if the dyeing apparatus were each working independently. The greater limiting factors are likely to have been the availability of the consumables and the customer demand. The source of fleece was seasonal, so a steady supply of fleece to a dye-works, and to each subsequent production process, depended upon sufficient storage and transport before dyeing and between each stage. This study has determined a theoretical maximum, but it is not yet possible to know how often all of the dyeing apparatuses were in use at once and how often supply of consumables or customer demand were limiting factors.

Experiment in context

This study was interdisciplinary and its significance goes beyond simply determining the size of the dyeing industry in Pompeii. Consideration must also be given to where this study is placed in the evolution of experimental archaeology, and to how its approach has added to each of the subjects upon which it draws. This study has incorporated disciplines that were not previously applied to archaeology, or classics, and has combined others in a way not seen before.

Experimental archaeology is a branch of archaeological science dating back to at least the 1960s, although experiments involving replication of archaeological artefacts were undertaken as early as the 19th century (Coles, 1973). Any experiment is undertaken in the paradigm of the subject, that is the framework of thought and theory that underlies the understanding of a subject. Since the 1960s there has been a marked paradigm shift in experimental archaeology. The original experiments undertaken by Coles and Heyerdahl, amongst others, were true experiments (Heyerdahl, 1971; Coles, 1973). They constructed full replicas and tested the capabilities and limitations of archaeological artefacts. They undertook experiments at the macro (gross) level. These experiments were allowed to fail, as it was through the method of failure that an artefact could be understood.

The most notable of these was the *Ra I*, Heyerdahl's replica ship that crossed the Atlantic. It was discovered that the original design had not been faithfully copied when the stern of the boat disintegrated 700km from the American coast. Repairs were abandoned due to shark infestation of the waters. The *Ra II* had a faithful design and crossed the Atlantic safely (Heyerdahl, 1971). At the start of this study, the nearest comparable replica was the *Matthew*, a copy of Cabot's original ship that crossed the Atlantic in AD 1497. Although this was publicised as a true replica, it was admitted on the *Matthew*'s own website that not only was it equipped with life-rafts and satellite navigation (to meet modern maritime regulations as it sailed alone), but that it also had an engine and fuel onboard. Although it can be argued that as times change new regulations have to be met, it must be asked how accurate a replica a ship can be if space and weight is taken up by such modern inventions. The reason for this change was that experiments, especially on the large scale, have increasingly moved from true experiments that test a replica and are allowed to fail, to experiences which involve a replica. At the end of this study a true experiment was launched that redressed this balance: the *Sea Stallion*, a replica Viking ship that successfully sailed from Roskilde to Dublin (*see* this volume). Like the *Ra I*, this had a support ship and so could be a true replica. That people were removed with hypothermia also supports its status as a true experiment.

Prior to this study no replica experimental work had been undertaken to understand the capabilities and limitations of a dyeing apparatus in Pompeii. The understanding of the output of the dyeing industry had come from calculations based upon theoretical assumptions applied to superficial measurements of the apparatus. This study not only constructed a replica, but used it as part of an experiment. The viability of the design was tested. The operating parameters were explored and defined. The consumables and cycle time were determined. These experiments not only provided a foundation for the understanding of dyeing in Pompeii, but the results can be transferred to inform other similar archaeological problems. Without experiments, theories cannot be tested, problems cannot be explored and answers cannot be gained.

The paradigm within which a discipline is based reveals itself in the approach to language used in the questions that each discipline asks. When an archaeologist asks how an artefact 'works' or is operated, they usually mean at the macro, gross level: how the whole artefact reacts. Very few archaeologists have attempted to explain why something happens in a reconstruction: the majority have reported the results of the experiments and then not explained the science behind what happened (for example Coles, 1973). Others did attempt to explain the physics, but did not go far enough (for example Speth, 1977; Cotterell and Kamminga, 1990, cited in Hopkins, 2007). When an engineer asks the same question, they may mean how an artefact 'works' at the micro level, how the materials that make up the artefact would react, break or how a force or change would propagate through it during its use. This simple difference in the definition of the word 'how' means a completely different approach to the question and so a different answer. Although either answer is accurate, neither is complete without the other. It was necessary therefore to combine these disciplines to provide a triangulated, fuller answer. When it was demonstrated that it was not possible to replicate a dyeing apparatus accurately without using a lead kettle, an understanding of the properties of materials meant that it was possible first to substitute a stainless steel kettle and then to construct a virtual replica to simulate the creep of a lead kettle. The experiment continued, albeit in an entirely different direction from that first envisaged and now provides answers to previously 'unanswerable' questions.

Each change or addition to the experiment was grounded in every discipline involved and now withstands the examination of each. Each stage of the vat experiment has been carried out separately, reviewed separately, and the mathematics have been determined separately. This has been to avoid circular arguments, or the application of a theory that fits neatly with one stage of the experiment to another stage, with the expectation that it will also fit. Each part of the experiment has been determined and confirmed through experimental, literary and theoretical engineering means, before moving on to the next stage.

The experiment was not allowed to become limited by preconception – as can be seen by the use of charcoal in the firebox. It was expected that the fuel used was charcoal and yet it was not possible to keep it alight, let alone achieve complete combustion.

This work differs also from previous projects in its sequential nature. It is possible to follow the understanding of the artefacts and the examination of their context from the original classical archaeological approach, through the different principles applied, the thermodynamics, the experimental replica, and ultimately to Finite Element Analysis of a virtual replica. This results in an understanding of both the artefacts and the principles applied. This means that what would otherwise be unknown principles to archaeologists are presented in a clear way with physical demonstration. It is then possible to take these principles and apply them to other artefacts. The ability to take the findings and new methodologies from the project, and use them in multiple disciplines to further understand each one, is a new development. The provision of a working example from an archaeological background through the application of a number of techniques more usually associated with engineering – in a way that makes each accessible to differing disciplines – is new. To focus not just on the original archaeological question, but also on the provision of explanation and examples that allow the transfer of techniques to other archaeological problems, is new.

Conclusion

This study provides an interdisciplinary approach to a previously 'unanswerable' question. It provides a maximum annual output for the dyeing industry of Pompeii and places this within the context of the requirements of the city's population. In so doing, it examined other factors that would have influenced the dyeing industry, such as the water supply, the fuel used and the role of the dyers themselves. The study involved disciplines and principles that were new to archaeology or new to this question, such as the application of ergonomics to an ancient population to test artefact use, or the application of Finite Element Analysis to an industrial artefact of more than one material. The study took each discipline back to first principles, showed a practical application of each, and travelled through each sequentially. This increased accessibility as it is possible to start at the known and work through the unknown. The triangulated approach means that the answers to each part of the study withstand the scrutiny of multiple disciplines. This study has answered a specific question regarding Roman dyeing, illustrated a new approach and provided a solid foundation for further work.

Acknowledgements

I would like to thank Dr Alexander Blustin for proofreading and assistance with editing.

Further reading

For a more detailed exploration of this work, please refer to Hopkins (2007). Please also refer to Hopkins, et al. (2005) and Watling (2004).

Bibliography

Allison, P. (1992): *The Distribution of Pompeiian House Contents and its Significance*, vols I and II (Sydney: University of Sydney).

Arthur, P. (1986): 'Problems of the Urbanization of Pompeii: Excavations 1980–1981', in *The Antiquities Journal*, 66 (1), pp. 29–44.

Bass, W. M. (1995): *Human Osteology: A Laboratory and Field Manual*, Special Publications No. 2, 4th edition (Springfield: Missouri Archaeological Society).

Bender Jorgensen, L. and U. Mannering (2001): 'Mons Claudianus: Investigating Roman textiles in the desert', in *The Roman Textile Industry and its Influence. A Birthday tribute to John Peter Wild* (P. Walton Rogers, L. Bender Jorgensen and A. Rast-Eicher, eds) (Oxford: Oxbow Books), pp. 1–11.

Brunello, F. (1973): *The Art of Dyeing in the History of Mankind* (Vicenza: Neri Pozza Editore).

Callister, W. D. (2000): *Materials Science and Engineering: An Introduction* (New York: John Wiley & Sons, Inc.).

Capasso, L. (2001): *I Fuggiaschi di Ercolano: Palaeobiologia delle Vittime Dell'eruzione Vesuviana del 79 d.C.* (Roma: L'Erma di Bretschneider).

Cardon, D. (2001): 'On the road to Berenike: a piece of tunic in damask weave from Didymoi', in *The Roman Textile Industry and its Influence. A Birthday Tribute to John Peter Wild* (P. Walton Rogers, L. Bender Jorgensen and A. Rast-Eicher, eds) (Oxford: Oxbow Books), pp. 12–20.

Clarke, D. L. (1978): *Analytical Archaeology*, 2nd edition (London: Methuen & Co. Ltd).

Coles, J. (1973): *Archaeology by Experiment* (London: Hutchinson University Library).

Cotterell, B. and J. Kamminga (1990): *Mechanics of Pre-Industrial Technology* (Cambridge: Cambridge University Press).

Crumlin-Pedersen, O. (1999): 'Experimental Ship Archaeology in Denmark', in *Experiment and Design Archaeological Studies in Honour of John Coles* (A. F. Harding, ed.) (Oxford: Oxbow Books).

Dalby, G. (1985): *Natural Dyes, Fast or Fugitive* (Mineheaed: Ashill Publications).

DeHaan, N. (1997): 'Water Use in Private Baths at Pompeii. The 98th Annual Meeting of the Archaeological Institute of America', in *American Journal of Archaeology*, 101, pp. 355–56.

Dreyfuss, H. (1967): *The Measurement of Man: Human Factors in Design* (New York: Whitney Library of Design).

Edmonds, J. (1999): pers. comm., Chiltern Open Air Museum.

Edmondson, J. C. (1989): 'Mining in the later Roman Empire and Beyond: Continuity or Disruption?', in *Journal of Roman Studies*.

Fagan, M. J. (1992): *Finite Element Analysis: Theory and Practice* (Harlow: Longman Scientific & Technical).

Greenfield, P. (1972): *Creep of Metals at High Temperatures* (Mechanics Engineering Monograph) (London: Mills and Boon).

Grierson, S. (1986): *The Colour Cauldron: History and Use of Natural Dyes in Scotland* (Perth: Mill Books).

Grieve, M. (1992): *A Modern Herbal* (Kent: Mayers of Chatham).

Grose-Hodge, H. (1944): *Roman Panorama*

(Cambridge: Cambridge University Press).

Groover, M. P. (1987): *Automation, Production Systems, and Computer-Integrated Manufacturing* (Upper Saddle River, NJ: Prentice-Hall).

Heyerdahl, T. (1971): *The Ra Expeditions* (New York: Doubleday), cited in J. Coles (1973).

Hodge, A. T. (1981): 'Vitruvius, Lead Pipes and Lead Poisoning', in *American Journal of Archaeology*, 85, pp. 486–91.

Hodge, A. Trevor (1992): *Roman Aqueducts and Water Supply* (London: Duckworth).

Hoffman, T. R. (1967): *Production: Management and Manufacturing Systems* (Belmont, CA: Wadsworth Publishing Co.).

Hopkins, H. (2002): 'Industrial dyeing processes in Pompeii – Practical research into their methodology and importance', undergraduate dissertation, Department of Archaeological Sciences, University of Bradford.

Hopkins, H. J. (2007): 'An investigation of the parameters that would influence the scale of the dyeing industry in Pompeii. An application of experimental archaeology and computer simulation techniques to investigate the scale of manufacture of the dyeing industry and the factors that influence output', PhD thesis, University of Bradford.

Hopkins, H., L. Willimott, R. Janaway, D. Robinson and W. Seale (2005): 'Understanding the economic influence of the dyeing industry in Pompeii through the application of experimental archaeology and thermodynamics', in *Scientific Analysis of Ancient and Historic Textiles, Informing Preservation, Display and interpretation* (R. C. Janaway and P. Wyeth, eds) (London: Archetype Publications).

Hopp, W. J. and M. L. Spearman (2001): *Factory Physics Foundations of Manufacturing Management*, 2nd edition (New York: Irwin McGraw-Hill).

Janaway, R. C. and D. J. Robinson (1994): unpublished, Lecturer and Postgraduate in Archaeology, University of Bradford.

Japan Management Association (eds) (1986): *Kanban Just-In-Time at Toyota. Management Begins at the Workplace* (trans. D. J. Lu) (Portland, OR: Productivity Press).

Jongman, W. (1988): *The Economy and Society of Pompeii* (Amsterdam: J. C. Gieben).

Lagercrantz, O. (1913): *Papyrus Graecus Holmiensis Recepte für Silbur, Steine und Purpur* (Uppsala, A.-B. akademiska bokhandeln).

Laurence, R. (1994): *Roman Pompeii Space and Society* (London: Routledge).

Mathieu, J. R. (2002): *Experimental Archaeology Replicating Past Objects, Behaviours, and Processes*, British Archaeology Reports, International Series, 1035 (Oxford: Archaeopress).

Moeller, W. (1976): *The wool trade of ancient Pompeii* (Leiden: E. J. Bril).

Monaghan, M. D. (2001): 'Coats of Many Colours: Dyeing and Dyeworks in Classical and Hellenistic Greece', PhD thesis, School of Archaeology and Ancient History, University of Leicester.

Monden, Y. (1993): *Toyota Production Systems. An Integrated Approach to Just-in-Time*, 2nd edition (London: Chapman and Hall).

Monteix, N. and M. Pernot (2005): pers. comm., 'Some features of the lead craftsmanship in Pompeii and Herculaneum', poster presented at 2005 conference, 'Summary of work to date undertaken at Centre Jean Bérard, Naples and Centre Camille Jullian, Aix de Provence'.

Mouritsen, H. (2001): 'Roman Freedmen and the Urban Economy: Pompeii in the first century AD' in *Pompei tra Sorrento e Sarno: Atti del terzo e quarto ciclo di conferenze di geologia, storia e archeologia, Pompei, gennaio 1999–maggio 2000* (F. Senatore, ed.) (Roma: Bardi Editore).

Muhlemann, A., J. Oakland and K. Lockyer (1992): *Production and Operations Management*, 6th edition (London: Pitman).

Murrel, R. F. H. (1957): 'Data on Human Performance for Engineering Designers', reprint. from *Periodical Engineering* (London: Newgate Press).

Peacock, D. (1989): 'The Mills of Pompeii', *Antiquity*, 63, pp. 205–14.

Pliny the Elder (1958–62): *Natural History*, 37 books in ten volumes (trans. H. Rackham, et al.) (London: Heinemann and Cambridge, MA: Harvard University Press).

Pliny the Younger (1969): *Fifty letters of Pliny, selected and edited by A. N. Sherwin-White*, 2nd edition (trans. B. Radice) (Oxford: Oxford University Press).

Pritchard, F. and C. Verhecken-Lammens (2001): 'Two wide-sleeved linen tunics from Roman Egypt', in *The Roman Textile Industry and its Influence. A Birthday tribute to John Peter Wild* (P. Walton Rogers, L. Bender Jorgensen and A. Rast-Eicher, eds) (Oxford: Oxbow Books), pp. 21–29.

Reynolds, P. (1999): 'The nature of experiment in archaeology', in *Experiment and Design. Archaeological Studies in Honour of John Coles* (A. F. Harding, ed.) (Oxford: Oxbow Books), pp. 156–62.

Rosetti, G. (1548): *The Plictho of Gioanventura Rosetti: Instructions in the art of the Dyers which Teaches the Dyeing of Woollen Cloths, Linens, Cottons, and Silk by the Great Art as Well as by the Common* (trans. S. M. Edelstein and H. C. Borghetty, 1969) (Cambridge, MA: MIT Press).

Ryder, M. L. (1990): 'The Natural Pigmentation of Animal Textile Fibres', in *Textile History*, 21(2), pp. 135–48.

Ryder and Stephenson (1968): *Wool Growth* (London: Academic Press).

Sahota M. K. and J. R. Riddington (2000): 'Compressive Creep Properties of Lead Alloys', in *Materials & Design*, 21(3), pp. 159–67.

Speth. J. D. (1977): 'Experimental Investigations of Hard-Hammer Percussion Flakes,' in *Experimental Archeology* (D. Ingersoll, J. E. Yellen and W. Macdonald, eds) (New York: Columbia University Press).

Storey, G. (1997): The Population of Ancient Rome, in *Antiquity*, 71, pp. 966–78.

Storey, J. (1978): *Dyes and Fabrics* (London: Thames and Hudson).

Temin, P. (2001): 'A Market Economy in the Early Roman Empire', in *Journal of Roman Studies*, 91, pp. 169–81.

Trotter, M. and G. C. Gleser (1958): 'A pre-evaluation of estimation of stature based on measurements of stature taken during life and of long bones after death', in *American Journal of Physical Anthropology*, 16(1), pp. 79–123, cited in W. M. Bass (1995): *Human Osteology: A Laboratory and Field Manual*, Special Publications No. 2, 4th edition (Springfield: Missouri Archaeological Society).

Vitruvius, P. (1960): *Vitruvius, The Ten Books on Architecture* (trans. M. H. Morgan) (New York: Dover Publications Ltd).

Watling, L. (2004): 'Fuel Consumption of Roman Dye Vats', undergraduate dissertation, University of Bradford.

Wild, J. Peter (1977): '"The Wool Trade of Ancient Pompeii" Review of Moeller, 1976', in *Textile History*, vol. 8, pp. 180–81.

Wilson, A. (2007): unpublished, Professor of Archaeology, Institute of Classics, University of Oxford.

FRASER HUNTER

The sound of Iron Age music

RECONSTRUCTING THE DESKFORD CARNYX

Introduction

ONE OF THE treasures of National Museums Scotland is the 2000-year old head of a carnyx from Deskford in north-east Scotland, found in a peat bog around 1816 (Fig. 3.1). A carnyx is an Iron Age animal-headed trumpet made from sheet bronze. Once widespread across Europe, only a handful of examples survive today, and Deskford is one of the finest. A reconstruction project in the 1990s attempted to reconstruct the carnyx, and is described here; full details can be found elsewhere (Creed, 2000; Hunter, 2000; Hunter, forthcoming; Kenny, 2000; Campbell and MacGillivray, 2000).

The project was the inspiration of John Purser, conceived in the course of creating his monumental radio series and accompanying book on Scotland's music (Purser, 1992, 2007). He obtained funding through a Glenfiddich Living Scotland award, and approached National Museums Scotland to collaborate on the project by providing both additional funding and archaeological expertise. Purser's key interest was to evoke the sound of the past; from a museum perspective, this had the potential to bring one of our key objects to

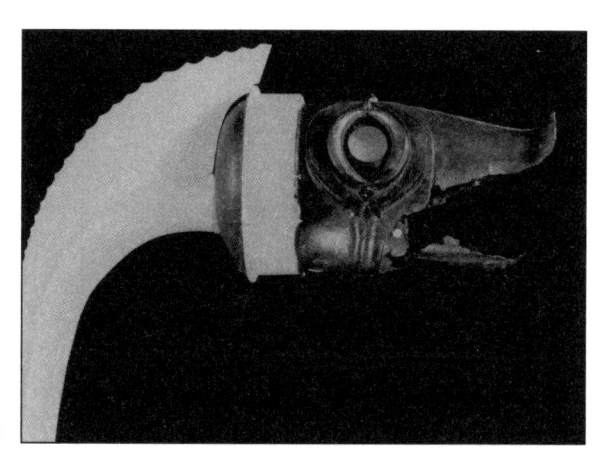

FIG. 3.1
The Deskford carnyx.
(© National Museums Scotland)

life (as a fragmentary object is always hard to understand) and gave the chance to engage in some sustained research into a sadly neglected field. The project team was completed by metalsmith John Creed and musician John Kenny; the fruitful exchange of ideas between the different specialisms was a crucial and highly stimulating part of the whole process.

Music-archaeology has a long, and not entirely honourable, tradition. It is a potentially dangerous area – literally so, in the case of Dr Robert Ball of Dublin, who in 1857 died of a burst blood vessel after attempting to blow an Irish late Bronze Age horn (Megaw, 1968: 347). But the dangers are intellectual as well; the instruments are generally incomplete (as in the current case), so the reconstruction is at best informed speculation, while even with intact instruments the actual music is wholly unknowable. These dangers are illustrated by another classic instance: the trumpets of Tutankhamen, played – with modern mouthpiece – by Bandsman James Tappern of the 11th Hussars in 1939 (Megaw, 1968: 346). The results, unsurprisingly, sounded not unlike a modern bugle, since this is what the musician was used to playing. Music is an entirely cultural construct, and we must guard against imposing our Renaissance-inspired ear on the sounds of other places and times.

The evidence – the object itself

The first port of call was the surviving fragments of the carnyx, a complex but very incomplete sheet metal construction in copper alloy (Fig. 3.2). Details are presented elsewhere, but the remaining portion gives considerable insights into the technology of the period (the late first to the early third century AD; Hunter, 2001, forthcoming). The key surviving components are a cylindrical head with repoussé decoration and two holes for the eyes, with a snout riveted to it, its upturned nose evoking that of a boar. This symbolism is carried onto the palate of the mouth, ribbed like a boar's. The rear of the palate was cylindrical and flanged to receive the tube; extensive repairs to this show that the instrument led a long and active life. An additional sheet riveted to the lower part of the head has two large holes which held the tapered lower jaw, the size of the holes suggesting the jaw was movable; this was confirmed by its back-projecting legs, angled to hold a counterbalance. Behind the head would have been a further cylindrical section with the ears attached to it; this was not recovered, but there was a saucer-like disc which formed the rear of the head, with a hole for the tube to pass through. The object had been deliberately dismantled before deposition in the peat bog. An account written some 20 years after its discovery records details now lost – notably 'a wooden tongue moveable by springs' (Innes, 1845: 66–67).

Scientific analysis threw further light on the find. The main joints had been

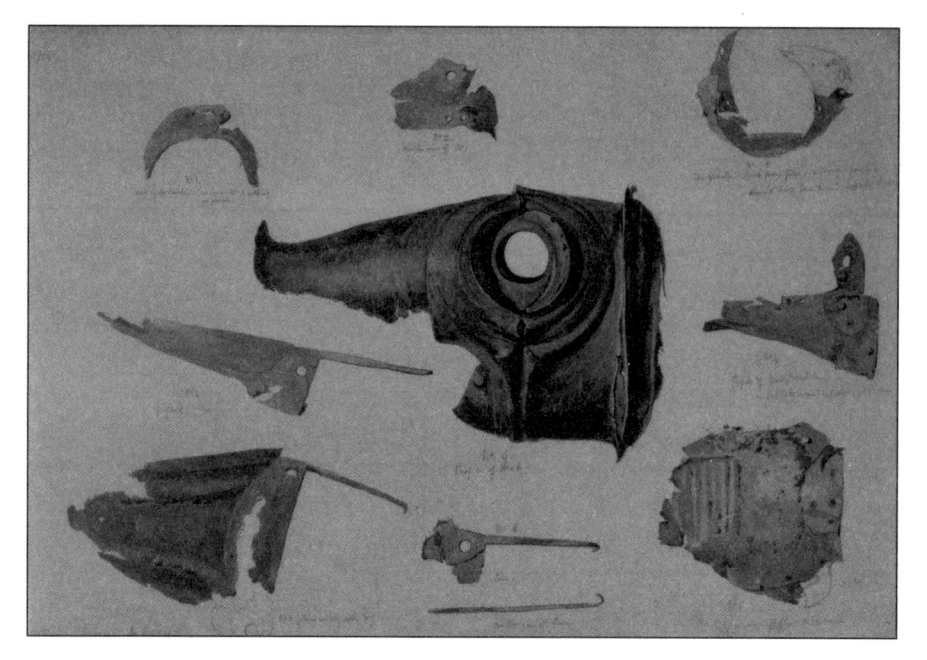

FIG. 3.2 Watercolour from c.1865–67 by Thomas Brown, showing the various components.

(© National Museums Scotland)

both riveted and soldered with a lead-tin solder, to ensure they were airtight. This analysis also showed that different alloys had been used for different components – bronze for the head, brass for the snout. When new, this would have created a striking colour contrast, and may also have had slightly differing acoustic effects. The results also help with the dating of the carnyx – the brass comes from recycling of Roman metal, showing that the instrument post-dates the first contact with Rome (Hunter, 2001: 78–80).

The evidence – parallels and analogies

Obviously much of the original instrument is missing in the case of Deskford, but other surviving examples and depictions provide analogies. The evidence is widely scattered, for such instruments were once widespread across Europe, and is of varying quality (Fig 3.3; see Piggott, 1959; Hunter, 2001).

The best evidence comes from other carnyx fragments. At the time of the reconstruction these were few and far between, but a major discovery at Tintignac in central France has both provided much fresh evidence (this one find doubled the known total!) and allowed fragments to be recognised in other finds (Maniquet, 2008, 2009; Hunter, 2009). This discovery also emphasises the variety of different styles, as it contains three different carnyx technologies,

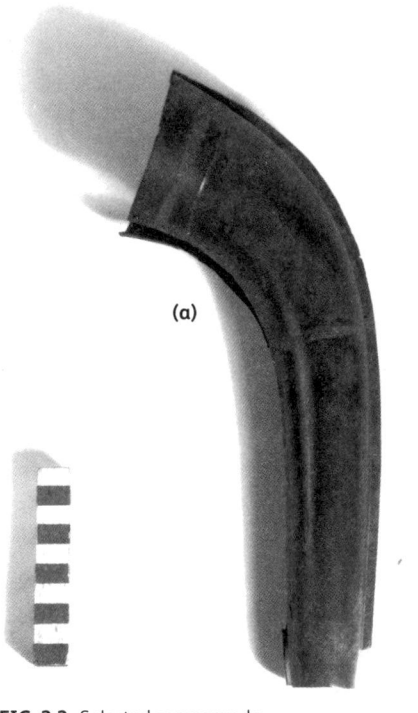

FIG. 3.3 Selected comparanda.

(a) Carnyx bell fragment, Kappel, Germany.
 (© Courtesy of Württembergisches Landesmuseum Stuttgart; Photograph by Fraser Hunter)

(b) Republican coins with carnyces.
 (© National Museums Scotland)

(c) Roman altar with carnyx and oval shield, Nîmes.
 (© Courtesy of Nîmes Museum; photograph by Fraser Hunter)

none particularly close to Deskford. This might be expected with such a widespread instrument, showing the development of local variants. The decoration on Deskford, for instance, is entirely in keeping with styles in its local area, and would look out of place elsewhere; we may assume the same is true of other aspects of the craftwork.

The other source of evidence is iconographic. Most reliable are depictions from groups who were actually using carnyces: they are illustrated, famously, on the silver cauldron from Gundestrup in Denmark, a product probably of south-east Europe where carnyces were in use, and are also found on Iron Age coinage in southern Britain and areas of France. Such depictions still require cautious use (for instance there are biases imposed by the small scale of many representations, while the artists were intending to create recognisable icons rather than technical drawings), but they carry considerable weight as they come from within cultures which used carnyces. More caution is required in using the few Greek and the abundant Roman depictions; these are almost universally in the context of propaganda pieces, commemorating the conquest of assorted tribes through display of their weaponry. As such, they were depicted as recognisable, visibly alien items, often exaggerated or misunderstood. Again, the medium imposed limitations; on coins, the knobs of the segmented tube are often emphasised, as these were easily conveyed in the limited available space, while in sculpture the gaping mouth is greatly exaggerated. Equally, for some sculptors these were clearly stock items taken from a pattern book, while others seem to have been drawing from life or personal experience; the information which can be gleaned from different depictions varies greatly.

The reconstruction

This broad data set informed the reconstruction effort, alongside the knowledge and specialities of the various participants. The archaeological evidence set a frame and a limit on the reconstruction, but there were many decisions and options where the surviving evidence was very incomplete, and here the various strands (archaeology, craft skills and musical potential) were considered together in reaching a decision. It must be remembered just how partial our knowledge is in the case of Deskford, where only parts of the head survive; the reconstruction is an exercise in informed speculation. For the head, the original was followed closely (including alloy choice), while clues to lost components were followed up. Thus the chamfer on the inside of the eye sockets suggests enamelled discs were once attached, a technique known in other metalwork from the area (e.g. Brailsford, 1975: 69–82), and enamelled eyes were restored in a style compatible with other finds. Ears were modelled on those of a wild boar, while a design for the nostrils was improvised from contemporary designs. For the 'wooden tongue moveable with springs', a leaf-spring solution was used, as coiled springs were apparently unknown in local metal-working traditions (Fig. 3.4). The construction of the bell drew on other examples, but it is the tube which is the greatest imponderable, and the greatest influence on the final sound. Clues to maximum possible diameter come from the rear disc, but length and construction are speculative, and draw on evidence from elsewhere. Both illustrations and examples indicate segmented tubes were typical, so bosses were used to join the three tube lengths together, modelled on those from Irish Iron Age trumpets as the nearest local parallels (Raftery, 1983; figures 202 and 204).

A further great imponderable was the form of the mouthpiece, another key element in the musical properties of the instrument. At the time no carnyx mouthpiece was known, and the musician John Kenny devised a range of possibilities based on other ancient examples (such as Roman ones) and his own experience. More contentious was the form of the tube at the base. In our reconstruction this was curved through almost 90°, based on our interpretation of the Gundestrup scene and musical arguments over player comfort. In fact, subsequent study made it clear that there are no convincing depictions of such

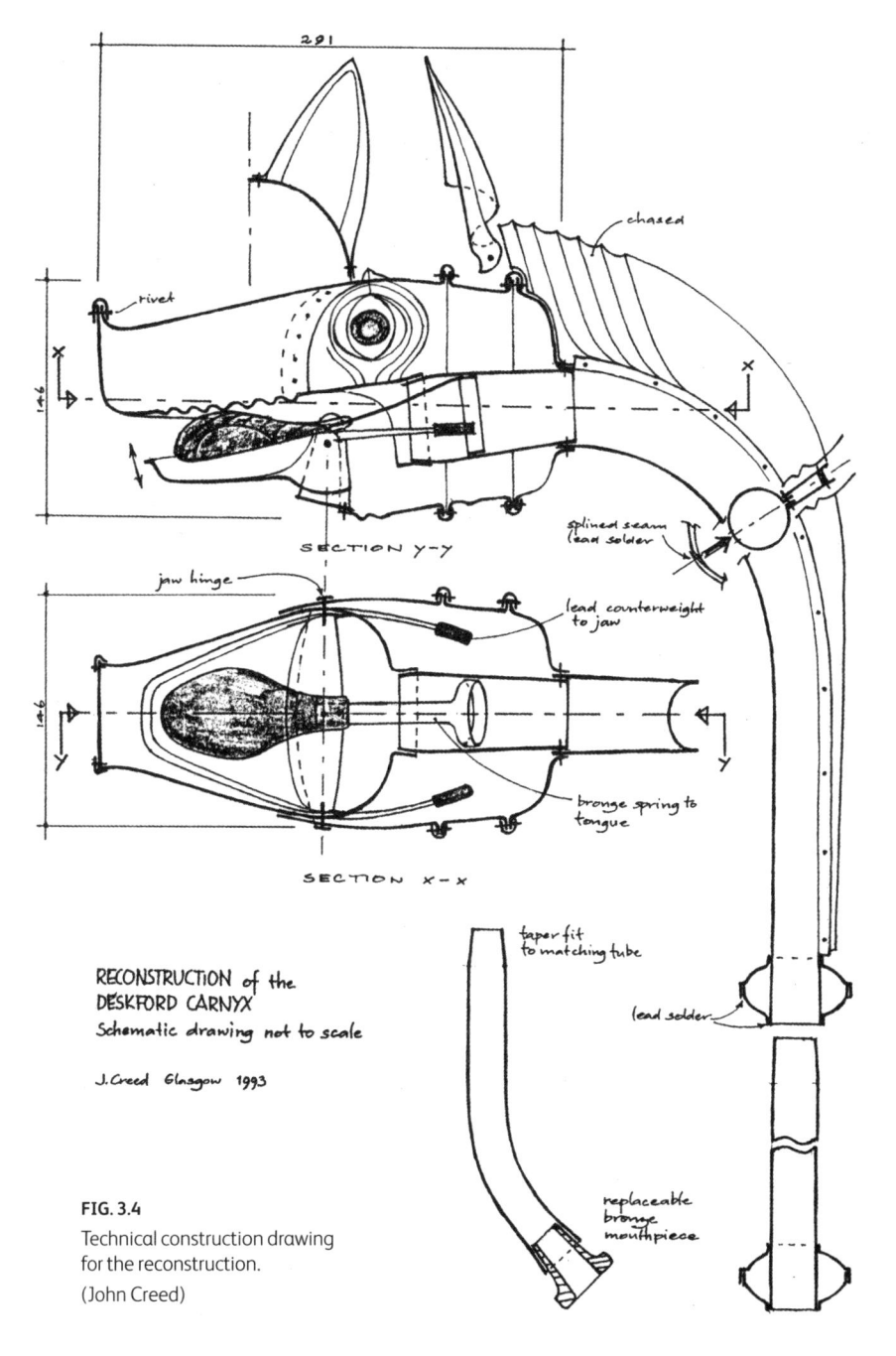

FIG. 3.4
Technical construction drawing for the reconstruction.
(John Creed)

FIG. 3.5
John Creed at work.
(© National Museums Scotland)

curved ends; the tube is shown as straight on all reliable depictions, and this was confirmed by the intact tubes from the Tintignac find. The comfort of the musician was clearly not seen as significant.

The reconstruction took 400 hours of labour, excluding the beating of ingots into sheet, which was done by machine (*see* Fig. 3.5 and Fig. 3 in art section). The surface was waxed to prevent corrosion, which dulls down (but does not remove) the colour contrast between bronze and brass; this prompted re-examination of ancient metalwork, where the frequent contrast between the good condition of the 'display side' and poorer condition of the reverse suggests similar surface treatments to enhance or protect the visible surfaces were used in prehistory.

Reconstructing ancient music?

From all the above, it should be clear that this is an exercise in possibilities rather than a secure replica of an ancient musical instrument. Not only are there great uncertainties over key elements such as the tube (a replica of the Tintignac carnyces would be much less uncertain), but there are also conceptual difficulties over the very idea of music – what classed as music in the Iron Age? Here the experience of the musician, John Kenny, was a valuable corrective to ingrained western traditions, as he is not only a concert trombonist but also plays the alphorn and the didgeridoo, both instruments which require a different technique from the western norm (circular breathing to maintain a constant note by manipulation of both lungs and stored air in the cheeks). The experience of the didgeridoo also showed something of the variety of music created in non-western areas, and this served to guide John's exploration of the instrument, along with attempts to make sounds evoking the boar which it depicted. The sounds themselves are entirely hypothetical (and for that reason are not discussed here in any detail; *see* Kenny, 2000; Campbell and MacGillivray, 2000, for this aspect), but it is clear it had the potential to be a tremendously versatile instrument. Its construction also acted to give it particular qualities, and we may suspect this was deliberate; sound radiated from the head, especially the ears, giving a great strength to the music, and this was augmented by the palate which acted to split the head into two separate chambers – one which the sound passed through (the mouth) and the other in which it resonated (the snout).

One upshot of the reconstruction was a desire to use the instrument in modern recitals. As it was also intended as a display piece, this led to the creation of a second instrument, thanks to generous funding from the Hope-Scott Trust and National Museums Scotland. The second instrument drew on the musical experience of the first, with the joints between tube sections being chamfered to ease smooth air flow and the gauge of metal being reduced from 0.65mm to 0.5mm. Both are within the range noted on the instrument, but the narrower gauge made the instrument lighter (and thus more easily held – 2.3kg rather than 3kg) and improved its acoustic qualities, the narrower walls reverberating readily and giving greater range and power.

Conclusions

The reconstruction is avowedly speculative, and there are dangers in such projects. A reconstruction has a physical presence which makes it seem real; its visual power is seductive, and most viewers will not know the caveats behind it. It can readily become an archetype rather than a possibility, closing off the prospect of other reconstructions rather than opening up the field. Indeed, just such a process can be seen in the current case, as the reconstruction is a recurring aspect of popular books and it features regularly in the reconstructions of artists and re-enactors; although thankfully the new French find is now providing a potent alternative model.

Yet these dangers seem to me to be outweighed by the benefits. It has brought a little-known object to life and made it comprehensible and accessible. It has raised awareness and discussion about the skills of the period among the general public. It has also become a modern icon, with the second reconstruction living an active life as a modern instrument for which pieces are composed in the unlikeliest of musical combinations (such as carnyx and double bass!). This is a good thing, giving the reconstruction a modern musical life which does not pretend to be a version of ancient acoustics. The process has also had powerful research benefits; working with the craftsman has clarified the role of certain technical details, such as a faint line on the head which can now be seen as a centring line for the decoration, or the question of waxing the surface noted earlier. It has also emphasised the power of sound in the all-too-silent past. Playing the reconstruction in the landscape has shown the carrying power of the instrument, and this would be all the more marked in the ancient countryside where there was no competition from combustion engines. While the music of the Iron Age remains unknowable, the reconstruction allows the significance of the instruments to be understood far better.

Acknowledgements

My partners in this project – John Purser, John Creed and John Kenny – made this a highly stimulating process with their enthusiasm, knowledge and commitment to a quality product. I am grateful to a wide range of archaeological colleagues for patient answering of a wide range of enquiries and enabling access to some highly obscure carnyx specimens.

Bibliography

Brailsford, J. W. (1975): *Early Celtic Masterpieces from Britain in the British Museum* (London: British Museum).

Campbell, D. M. and T. MacGillivray (2000): 'Acoustics of the carnyx', in Hickmann, et al. (eds) (2000), pp. 357–63.

Creed, J. (2000): 'Reconstructing the Deskford carnyx', in Hickmann et al. (eds) (2000), pp. 347–50.

Hickmann, E., I. Laufs and R. Eichmann (eds) (2000): *Studien zur Musikarchäologie II: Musikarchäologie früher Metallzeiten* (Rahden: Verlag Marie Leidorf).

Hunter, F. (2000): 'Reconstructing the carnyx', in Hickmann, et al. (eds) (2000), pp. 341–45.

Hunter, F. (2001): 'The carnyx in Iron Age Europe', in *Antiquaries Journal*, 81, pp. 77–108.

Hunter, F. (2009): 'Une oreille de carnyx découvert à La Tène', in *Le site de La Tène: bilan des connaissances – état de la question, Actes de la Table ronde internationale de Neuchâtel, 1–3 novembre 2007* (M. Honegger, D. Ramseyer, G. Kaenel, B. Arnold and M.-A. Kaeser, eds) (Hauterive: Office et musée cantonal d'archéologie de Neuchâtel (Archéologie neuchâteloise 43), pp. 75–85.

Hunter, F. (forthcoming): *The carnyx in the European Iron Age* (Edinburgh: NMS Enterprises Limited – Publishing).

Innes, G. (1845): 'Parish of Deskford', in *New Statistical Account of Scotland*, 13, pp. 63–78.

Kenny, J. (2000): 'The reconstruction of the Deskford carnyx – an ongoing multidisciplinary project', in Hickmann, et al. (eds) (2000), pp. 351–56.

Maniquet, C. (2008): 'Le dépôt cultuel du sanctuaire gaulois de Tintignac à Naves (Corrèze)', *Gallia* 65, pp. 273–326.

Maniquet, C. (2009): *Les guerriers gaulois de Tintignac* (Limoges: Editions Culture et Patrimoine en Limousin).

Megaw, J. V. S. (1968): 'Problems and non-problems in palaeo-organology: a musical miscellany', in *Studies in Ancient Europe: Essays presented to Stuart Piggott* (J. M. Coles and D. D. A. Simpson, eds) (Leicester: Leicester University Press), pp. 333–58.

Piggott, S. (1959): 'The Carnyx in Early Iron Age Britain', in *Antiquaries Journal*, 39, pp. 19–32.

Purser, J. (1992): *Scotland's music* (Edinburgh: Mainstream Publishing).

Purser, J. (2007): *Scotland's music* enlarged version (Edinburgh: Mainstream Publishing).

Raftery, B. (1983): *A Catalogue of Irish Iron Age Antiquities* (Marburg: Veröffentlichung des Vorgeschichtlichen seminars Marburg-Sonderband 1).

SØREN NIELSEN

The *Sea Stallion* from Glendalough

RECONSTRUCTING A VIKING-AGE LONGSHIP

The real thing[1]

THE CELTIC SEA. Sea Stallion is sailing with a 'half wind' (beam reach) on starboard tack (the wind is coming across the ship from the starboard side). The wind is strong, 11–13 m/s, and the sea is between 2½–3½ m high. We are sailing with two, then three reefs (out of a total of five) in the sail.

Well over half the crew is seasick, and a handful of them are inactive. Sea Stallion is doing well despite the swells and waves not coming from the same side, so it's a strange restless sea on which to sail. But we are taking a lot of spray over the side, especially in the bow, just in front of the tack (the front corner of the sail), and the crew in the bow is getting a lot of water down their necks. Spray also is coming on board just behind the mast shrouds. Some waves are coming in over the stern, up in the navigation box and VHF radio.

Once the sail is reefed (reduced), we tighten the parrel rope hard and set it farther back than usual to relieve the drag of the yard on the middle of the mast, causing it to sway forward in an arc. The parrel rope broke when the ship came down hard on a wave. We choose to proceed without the parrel rope because it would be critical to take the sail down across the ship in order to mount the parrel rope again in this weather. We'll get the sail down when it should, with the help of the braces and clue lines.

I'm sitting aft on the starboard navigation box (opposite the navigational equipment) and wondering if I should propose to the skipper to take in one more reef, but the ship is doing fantastically as it is being hit by waves and twisting in the sea. Through the many miles we have sailed, we have become accustomed to the fact that Sea Stallion is a very flexible ship. When the ship sails out of a wave and immediately knocks down the next, the whole bow moves 20 to 30 cm back and forth. If one sits at the mast and looks fore or aft, you see Sea Stallion move up and down and to both sides. Even with the reinforcements we set in the ship before the trip to Ireland, it is still flexible. The new reinforcements amidships on the gunwale that were added in Dublin before leaving [*see* podcasts from 28 April], also seems to hold.

The ship that we have reconstructed holds on without problems in relationship to these issues, but it is also very worn out after such a trip where it constantly twists and is hit. Based on calculations of when the *Sea Stallion's* predecessor, the Skuldelev 2 find, was built, the ship ended her days as part of a blockade at the bottom of Roskilde Fjord, and was around 20–25 years old. And as the *Sea Stallion* has given so much during this voyage, it is easy to understand that such ships quickly wore out.

The wind has increased and the waves have grown. We are really pleased that we decided to reconstruct the *Sea Stallion* with a loose extra strake (the red top splash board) for open-sea sailing.[2] Several times it has saved us from getting a wave in the ship. But later that night, we found out that the splash board was not always enough. Some waves are significantly higher than average, and when such a wall of water comes across a ship, hundreds of litres of water will pour over the side and drench the crew sitting to windward, keeping the vessel balanced. After which the wave turns the ship towards port, and a load of water will come in over the opposite side. Besides spray and waves, many litres of water also come in under the flaps that cover the oar holes and the holes where all the rigging is attached to the sides of the hull. In the 21 hours it took us from leaving the shelter of Ireland to reaching Lands End, and could sail with the sea behind us, we pumped about 18,000 litres of water from the *Sea Stallion*, with 4-6 pumps working almost constantly.

The rudder withy is also on trial now. On starboard tack the rudder is drawn away from the hull, so there is a steady hard pull on the birch withy. However, the withy is doing really well: it has been pulled a little further out from the ship, so it now sits 5–7cm away from rudder boss instead of 3–4cm. We should tighten and inspect it when we come into port again. The withy still looks good, however, and it'll keep for the next day's sail.

The five Skuldelev ships

At the Viking Ship Museum, experimental archaeology began in 1982–84 with the reconstruction of Skuldelev 3 ship-find, the remains of a small trading vessel from the Viking Age.[3] Skuldelev 3 was the most well-preserved of the five Viking ships excavated near Skuldelev in Roskilde Fjord in 1962,[4] and this ship was therefore the most obvious candidate for a reconstruction project. Before the building of the reconstruction of the large warship Skuldelev 2 began in 2000, three other vessels from the Skuldelev ship-find had been built: the small warship Skuldelev 5 (reconstructed in 1992), the fishing boat Skuldelev 6 (in 1998), and the large merchant ship Skuldelev 1 (in 2000). The original Skuldelev 3 and Skuldelev 5 ships were both built in southern Scandinavia. Skuldelev 6 and Skuldelev 1 were built beside the Sognefjord in western Norway. Skuldelev 2, was built in Ireland, on the outskirts of Dublin.

FIG. 4.1
Reconstructions of the five Viking ships from Skuldelev, Denmark. Left to right: *Kraka Fyr* (1997), *Roar Ege* (1984), *Helge Ask* (1991), *Sea Stallion from Glendalough* (2004) and *Ottar* (2000).
(Werner Karrasch/The Viking Ship Museum, Roskilde)

The traditional Nordic boat

As well as experience in reconstructing archaeological finds of shipwrecks, the Viking Ship Museum maintains a collection of *circa* 30 boats. The collection consists of traditional clinker-built boats from all over Scandinavia: Denmark, Sweden, Norway, Finland and the Faroe Islands. These boats represent the last step in a development that started before the Viking Age. It is the maintenance of this collection and the construction of traditional boats that gives us the experience of how to think like a traditional boatbuilder. An example of this mindset, which exists in only a few places in Scandinavia today, was demonstrated at a meeting of boatbuilders by the experienced western Norwegian boatbuilder Harald Dalland from Os, near Bergen, Norway. At one point, Harald questioned whether we are simply able, by looking at the shape of an old boat, to determine for what purpose the boat was built. Was it a sailboat, a rowboat, a boat for transport, or a boat for fishing? Was the boat built to sail into the North Sea or for coastal navigation? What is special about traditional boatbuilding is that the boats are never exactly alike. They were built to follow the buyer's specific wishes, but still within the local tradition. Similarly, during the Viking Age, boats were built for specific purposes following a local tradition. If a modern boatbuilder knows how a boat must be designed in relation to special needs, he is also able to assess what purpose a boat was originally built for, perhaps even if it is from the Viking Age.

Experimental archaeology

At the Viking Ship Museum, experimental archaeology serves as a research and analytical method of studying archaeological ship-finds.[5] As various scientific methods can illuminate an archaeological find from different angles, so does the methodology of experimental archaeology adds information about archaeological finds that is otherwise impossible to produce using other methods.

The method's tangible result is a reconstruction – our interpretation of how the original ship looked when it was built – but this result also includes the significant knowledge of the crafts such as boatbuilding, rope making, sail-making, tar production and sailing that contributed to the manufacture of the ship. Analyses of archaeological material often give other types of results, when the focus is on wood quality, tool marks and wear marks, among other things. These assessments serve as the primary source for selecting the proper trees in the forest, and the proper tools for construction. Furthermore, subsequent analysis of our consumption of human and material resources gives us insight into the original ship's resource requirements that ultimately can illuminate contemporary organisation and society.

The ships are reconstructed and built to the same size, shape and appearance following our interpretation of the original ships. All well-preserved parts of a ship are copied, and the missing parts are reconstructed. If there are well-preserved sections that are not well understood, or do not seem logical or strong enough, we will never change the construction. Rather, we see it as a challenge to discover why things were made as they were. The sail and rig are manufactured by the same principles as the rest of the ship.

Reconstruction must work in practice, and the construction, rigging and sailing of a ship require concrete solutions. If we have reconstructed incorrectly, our mistake will be demonstrated when a ship is under sail. This is where we get concrete answers on 'hypotheses', interpretation and reconstruction. It is also where we reach beyond the desktop restrictions, asking crucial questions and providing answers to those still unresolved. It is only due to being at sea and experiencing the wind, tides and the nights, with cold, with fatigue and with the other crew members, that some new obvious questions arise and other previously neglected issues become important. If the ship or part of the ship is not working properly, it may be due to erroneous interpretations of the original findings. In such cases we go back to square one: the archaeological material, iconographic material, ethnological material and experiences from other reconstructions to investigate whether there are other interpretations than first assumed. If the ship or parts of the ship do not work, this can also be due to incorrect use of the ship, resulting in a modification of the handling and manoeuvring of the ship.

FIG. 4.2
Skuldelev 2's aft stem.
(The Viking Ship Museum, Roskilde)

The prerequisite for a successful experimental archaeology project is that it is well-defined. The area of ship construction must be given priority. If logging is not the focus of the research project, for example, cutting down the trees for a ship's timber with a saw will have no impact on the final quality of the planks. Otherwise using a chainsaw will have a decisive impact for cutting the planks for a Viking ship, since the properties of a sawn plank and a split plank are very different. If a ship is built with other types of materials and other manufacturing methods than those possible for the original ship, it can also be crucial in determining whether the experience is being collected under the building and sailing, thereby illuminating the original ship and the society to which it was a part.

If the goal is to describe the craft and the society that gave rise to contemporary shipbuilding, navigation and organisation, it is also important to use professionals for different parts of the project: experienced boatbuilders for construction, sailors for sea trials, historians and archaeologists for interpretations, and comparisons of historical and archaeological sources.

For 25 years, the Viking Ship Museum has reconstructed ancient and historical vessels. Our background is interdisciplinary, and includes archaeologists, historians, boatbuilders, sailors, loggers, technicians, and so on. We realised that we can only work in such an interdisciplinary manner and must respect each other's knowledge and skills in order to undertake a meaningful reconstruction project. To assess, reconstruct and build a vessel from archaeological finds and other comparable archaeological and historical sources, the craftsmen and sailors require – as well as skill – the broad knowledge and understanding similar to academics such as archaeologists and historians. Only through close co-operation can results from the experimental archaeological reconstruction projects approach the quality of those of the original vessels that we are attempting to reconstruct.

RECONSTRUCTIONS

Primary and secondary sources for reconstruction

The sources that form the basis for our reconstruction work are comprised of:

- *The original ship-finds*: the primary source for the reconstruction;
- *Other comparable archaeological finds*: if the original ship-find is not enough to determine essential components for the reconstruction, information from other archaeological finds that originate as close as possible from the same period and same geographic/cultural area are used;
- *Iconographic sources*: images carved on stone, graffiti on wood, murals, etc.;
- *Historical sources*: written sources;
- *Ethnological sources*: traditional Nordic boatbuilding techniques; living tradition bearers;
- *Natural science sources*: wood technology, pollen analysis, etc.

The process of reconstruction

Documentation

The first challenge for the reconstruction of a ship is to reconstruct the original hull shape. Centuries of submersion in water or soil not only lead to varying degrees of corrosion of iron and wooden elements, but also to deformation in a ship's general shape. To reconstruct a ship's original form and design, it is necessary to document thoroughly the find immediately after excavation and cleaning of the individual parts. This is important to do before any conservation is attempted, as the process often causes shrinkage or other changes in the wood.

FIG. 4.3
Tool marks on the remains of the Skuldelev 3 ship-find.
(The Viking Ship Museum, Roskilde)

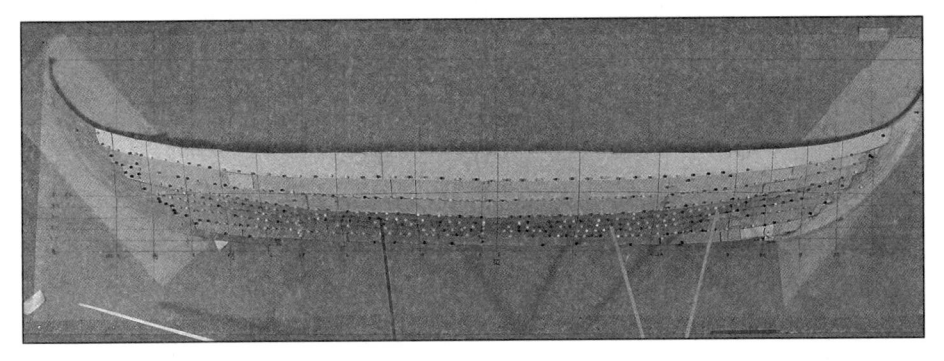

FIG. 4.4
Cardboard model of the hull of the Skuldelev 3 ship-find.
(Lars Kann-Rasmussen/The Viking Ship Museum, Roskilde)

In the case of the Skuldelev ship-finds, the parts were drawn manually in 1:1 scale using a special technique of clear folio; today ship-finds are drawn digitally in 3D.

Cardboard model

The 1:1 scale drawings are reduced to 1:10 scale, printed out and glued on to strong cardboard. Each ship part is cut out and assembled using pins through the markings that indicate the location of nail holes in each strake. This allows for the reconstruction of the original ship form from the excavated parts, and the resulting cardboard model serves as the basis for the production of line drawings and reconstruction drawings of the ship (Fig. 4.4). Eventually it may be possible to reconstruct the form of vessels using only digital means.

Wooden model

Subsequently, a wooden model of the entire vessel, also in 1:10 scale, is built. The sections of the ship that were not preserved are reconstructed through the above-mentioned secondary sources. All the details of the ship and objectives of the project are reviewed to form a solid basis for building a full-scale reconstruction.

Full-scale construction

Based on the background of the original hull form established from the cardboard model and then the reconstruct hull form established from the wooden model, the reconstruction begins in full scale. During this process, the focus is on the craftwork and the technologies that were used in the construction of the original ship. The preserved part of the original ship are examined for traces of tool marks that can provide answers to tool use, material quality and manufacturing methods.

RECONSTRUCTIONS

Analysis of the original ship parts

Wood species and quality

The original ship-find indicates what materials were used in its construction: the nature of the wood, dimensions of the framing system, quality and location of the element in the original tree. In Skuldelev 2 both the planks and frames are made of oak (*Quercus robur/Quercus petraea*). The treenails that attach the frames are manufactured of willow (*Salix caprea*). Closer analysis also reveals the quality of the materials used. Knots and tree growth are of crucial importance for wood quality. The pith ray pattern of the original oak planks reveal whether the trees used were straight-growing. If an oak trunk has an irregular and twisted growth, the fibres will cross the planks that are made from it; when corrected (or straightened), the planks thereby lose their strength. In such cases the pith ray pattern looks like tiger stripes on the surface of the planks (radial cleaved oak planks are also called 'tiger figured oak'). The fibre's position and location of the pith in the planks reveal that they were made of curved oak or of trunk-branch elements. Observation has revealed that the smaller frame elements (knees and the like) were also obtained from quarter sawn curved oak trees (Fig. 4.5).

It requires considerable experience to assess how an oak tree standing in the forest and covered with bark looks underneath and in the pith (Figs 4.6 and 4.7). It can be hard to tell if the tree is straight-grown and knot-free. Unblemished wood can be converted into many planks, depending on the required width. A tree 1m in diameter gives an average of 16 planks with a width of approximately 30cm. It requires far fewer hours to fell a fine straight-grown tree and cleave it into planks to fit a ship, than to do the same with an irregular, twisted tree. However, it is not so important to use exclusively oak of very fine quality. Using a slightly lower quality oak might result in the production of only 8–10 planks. Unlike today, the number of hours spent to make each plank before it was added to a hull of a ship was not very important in the past.

Radial cleaving

The Vikings did not apparently use saws in boatbuilding; planks were cleaved and not sawn out of big oak trunks. To split an oak tree over 1m in diameter, with a length of up to 8m, requires a specific quality of wood. Cleaved planks have a very high working quality, which means that the dimensions can be minimised without the planks losing their strength. Perhaps using the cleaving technique is the reason why the Vikings could build such long, light and simultaneously strong warships. In the Middle Ages, planks began to be manufactured by sawing, and this technique meant that planks could be made from very inferior quality oak. This resulted, however, in the increase in the dimensions

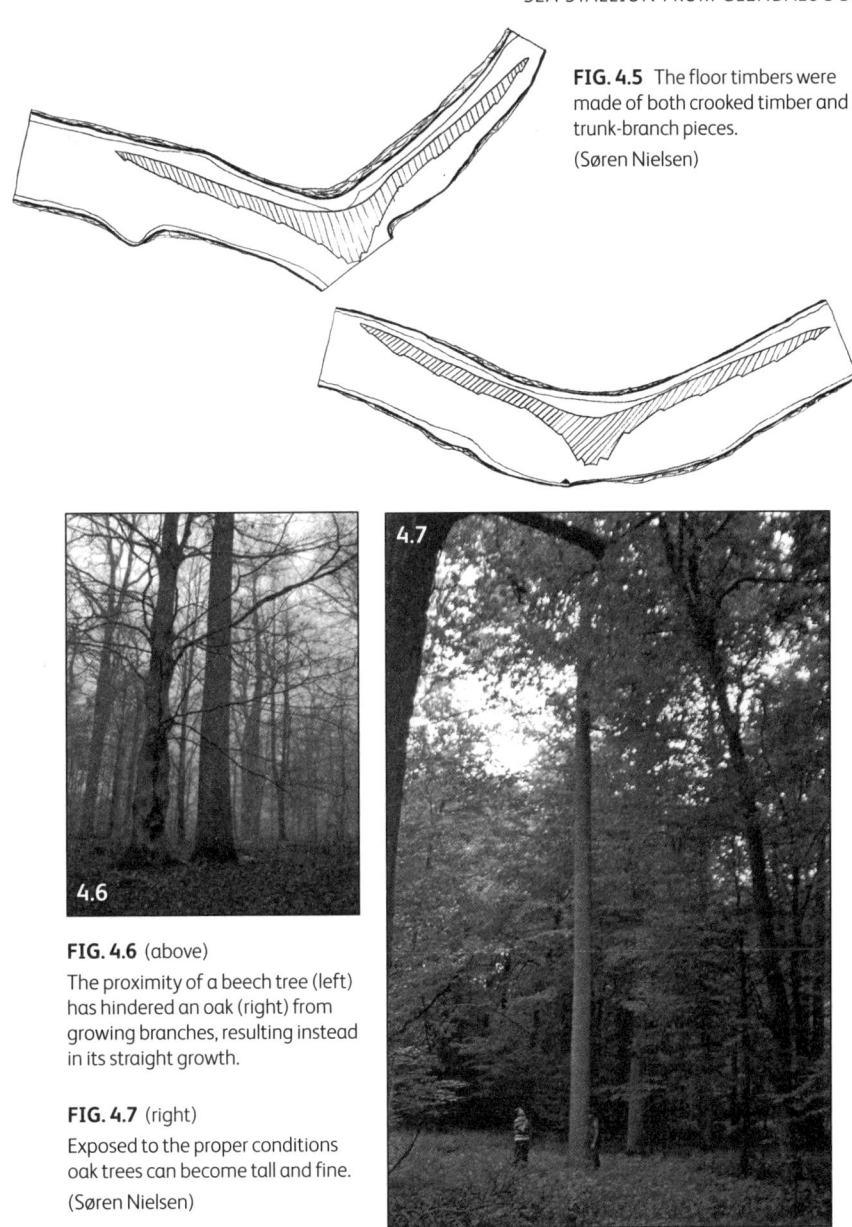

FIG. 4.5 The floor timbers were made of both crooked timber and trunk-branch pieces.
(Søren Nielsen)

FIG. 4.6 (above)
The proximity of a beech tree (left) has hindered an oak (right) from growing branches, resulting instead in its straight growth.

FIG. 4.7 (right)
Exposed to the proper conditions oak trees can become tall and fine.
(Søren Nielsen)

of both planks and frames, leading to the building of stronger, heavier and less elegant ships.

During the Viking Age cleaving was done in the forest, where the trees stood. We chose to carry most of the oak to the museum's boatyard to disseminate the process of cleaving to interested audiences. An oak trunk is cleaved using small wooden wedges starting at the base of the cut trunk. Small wedges are replaced gradually with larger wedges as cracks open up following the tree's natural fibres. The first cleaving, where the tree is halved, takes the longest time. A good oak can be split into two parts in about 2½ hours. Then it takes around half to one full hour to split the two halves into quarters and then again into eighths. Depending on how wide the planks should be, and depending on the wood quality, it can also be chopped into sixteenths and in rare cases, 32 parts.

The irregular part of the cleaved piece, which lies up against the pith, is then chopped away and the sapwood and bark are removed. The cleaved piece is rough hewn on one side with an ordinary axe and then hewn down with a broad axe to the desired plank thickness, about 28 mm.

If planks are not going to be used immediately after cleaving, they are stored in water until needed. It is important that planks are moist when they are used, as it is easier to twist them into the desired shape, and easier to manipulate with hand tools, axes, planes and spoon augers. Dry oak becomes rock hard, and it is almost impossible to bend or cut with hand tools.

FIG. 4.8
Radial cleaving.

FIG. 4.9
An oak trunk ready for cleaving.

FIG. 4.10
A strong radial cleaved oak plank, ready to set on the hull.

(Werner Karrasch/The Viking Ship Museum, Roskilde)

FIG. 4.11
Copies of tools from the Viking Age. The Vikings also used planes, spoon augers, hammers, chisels, etc. (Werner Karrasch/The Viking Ship Museum, Roskilde)

Tool-marks

Archaeological finds are cited as the direct and primary source for building reconstructions. The original ship parts are not only fragments of the original product, they also reflect contemporary technology such as wood-splitting techniques and traces of the tools that boatbuilders used. The tool marks that can be found on various original ship components are the basis for the choice of tools with which the reconstruction is built.

On the original wood elements were found traces of cuts from various axes in different sizes, different types of planes, scrapers and chisels. Reconstruction projects have resulted in an increased awareness of tool marks, not just from elements of shipwrecks but also other archaeological finds. The focus on tools marks emphasises the importance of careful documentation in connection with the excavation of archaeological ship-finds.

Sometimes specific tools are known to have existed, but no traces of them remain. Very characteristic broad axes are known from Viking-Age archaeological sites, but there is not yet a trace of the use of this axe on Viking-Age ship-finds. The broad axe has probably been used in ship construction, although the marks subsequently have been planed away. This type of axe was also used for the reconstruction of the Skuldelev 2 ship-find.

Building the *Sea Stallion from Glendalough*

How were these ships built and in what order were they put together?[6]

Skuldelev 2's oak keel is comprised of three pieces, as are the stems. The keel and stems are the ship's backbone, and the first raised. On each stem are three steps (Fig. 4.12) which are attached to a multiple hooding end (Fig. 4.13). To this are attached two to three planks. The ship has twelve strakes.

After the first seven strakes, the ship's bottom is built up, and floor timbers are set in order to stabilise its shape. Then the keelson is added, on top of the floor timbers (Fig. 4.14). The purpose of the very long keelson is to give the

RECONSTRUCTIONS

ship longitudinal stiffness and therefore it passes over all the floor timbers and then is fastened with knees. After the floor is laid, the last strake is fastened, and two layers of beams with knees over the floor timbers, as part of the frames. In addition to the keel and keelson, the ship has four longitudinal stringers that are attached to the inside of the hull. All these elements must ensure that the long, light ship keeps its shape and not collapse when it is writhing like a worm on large, long waves.

The ship's 30 pairs of oars pass through holes in the second highest strake, which is made of ash (*Fraxinus excelsior*). In contrast to oak, of which the other strakes are made, ash does not split so easily. This property is required of wood when oars accidentally are stuck in the hole when rowing. The mast, yard and 60 oars are of pine (*Pinus sylvestris*).

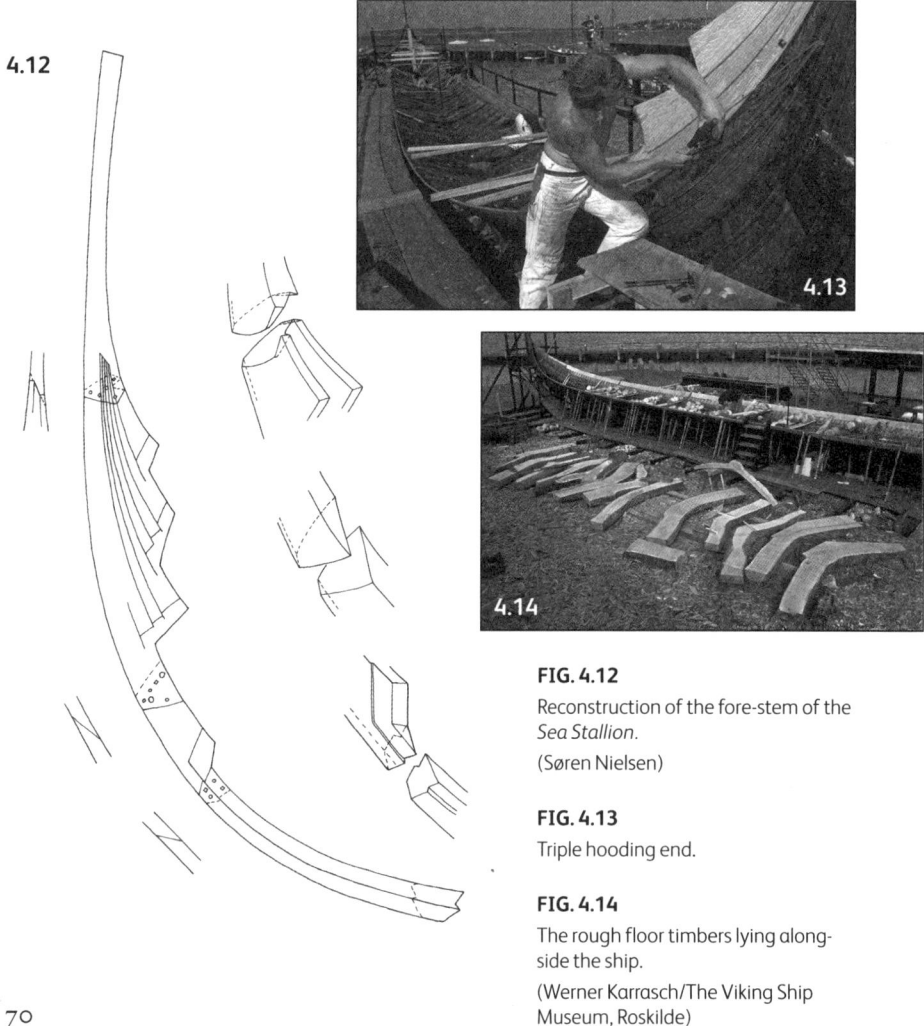

FIG. 4.12
Reconstruction of the fore-stem of the *Sea Stallion*.
(Søren Nielsen)

FIG. 4.13
Triple hooding end.

FIG. 4.14
The rough floor timbers lying alongside the ship.
(Werner Karrasch/The Viking Ship Museum, Roskilde)

Resources

Materials

Four smaller oak trees were used for the *Sea Stallion's* keel and stems, 14 large oak trees for the planks, two medium oak trees for the keelson and mast fish, *circa* 285 pieces of crooked timber of oak for floors and knees, four ash trees for the oar-hole planks, three long, medium oak trees for the longitudinal reinforcement, two pine trees for the mast and yard, 35 pine trees for oars (each tree makes two oars), and ten willow trees for the treenails.

In modern Danish forests it is not very common to find large straight oaks with a diameter of one metre, despite the fact that oaks have been cultivated for over 200 years and it is precisely these types of trees that are coveted. This is largely due to the fact that forest managers now consider trees to be mature for harvesting when they are 85–90 cm in diameter. If a tree is much larger, it is difficult to handle in sawmills and the risk of diseases in the root increases. Viking-Age forests must have had numerous large trees. From archaeological finds we know that even small boats used big oak trees, as evidenced by the broad planks with which they were built.

As such a large number of clinker-built boats and ships were built during the Viking Age, the question therefore arises: what did contemporary forest look like? The forest coverage must have been larger than today with more trees from which to choose. There must also have been rules for who cultivated the trees and had permission to fell the large, older oak trees. Not everyone could have been able to fell a tree at will. It would be interesting to conduct research with forest historians and foresters on how forests appeared during the Viking Age. This issue is just one of many interdisciplinary subjects that stems from the *Sea Stallion from Glendalough* experimental archaeology project.

Use of time

During the construction of the *Sea Stallion*, the duration of each task was carefully recorded. Against this background, we can speculate how long it may have taken to build a 30 m-long ship in the Viking Age: there was work for about ten boatbuilders, if the construction was to be as efficient and effective as possible. The boatbuilder's work was to form the vessel's shape and adapt the rough-cut planks and rough-cut elements into frames, knees, etc. With the finishing of pieces and aligning of planks and floor timbers, knees, and so on, ten men would have been able to build a longship such as the *Sea Stallion* in approximately seven months.

This could happen, however, only if there were many other people who could supply parts for the ship. Ten men would be needed to fell trees and rough-cut planks in the forest in three months. People would also be required

RECONSTRUCTIONS

FIG. 4.15
Tar burners from Kuhmo, Finland, burn tar in the same way as the Vikings did.

FIG. 4.16
It takes two hours to fell an oak tree of this size.

(Werner Karrasch/The Viking Ship Museum, Roskilde)

to transport the timber from the forest to the construction site. Two men would be needed to rivet the planks and other ship parts. Bog iron would have to be collected and other iron sources mined, which would then have to be transported to the blacksmith. A blacksmith could forge the approximately 8000 iron nails in about eight months. Four tar burners would have to spend two months felling trees, digging up roots and establishing pits in order to produce the approximately 500 litres of wood tar needed for the ship both during construction and afterwards. Others would have made shields, and some people would have carved the mast and oars. The making of the sail would have required about twelve people to spend six months weaving the individual pieces, and it would have taken two people one month to sew the pieces together. In six months, five people could produce the bast and horsehair for the ropes, and a further six months would be needed for the rope maker to make the ropes. Finally, a 'stem-smith' was in charge of the large building project.

For comparison, the longship *The Long Serpent*, according to the Icelandic sagas, was built during one winter.[7] Our experience suggests that the building could take place in around seven months with a 'stem-smith' and ten boat-builders, in addition to approximately 50 others who would contribute with specialised tasks.

Launch

Dendrochronological studies reveal that Skuldelev 2 dates to 1042 and was built from trees that grew in the area south of Dublin.[8] At the launch of the full-scale reconstruction on 4 September 2004, the ship was named the *Sea Stallion from Glendalough*. The sagas mention ships as horses of the sea.[9] The ship is built according to the Nordic tradition of Scandinavian Vikings who had settled in the Dublin area by the 9th century.[10] It is unclear why the ship ended her days in Roskilde Fjord in the second half of 11th century, but there was very extensive contact between Ireland/Dublin and Denmark during this period.[11]

Objectives and principles of the sailing project

If the Vikings could sail from Dublin to Roskilde, we could also sail the reconstructed ship in the opposite direction, back to the waters for which the original ship was built. We would be able to test our reconstruction of this seagoing warship under realistic conditions in the waters around Scotland, England and Ireland and in the seas of southern Scandinavia.

Concerning the trial voyage the research plan stated:[12]

> *The task*: The reconstruction will be tested under realistic conditions in the seas for which she was built, the North and Irish Seas. The investigations will pursue three overall objectives:
>
> - To test the reconstruction – our hypothesis on the original ship;
> - To test the sailing characteristics of the reconstruction, thus also testing this type of vessel in general;
> - To investigate the functions, organisation and logistics of the ship and her crew
>
> *The challenge*: The project's main challenge is to create representative results capable of contributing to our knowledge of Skuldelev 2, Viking longships in general and Viking society.

Ships of this type are unparalleled today and the sailing experience and seamanship required for handling the original ship and its many functions have been long since lost. Part of the challenge is therefore to train a present-day crew to sail the ship in a way that allows representative quantification of the vessel's sailing characteristics. The credibility of the results depends on the voyage being completed under realistic conditions, using the original means of propulsion – sail and oars. Sending 65 men to sea in an open boat with no motor is a dangerous undertaking. Completing the archaeological experiment with a high degree of authenticity while ensuring the safety of the ship and her crew will therefore be a major challenge. It will also be a challenge to accomplish and document the planned investigations under conditions that affect both the people participating in the experiment and those performing the investigations. And finally, it will be a challenge to safely store the recorded data under all possible conditions.

RECONSTRUCTIONS

FIG. 4.17
Queen of Denmark, Margrethe II, names the reconstruction *The Sea Stallion from Glendalough*, 4 September 2004.
(Werner Karrasch/The Viking Ship Museum, Roskilde)

FIG. 4.18 (opposite page)
The route of the *Sea Stallion* 2007–08. (Mette Kryger/The Viking Ship Museum, Roskilde)

The trial voyage

In the summer of 2007, the *Sea Stallion* sailed from Roskilde (Fig. 4.18) to Norway and from there to the Orkney Islands, further down the Scottish coast, through the North Channel and via the Isle of Man to Dublin. During the winter the *Sea Stallion* was on display at Collins Barracks, at the National Museum of Ireland in Dublin. The following year, in the summer of 2008, we sailed from Dublin south through the Celtic Sea and into the English Channel, where we berthed in Torquay and Portsmouth, among other ports. In Lowestoft, on the east coast of England, the ship had to wait for good winds to cross the North Sea, and after many days the *Sea Stallion* sailed to Holland and then to northwest Denmark and back to Roskilde. In total, the ship sailed 2482 nautical miles and the crew spent 495 hours at sea.

Selecting and training a crew

To obtain reliable results, the crew initially had to be trained to sail the ship. The crew was chosen based on the requirements that we believe were necessary for making the ship and crew functional: experience sailing of vessels with square sails, experience of travelling at open sea, experience of sailing in general, social, physical and recreational skills, and those with special skills such as nurse, cook, mate and skipper.

After the crew was selected, we planned 2½ years of training voyages with the *Sea Stallion*. In 2004 we sailed only in Roskilde Fjord; in 2005 we sailed in Danish waters; in 2006 we sailed into the Kattegat along the west coast of

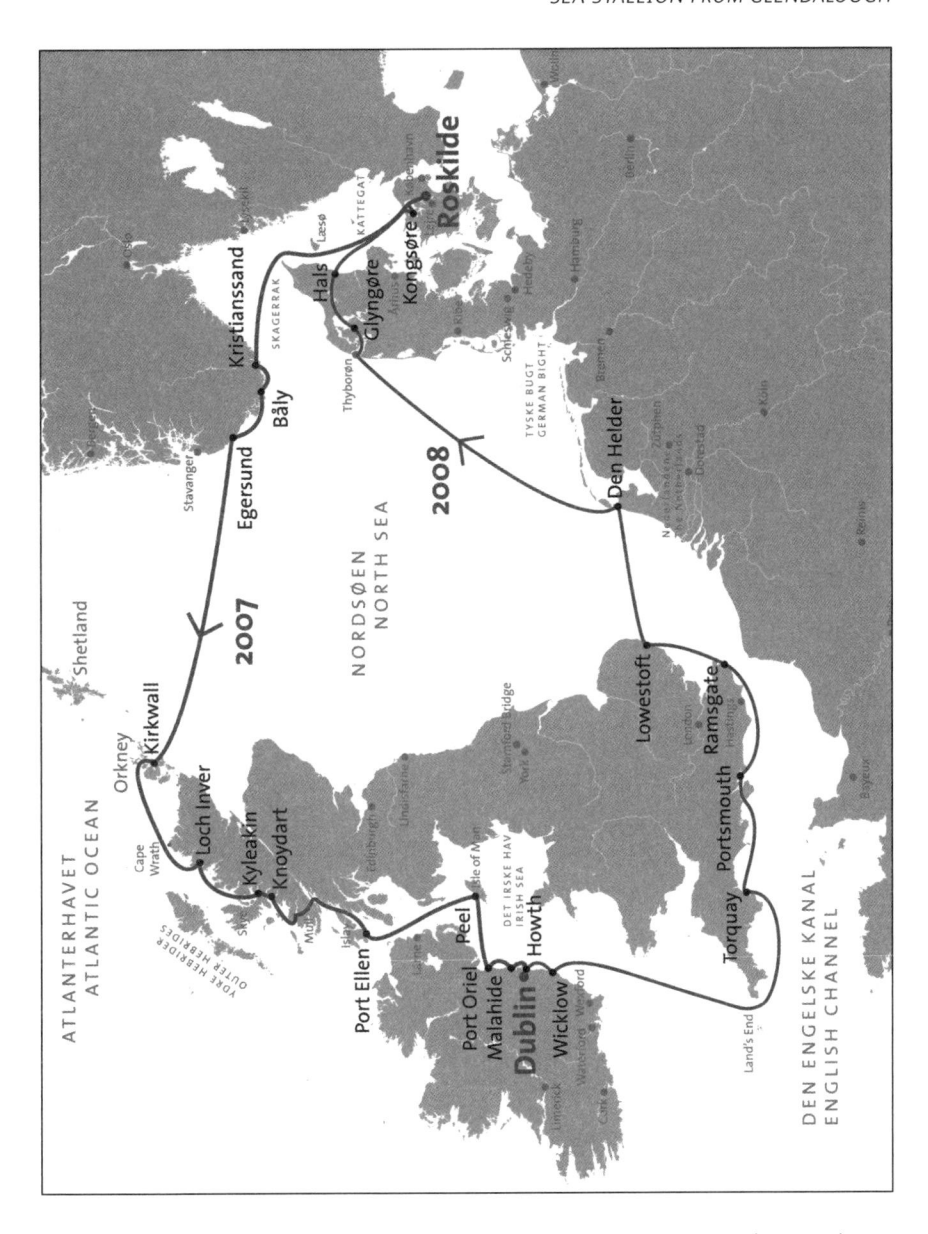

Sweden to Norway, along the eastern coast of Norway, out into the North Sea and back to Denmark through the Limfjord east to Roskilde.

The ship's balance (hull, displacement and rig)

In addition to a highly-trained and hundred per cent ready crew, the ship also had to be ready. Its basic trim had to be established before any conclusions could be made regarding the ship's general properties. Rig, hull, ballast and crew had to be placed correctly in order for the ship to be balanced.

A ship is balanced when it sails straight ahead without correcting its heading with the rudder or through weight distribution (the crew) (*see* Fig. 4b in art section). A ship should not turn up into the wind or away from the wind during a voyage. It is in this context important to have control over the weight of the ballast (stones), personal gear, the ship's equipment, such as extra rope and anchors, drinking water, food, crew, etc., in order to establish the ship's basic trim.

The ship weighed 8.3 tons empty. Stone ballast weighed 3.6 tons. Sail, rigging and other equipment weighed 6.4 tons and its crew weighed 4.7 tons. Finally, food and water weighed around 1.5 tons. In total during the sea trials, the *Sea Stallion* weighed around 24.3 tons.

Crew weight is important as the crew represents a very important factor when the ship sails. In addition to being used to steady and manoeuvre the ship, the crew must also act as 'live ballast'. They must move from side to side, depending on the direction of the wind. Weight placed on the railing on the side to windward is much more efficient than weight at the bottom of the ship (according 'force x arm' principle).

Safety

The safety of the crew was a priority for the sailing project. On board the *Sea Stallion* we had life rafts and personal life jackets and survival suits. We chose also to have an escort ship during the trial voyage. In the event of capsizing, we reviewed and practised safety procedures. Should we capsize in rough weather so that we could not be rescued, the escort ship was to contact land-based life-saving services.

For security reasons, we had installed modern navigational equipment on board the *Sea Stallion*. It was important that other large ships could see and contact us. It was also important that we always knew where we were, so we could take our precautions for the onward voyage. The Viking Ship Museum was responsible for all the 60-person crew (male/female) on board to arrive safely back in Roskilde again.

There is no doubt that modern equipment influenced how we undertook the voyage. We did not navigate on Vikings terms; instead we had charts and forecasts of wind and tidal conditions, which the Vikings did not have. As a result, Viking navigation methods were not a focus of research of this project. But we sailed the same waters, in the 'same' ship, and through the same navigational hazards as the Vikings did 1000 years ago, and it was the ship, as the embodiment of technology, that was our research project.

FIG. 4.19
Training: A rescue helicopter lowers a doctor and hoists a crew member.
(Werner Karrasch / The Viking Ship Museum, Roskilde)

Research and dissemination

The project was a combined research and dissemination project. The goal was to disseminate the research we conducted underway, directly via the project's website, documentaries and in various media. Along the way we wrote different sailing log books. These served both as documentation of the voyage and real-time dissemination on the project website.

The log books consisted of:

- The ship's log book: the traditional log book on the voyage, kept by the skipper and the mates;
- Life on board: the daily routines as related by the crew;
- Research and experience: diary on technical observations of the ship, the rig and the sail, kept by the boatbuilders on board.

The log book entries, photos and film clips were posted daily on the project's website. In addition, the current electronic data (weather and wind conditions, sails and ship's speed) were updated on the website every ten minutes, so the ship could be followed directly on a Google map of the entire trip from Roskilde to Dublin and back again.[13]

Under the entry 'Armchair comments', various specialists outside the project also wrote their criticisms, comments and questions about the voyage and the project as a whole. In some cases, on board the *Sea Stallion* we could test ideas and theories or answer questions along the way which hitherto had been only thoughts or theoretical considerations.

The voyage was also followed daily by the BBC (in 2007) and *Politiken* (one of the largest Danish daily papers), as well as by several Danish and foreign news media outlets.

Interest in the trial voyage was overwhelming. Everywhere we went, people would flock together to watch the ship, even in the smallest villages in northwest Scotland. Up to 10,000 people witnessed our arrival in Dublin, and the arrival was announced on the big billboard on Times Square in New York. In Denmark, SMS messages were sent around, so people were lined up in ports and along the shores to welcome the *Sea Stallion*. In 2007 the project website (www.havhingsten.dk) was viewed 1,876,985 times, and in 2008 the website was viewed 1,579,658 times.

Preliminary results

The side rudder, which was fastened with a rudder rope or rudder withy to the ship's starboard side aft, caused problems. Archaeological ship-finds reveal that Viking ships' rudders were positioned on the starboard side. It was not until the 1300s that a new rudder system situated on the stern was invented, as is known today.

For the first part of the sea trial, the *Sea Stallion's* side rudder was held fast with a hemp rope with a circumference of *c*.50 mm. The rope could not withstand the pressure to which it was exposed when the ship sailed from 8–10 knots in a swell. After about ten hours, with the wet hemp rope being alternately stretched and tightened again, the fibres of the rope began to fray and eventually broke. A solution had to be found before the journey back to Roskilde and within the framework of the extant archaeological and historical source material.

Rudder withies of wood (Fig. 4.20) are known from archaeological finds and discussed in sources from the Middle Ages. In find material from the excavation of the Nydam ship, from AD 320, a piece of lime bast was found with the rudder; the remains of a pine withy were found in the helm of the Oseberg ship from Norway, and the remains of a rudder frame with oak withy were found at Fribrødre Å on Falster.[14] Rudder withies can be made of oak, birch, hazel and juniper. The young trees are selected and twisted in the spring when they are filled with sap. The stem, at about 50 mm thick at root, must be free of larger branches, and small twigs must be cut carefully without damaging the fibres of the trunk. The roots on the one side are cut *c*.20 cm from the trunk, after which the tree is laid down. The tree is twisted by the outer small branches down to *c*.30 cm from the root. The root is trimmed and used as an exterior stop-block when the withy is mounted in the rudder. It takes two to three people to wring a withy of this size. Twisting the tree causes the longitudinal fibres to detach from each other, and makes the tree easier to bend.

On the return journey to Roskilde, the *Sea Stallion* used a withy of birch (Fig. 4.21). After 437 nautical miles, six days and a tough trip across the Celtic Sea, the rudder withy was changed in Portsmouth. The withy was not worn, but it was not in good shape: it was too weak to act as a stop-block on the outside of the rudder, while at the same time it was violently wearing against the rudder's exterior. A new rudder withy was put on. It had a worse stem but a better root, which was protected with a leather lining. This withy held all the way to Roskilde, 829 nautical miles and ten days.

Flexibility

The longship Skuldelev 2 was built to carry many warriors from one place to another. The ship would not only sail, but it would be dragged ashore on foreign beaches. It should be at once strong and heavy enough to carry her sail, and light and sufficiently long enough to be rowed by as many men as possible. The longship was a compromise between strength and lightness, and the result was an extremely flexible ship.

The strength of a strake or frame, as mentioned above, depends on whether the wood fibres are followed. Planks in Viking-Age ships were cleaved out, and trees with curved growth were carefully selected for the frames. In both cases, this means that the wood fibres follow the finished ship parts, making it possible to minimise the dimensions of the planks and frames. This was the basis for building a light and simultaneously strong ship. However, the thin and light design resulted in extreme flexibility, and Viking-Age boatbuilders must have known the limit beyond which the flexible structure would collapse. The voyage of the *Sea Stallion* was an ultimate test of the reconstructed ship. Would the ship be flexible? Was the Skuldelev 2 ship-find reconstructed correctly?

Before the voyage to Dublin, the ship's flexibility was tested in different ways. Rivets in the ship's side broke, the mast fish moved from side to side,

FIG. 4.20
On the ship rudder, withies of birch and oak were ready for use if necessary.
(Werner Karrasch/The Viking Ship Museum, Roskilde)

FIG. 4.21
The rudder withy is inserted through the rudder from the outside where the root end works as a stop-block.
(The Viking Ship Museum, Roskilde)

wedges worked themselves out of treenails, and joints in the longitudinal stringers weakened. The stems moved 50cm in relation to each other – up and down and side to side. When the bow was on its way down the side of a wave, the stern moved in the opposite direction up the side of the same wave. Ropes from the stem and stern to the top of the mast were tight as a steel wire, and the next moment hung loose when the ship hung between two waves. The construction writhed and gave way to everything.

After the first sea trials, concerns led to the decision to reinforce the vessel in order to restrict its flexibility before the voyage to Dublin. Based on traces found on the original ship, the *Sea Stallion* was strengthened amidships. And based on other ship-finds, the floor boards along the ship's sides were nailed to the hull to strengthen the ship diagonally.[15] In addition, the rig was tightened to strengthen the hull and make it stiffer.

Wind roses and beating to windward trials

The analysis of the data from the sea trials is still being undertaken. The goal is to summarise the data in such a manner so that it can be compared with measurements of other square-sail vessels, and other reconstructions as other traditional square-rigged boats.

A wind rose describes a ship's speed potential in a certain type of weather such as wind speed and wave height. A beating to windward trial is conducted to detect how close the ship can sail to the wind under certain weather conditions and makes it possible to measure the effective speed directly against the wind. For a longship, a beating to windward trial is comparable to a rowing test if performed in the same weather and sea conditions. The aim is to investigate whether it is worth sailing against the wind or rowing directly against the wind. These experiments are important when considering the ship's potential under different conditions (was the ship built for sailing or rowing?).

Conclusions so far

Experience has shown that in cases of extreme sea, it is virtually impossible to row the *Sea Stallion* against the wind and waves. In calm seas and little wind, the *Sea Stallion* can be rowed at an average speed of approximately 2.5–3 knots and a top speed of 4–4.5 knots. Under sail, the *Sea Stallion* has sailed up to 13 knots and reached an average speed over a longer stretch of about 10 knots. The ship can sail around 60 degrees against the wind. On the voyage to Ramsgate, in the mouth of the Thames in a west wind, 12–15 m/s, we sailed over 10 knots![16]

The ship's hull and rig are in equilibrium, and we have found the *Sea Stallion's* basic trim. The ship has functioned satisfactorily in all situations it has been exposed to on the voyage. It has been possible for the 60-man crew to handle the ship in 20–25m/s and swells of 4–5m. The top speed is not yet known, but the ship has the potential to sail above 12 knots over short distances. Our overall experience with the *Sea Stallion* is that it is a highly manoeuvrable and seaworthy ship that feels safe to be on board, including sailing in a fresh wind.

We consider the reconstruction of the Skuldelev 2 ship-find, the *Sea Stallion from Glendalough*, to be representative of a Viking-Age longship. At the same time, we believe that the experience and results of the trial voyage may contribute to new insights and understanding of Viking-Age society. We consider the voyage, as a long-distance trip and with a heavily-laden ship, under the weather conditions to which we were exposed, as being representative of a Viking voyage. However, a Viking skipper and crew would probably be willing to press themselves and the ship further and take more chances if they had a bloodthirsty fleet in hot pursuit ...

The results of parts of this project are already published in various books from the Viking Ship Museum, including *Welcome on board – The Sea Stallion from Glendalough. A Viking long ship recreated.*[17] The next book, *Thoroughbred of the Sea. The Sea Stallion from Glendalough. Trial voyage with a long ship* is under preparation in this series. *The Skuldelev Ships I*, in the series 'Ships and Boats of the North', volume 4.1,[18] deals with the archaeological ship-finds from Skuldelev, and the forthcoming volumes, *The Skuldelev Ships II and III*[19] will present the reconstruction, construction, sea trials and experience with the five Viking ships, and the value of these experimental archaeology reconstructions will be presented.

Acknowledgements

This project is planned and implemented in response to an interdisciplinary collaboration between theoreticians and practitioners at the Viking Ship Museum in Denmark. Neither party could have completed the project alone and obtained the same results. This collaboration applies even to the last comma in this article.

References

1. Nielsen (2008): 'The Sea Stallion in rough weather', in the log book *Research and Experence*, 8 July 2008. Access at: www.vikingeskibsmuseet.dk
2. See Andersen, 'Hlyða. A detachable wash strake', in *Armchair comments*, 8 August 2007. Access at: www.vikingeskibsmuseet.dk
3. Andersen, Crumlin-Pedersen, Vadstrup and Vinner (1997).
4. Crumlin-Pedersen (2002).
5. Nielsen, (2003).
6. Bill, Nielsen, Andersen and Damgård-Sørensen (2007).
7. Snorre Sturluson (1997): 'Olav Tryggvasons saga', in Hødnebø and Magerøy (eds) (1997), ch. 88 (Den norske Bok-klubben), p. 192.
8. Crumlin-Pedersen (2002), pp. 66–67.
9. Snorre Sturluson (1997): 'Håkon den godes saga', in Hødnebø and Magerøy (eds) (1997), chs 5 and 6, pp. 89–90

10 Ó Corráin (2001), pp. 19–21.
11 Holman (2007), pp. 104–107; Crumlin-Pedersen (2002), pp. 328–30.
12 Damgård-Sørensen (2006). Access at: www.vikingeskibsmuseet.dk
13 Access at: www.vikingeskibsmuseet.dk
14 F. Rieck, pers. comm. (1917): Brøgger, Falk and Shetelig (1917), vol. 1, pp. 309–10; Skamby Madsen and Crumlin-Pedersen (1989), pp. 18–19 and p. 31, note 10.
15 Brøgger, Falk and Shetelig (1989): (14), pp. 312–13.
16 Damgård-Sørensen (ed.) (forth. a).
17 Bill, Nielsen, Andersen and Damgård-Sørensen (2007).
18 Crumlin-Pedersen (2002).
19 Damgård-Sørensen (ed.) (forth. b); (forth. c).

Bibliography

Andersen, E., 'Hlyða. A detachable wash strake', in *Armchair comments*, 8 August 2007.

Andersen, E., O. Crumlin-Pedersen, S. Vadstrup and M. Vinner (1997): *Roar Ege. Skuldelev 3 skibet som arkæologisk eksperiment* (Roskilde: The Viking Ship Museum).

Béat, A. (1999): Altaripa: archéologie expérimantale et architecture navale gallo-romaine, in *Archéologie Neuchâteloise* 25 (Neuchâtel: Musée cantonal d'archéolgie).

Bill, J., S. Nielsen, E. Andersen and T. Damgård-Sørensen (2007): *Welcome on board! The Sea Stallion from Glendalough. A Viking longship recreated* (Roskilde: The Viking Ship Museum).

Bischoff, V. (forthcoming): *Reconstructing the Oseberg ship – Evaluaction of the hull form*. Between Continents, Proceedings from the 12th International Symposium on Ship and Boat Archaeology, 12–26 October 2009, Istanbul.

Brøgger, A. W., H. Falk and H. Shetelig (1917): *Osebergfundet* (Kristiania).

Crumlin-Pedersen, O. (2002): *The Skuldelev Ships I. Topography, Archaeology, History, Conservation and Display*, Ships and Boats of the North, vol. 4.1 (Roskilde: The Viking Ship Museum).

Damgård-Sørensen, T. (2006): 'Research Plan', Project: Thoroughbred of the Sea, The trial voyage to Dublin (Roskilde: The Viking Ship Museum).

Damgård-Sørensen, T. (ed.) (forth.a): *Thoroughbred of the Sea. The Sea Stallion from Glendalough. Trial voyage with a long ship* (Roskilde: The Viking Ship Museum).

Damgård-Sørensen, T. (ed.) (forth. b): *The Skuldelev Ships II*, Ships and Boats of the North (Roskilde: The Viking Ship Museum).

Damgård-Sørensen, T. (ed.) (forth. c): *The Skuldelev Ships III*, Ships and Boats of the North (Roskilde: The Viking Ship Museum).

Hødnebø, F. and H. Magerøy (eds) (1997): *Norges konge sagaer* (Den norske Bokklubben)

Holman, K. (2007): *The Northern Conquest. Vikings in Britain and Ireland* (Oxford),

Nielsen, S. (2003): 'Experimental archaeology at the Viking Ship Museum in Roskilde', in *Connected by the sea. Proceedings of the Tenth International Symposium on Boat and Ship Archaeology* (L. Blue, F. Hocker and A. Englert, eds) (Roskilde: The Viking Ship Museum).

Nielsen, S. (2008): 'The Sea Stallion in rough weather', in the log book, *Research and Experence*, 8 July 2008.

Ó Corráin, D. (2001): 'The Vikings in Ireland', in *The Vikings in Ireland* (A.-C. Larsen, ed.) (Roskilde: The Viking Ship Museum)

Skamby Madsen, J. and O. Crumlin-Pedersen (1989): *To skibsfund fra Falster* (Roskilde: The Viking Ship Museum).

Sørensen, A. C. (2001): *Ladby. A Danish Ship-Grave from the Viking Age*, Ships and Boats of the North, vol. 3 (Roskilde: The Viking Ship Museum).

Ravn, M., Bischoff, A. Englert and S. Nielsen (forthcoming): 'Recent Advances in post-Excavation Documentation, Reconstruction and Experimental Maritime Archaeology', in *Oxford Handbook of Maritime Archaeology* (A. Catsambis, D. Hamilton and B. Ford, eds) (Texas: Oxford University Press).

MICHAEL R. BAILEY

Learning through replication

THE *PLANET* LOCOMOTIVE PROJECT

The third resource

THE STUDY OF the history of engineering and technology relies largely on archival material as a primary resource. A glance through the papers presented to the Newcomen Society since 1920 reveals a rich bibliography and reference source, second to none in this discipline. However, engineers and technologists of yesteryear have often been sparing in the records of their progress, their patents and published works usually reflecting a successful outcome to their work. But what of their experiments, often chasing flawed concepts or uneconomic designs? And what of their approach to design and understanding of materials and construction methods? Unlike scientists, who are required to supply full details of their research in order that their peers can evaluate the validity of their achievement, technologists rarely recorded full details of their research and development, for reasons of reputation, commercial concern, or simply lack of systematic approach. Indeed, there are dangers for the historian of technology from an over-reliance on documentation, which may hide important detail, and make conclusions suspect.

Only to a much lesser extent has the historian of technology turned to the second resource, namely surviving structures and artefacts. From these may be learned more, not just about design, but about materials, methods of manufacture and component assembly, which were once the preserve of the tradesmen. Their knowledge and skills were rarely recorded, but were transmitted from master to apprentice, each generation building on and developing the skills of its forebears. Indeed, there is much opportunity for the historian to examine and learn from artefacts in science and technology museums around the world, which have for so long been repositories for collections rather than centres of study.

Experimental development has been an essential part of the history of technology, and there are many examples to counter the theory, popular among scientists, that scientific advance precedes technological progress. After experimental development, however, scientific interpretation of technological progress was usually left to others. The extraordinary development of the steam locomotive by Robert Stephenson and his team in the late 1820s/early 1830s, the result of which is the subject of the case study discussed in this paper, is a good example. It was left to Chev F. M. G. De Pambour, Rev. Dr Dionysius Lardner and M. Navier to determine the dynamics and efficiencies of the new technology that was to have such a dramatic effect on popular travel.[1]

Our understanding of such technological progress can, therefore, be significantly enhanced through practical experience and experimentation, to complement the usual historical research methods. Such practical experience, the 'third resource', can lead to a much better understanding of design philosophies and concepts, as well as material characteristics and strengths, and construction techniques. The research carried out in Manchester by the University's Historic Structures Group, enquiring into the structural characteristics of cast iron and other materials, is a good example of 'practical history'.[2] For machinery, in addition, experimentation could lead to a better understanding of operating techniques, component reliability and maintenance regimes, together with a more rigorous understanding of performance capabilities and efficiencies. Papers on the history of engineering and technology are rarely based on experimentation, however, although machinery, both static and transport related, was itself usually experimentally developed. The only paper presented to the Newcomen Society based on experimentation was by Dr R. L. Hills, who headed the University of Manchester Institute of Science and Technology (UMIST) team which made the third-scale working model of Thomas Newcomen's 1712 steam pumping engine.[3]

Operation of historic machinery for the purposes of experimentation is, however, controversial, as it raises issues of long term wear and tear, with the risk of damage or even loss. The debate about the merits of keeping a machine working to aid the conservation of its components, against the perceived risks, will remain a preoccupation for technology museums for many years to come. Also, an incomplete machine may require unacceptable modifications to return it to operable condition and, even when a machine is complete, it has often been rebuilt or modified during its life, and is not necessarily representative of its original technology.[4]

Replication

For the purposes of historical research, replication is the most likely means of providing the third resource, as it can offer not just hands-on experience but the opportunity to understand the thought processes that lay behind technological advancement through experimentation, design, construction and operation.

Introducing replication as a concept directly leads to consideration of the recreation of any structure, machine, craft, vehicle or tool that played a part in the history of technology. Such projects need not always relate to known successes, but could include some of the failures. Indeed, replication could confirm or deny the very claims made for long-past technology. What would replicas of Savery's 1699 steam engine, or of Davidson's 1842 battery-electric vehicle tell us?[5]

Replication provides the opportunity for relearning lost manufacturing skills and for the reuse of superseded materials. Replicated machinery, sea and air craft, and rail and road vehicles, allow us to relearn, or keep alive, operating and handling skills and maintenance practices, as well as revealing any design weaknesses, recreating the experience of the original operators and maintenance teams. Furthermore, experimentation with replica machines, through performance trials, can provide a much better understanding of the dynamics, efficiency, power and fuel consumption of the original machines.

A 'replica' is defined in the *Oxford English Dictionary* as being an 'exact copy', which few of them could claim to be. It has been argued that a 'reproduction' would be a more accurate term, defined as 're-manufactured from raw materials'.[6] The author certainly subscribes to this view and has often used the term accordingly, but it has to be accepted that the distinction is popularly perceived to be too subtle, and the term 'replica' is now more usual.

Replication has been practised throughout history. Ancient Greek 'trireme' warships were built by shipwrights of the Roman Empire, another was built for the French Emperor, Napoleon III, and yet another was completed just a few years ago.[7] The modern era of replication to recreate and demonstrate long-past technology can be traced back to 1893, when several were made for the World's Columbian Exposition in Chicago, USA. The *Viking*, a 24-metre long replica Gokstad long-boat, sailed from Norway to be exhibited at the Exposition. Replica machinery was also exhibited, including the oldest operable replica steam locomotive, the *De Witt Clinton*. Built 61 years after the original locomotive, it is itself now a revered, 116-year old museum artefact.[8] Indeed, it pre-dates by a decade the beginning of the aircraft industry, itself now much the focus of replication projects.

Most replica machines are transport 'prime movers', with planes, ships and railway locomotives forming the large majority, and there are now several hundred operable replicas in several countries. Quite apart from the heightened interest in these objects, the need for replication is enhanced – by comparison with fixed machinery and structures – as the very characteristics of the originals meant that they were more quickly superseded, with a consequential lower survival rate. Replica projects in recent years, however, included buildings, notably the Shakespearean Globe Theatre in London completed in 1997. A decade earlier saw the completion of the full-scale, operable 1712 Newcomen designed *Dudley Castle* steam pumping engine erected at the Black Country Museum in Dudley, and itself housed in a replicated engine house.[9]

There are some fine examples of replica sea-going craft, such as the 1947-built balsa wood raft *Kon Tiki*, which sailed from Peru to Polynesia to demonstrate the feasibility of emigration in AD 500,[10] and the 1976-built leather hulled curragh *Brendan*, which sailed from Ireland to Newfoundland also to demonstrate the feasibility of navigation from the same era.[11] Both craft were dependent upon precise replication to prove their respective theses. At the other end of the scale, the 600-ton Dutch East Indiaman *Batavia*, of 1682, is perhaps the largest replica vessel to have been completed so far.

There are many dozen small and large replica aircraft in actual or potential flying condition in museums and private collections around the world. One example is that of the Bleriot XI flying machine of 1909, which recently flew across the English Channel on the centenary of Bleriot's momentous flight. Another is the 1994-built Vickers *Vimy* twin-engined bomber aircraft of 1919 which re-enacted, in that 75th anniversary year, Ross and Keith Smith's epic flight to Australia.[12]

Over 30 replica steam locomotives have been made since the *De Witt Clinton* in 1893. The majority of them represent mid-nineteenth century designs, but the biggest project that has recently been undertaken is the building of *Tornado*, a 1947-designed, London & North Eastern Railway A1-class Pacific locomotive.

Technology and transport museums around the world are increasingly turning to replicated exhibits to fill in gaps in their collections. They range from small hand tools to large machines. But what constitutes a replica and how accurate should it be? The natural inclination of the curator is to strive for accuracy, but commercial and safety considerations may dictate otherwise.[13] The options range between operable replicas and general representations providing only an indication of the appearance of the original type. Within that range there are various degrees of replication, such as an original machine with replacement components, such as George Stephenson's *Hetton* locomotive of 1822[14] and a dummy replica with original components, such as Braithwaite and Ericsson's *Novelty* locomotive of 1829.[15] And, of course, there are large scale models such as the UMIST Newcomen engine.

Consideration of replica aircraft includes the added dimension of their ability to fly. Some fully-flying replicas have original engines, whilst others, such as the *Vimy*, have latter-day engines which are perceived to be more reliable. Others are 'semi-flying' with the ability to fly at low altitude within the confines of an airfield, but many good replicas are only intended to taxi or, indeed, to remain on static display, even though they might potentially have flown.

Replication should be much more than the opportunity to exhibit or demonstrate a long-redundant structure, craft, machine or tool to an interested audience, important as that is. The essential matter is not what is, or is not a good replica, but what has been learned during the project to make and operate it. It is unfortunate that, over the last century, most operable replica locomotives have been made for the purposes of demonstration alone, usually associated with a major event, such as a centenary or a 'World's Fair', the motivation for which has usually been promotional. Lessons learned during their design, construction and operation, have not been written up for the benefit of learned debate and, before now, there are no known examples of replica locomotives being subjected to performance trials.

Replicas are, of necessity, compromises between historical accuracy and practical manufacture. Commercial pressures apart, all project teams naturally strive for accuracy as, without such motivation, the project is of limited historical value. Two things serve to prevent fulfilment of that objective, however. First, correct materials and components are not always available, and, second, safety and certification standards have evolved to become more rigorous with each generation. Although, for the most part, replicas represent the best endeavours of their project teams, each example reflects the availability of materials and the safety standards of the era in which it was made. A striking example may be seen with George and Robert Stephensons' *Rocket* locomotive, the two operating replicas dating from 1929–31 employing different materials and operating specification from their 1979 successor.[16]

Use of materials comparable with those of the original artefacts is a particular frustration for the replica-builder. Although, for example, it has often been possible to employ the same form of timber, replication may not even then be precise if treatment of that timber has not been similarly undertaken. Certain man-made materials, such as wrought iron plate, are now unobtainable. Even such base materials as copper and brass are very different from, say, 150 years ago, the purity of today's supplies reflecting the needs of the telecommunication industries.

Those fortunate project teams who have had the opportunity to replicate a vehicle, craft or structure with precisely the correct materials and fabrication methods, such as the *Sea Stallion* Viking longboat and the *Brendan* curragh, have been able to satisfy that quest for precision. Their replicas have, however, been limited to expeditions in which only the project teams themselves could participate, and with which they could demonstrate the feasibility and endurance of the original craft.

The second major influence on replica development relates to today's tightening standards of safety which, in an increasingly litigious society, often constrain the replicator's level of precision. As soon as application is made for

'Health and Safety' approval, or for a 'Marine Certificate', or for a 'Permit to Fly', themselves prerequisites for insurance certification, and necessary for public demonstration and participation, the safety practices of the present day override the standards of yesteryear. Thus begins the compromise to fulfil those safety requirements with minimum influence on the original design. It is of course a sterile argument to veto a replica project on the grounds that precision is compromised. We learn more from a compromised project than from no project at all.

The full benefits of a project, apart from the pleasure in recreating extinct technology, come with the presentation of its findings. It has to be said, however, that, popular articles apart, the findings from the design, development, construction and operation of replicas mostly remain unpublished.[17] Exceptions include the welcome books about the *Kon-Tiki* and the *Brendan*, whose authors have provided detail about the materials and construction methods of the craft.

The *Planet* locomotive project

The project, which commenced in 1986, undertook a programme of design and construction of a replica that would lead to a better understanding of the technology of the first class of main-line railway locomotives developed for use on the Liverpool and Manchester Railway (Fig. 5.1). The class prototype, *Planet*, was the culmination of an extraordinary period of research and development by Robert Stephenson and his close colleagues in the 33 months between January 1828 and September 1830. Some 40 members of the class were subsequently built, together with a similar number of their sister four-coupled *Samson* class, by the Stephenson Company at its Forth Street, Newcastle

FIG. 5.1
Prototype *Planet* at Rainhill Bridge, Liverpool and Manchester Railway.
(Drawn and engraved by I. Shaw, published in 1831)

FIG. 5.2
Replica *Planet* at Liverpool Road, Manchester, the terminus of the Liverpool and Manchester Railway, and now the Museum of Science & Industry. The 1830 station building is to the left and the warehouse in the centre.
(Michael R. Bailey)

upon Tyne, works. Further examples were built by other companies in Britain and the United States.

The replica project was undertaken by the Friends of the Museum of Science & Industry in Manchester, a team which included several members of the Newcomen Society. The Museum site includes the Liverpool Road terminus of the Liverpool and Manchester Railway, the surviving 1830-built station and warehouse buildings being the oldest mainline railway buildings in the world (Fig. 5.2). The objectives of the project were five-fold:

1. To design and build an operable replica, as close as possible to the original *Planet* type, but within the constraints of material availability and prescribed safety standards;
2. To understand fully component design and development by Robert Stephenson and Co. in the early 1830s;
3. To understand better the materials employed and methods of manufacture in those years;
4. To conduct trials, for comparison with the recorded achievements of the original locomotives, and to understand better the performance characteristics and efficiencies of the class;
5. To provide visitors to the museum of with the opportunity of seeing the locomotive working within the 1830 station/warehouse complex and travelling behind her in period carriages as part of a wider visitor 'experience'.

For the historian with an interest in early locomotive development, the *Planet* and *Samson* classes have been well written up in contemporary books, and a review of their design characteristics and component detail is well presented by Warren in his history of the Stephenson Company.[18] This account adds to that knowledge by presenting the findings of the replica project team regarding design, construction, operation, maintenance, performance and efficiency.

Specification and design

George Phipps, the Stephenson Company's first draughtsman, introduced and developed design practices into the company after his appointment in 1828. Initially only general arrangement drawings, in terms of layout and centre dimensions, were prepared. Component manufacture was undertaken under the supervision of the works manager, William Hutchinson, by the Forth Street tradesmen to fit the general arrangements, their work being based on their experience rather than on working drawings. The rapid evolution of the *Planet* class, within a basic configuration that had very little room for driving and valve motion, led to the preparation of the first working drawings, all too few of which have survived.[19]

However, such was the interest in the new main-line railways and locomotives in the 1830s that a number of publications produced creditable illustrations of the *Planets* and *Samsons*.[20] The illustrations and attendant descriptions confirm that each member of the class, built between 1830 and 1837, incorporated improvements in materials, arrangements and component detail, benefiting from the experience being gained. The replica's design was based on Phipps' few surviving arrangement drawings and the several published illustrations and descriptions. It is, therefore, a 'typical' rather than a specific member of the class, incorporating features that would generally reflect the practice of 1832.

The key design features of the *Planet* class include:

- 2-2-0 wheel arrangement with outside 'sandwich' frames supporting the boiler, and four independent inside frames stretching between the smokebox and firebox supporting the crank-axle bearings and crosshead slide bars (Fig. 5.3);
- Inside horizontal cylinders, housed in the bottom of the smokebox, with slide valves driven by slip eccentrics, via rocking shafts in front of the cylinders. An eccentric cluster, which is free to rotate on the crank axle, is normally engaged by one of two drive 'dogs' (or 'drivers' as they were known) located on either side of the cluster, for forward or reverse movement;
- Provision to shift the cluster laterally, to engage the required 'dog' using a 'yoke' operated indirectly from a pedal on the footplate;
- Provision to move the slide-valves manually, with the eccentrics disengaged, using levers on rocking shafts mounted on the boiler backplate, and linked to the front rocking shafts via side rods.

The project's specification stage was the forum to debate the compromises that would be necessary to build a replica that was as close as possible to the original, but incorporating requisite changes to meet specified safety require-

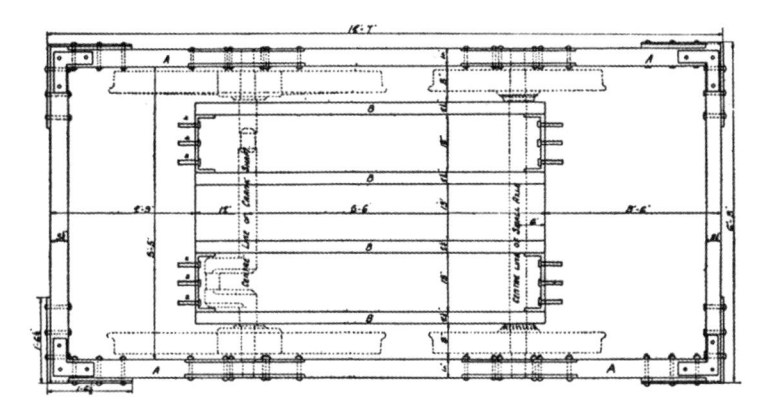

FIG. 5.3

Plan view of an early *Planet* frame arrangement drawn by George Phipps, illustrating the limited room available for the crank axle. (Traced from the original drawing in the Science Museum, R. Stephenson & Co. collection [Inv. 1924–159], drawing no. 15)

ments. The Department of Transport's Railway Inspectorate, British Railways and the British Engine Insurance Company all played important roles in identifying those requirements, including confirmation of the materials that would be acceptable in the early 1990s. The late Michael Satow, who had much experience of replica production, also offered helpful advice. The main specification variations from the original locomotives were:

- Substitution of a water-feed injector for the right-hand crosshead boiler feed-pump;
- An increase in boiler pressure from a nominal 50lb/in^2 to a working pressure of 100lb/in^2, a requirement deemed necessary for the stable working of the injector;
- Decrease in cylinder diameter from 11in to 8in, being in direct proportion with the increase in boiler pressure for comparable performance;
- Inclusion of leading and central cross members to link inner and outer frames to provide a coherent running chassis. The drawbar bracket was welded to the rear of the chassis in order that the longitudinal drag forces could be taken through the frame and not through the boiler as was originally the case;
- Insertion of a mechanism to indicate to the driver the required positioning of the slide-valves in order to achieve forward or reverse running;
- Use of welded steel instead of riveted wrought iron construction;
- Inclusion of an air brake system.

The *Planet* project team prepared arrangement drawings, with the overall objective of a locomotive that came closest to the original class in terms of both appearance and performance. About 160 working drawings were prepared specifying dimensions, machining tolerances and materials. To fulfil the requirements of insurance certification, the boiler was designed by a specialist firm, in consultation with the project team. The detailed design and construction process was a far cry from the practice of 1830!

Fitting up

The learning process began almost at once. The general layout is very cramped, with the short (5ft 2in) wheelbase determined by the boiler's length. The Stephenson boiler-makers were limited by a 4-ton axle load limitation and a maximum width of boiler plate (2ft 6in) that it was then possible to roll. The boilers were made of three circumferential plates with two intermediate lap seams. Within less than seven feet therefore, between the back-plate of the smokebox and the front-plate of the firebox, had to be fitted the slide bars, connecting rods and crank throw. The horizontal and vertical clearances for the cranks were just fractions of an inch (Fig. 5.4).

The fitting and operation of the pedal-operated slip-eccentric change mechanism was constrained by the layout of the footplate, firebox, inner frames and cranks, which was difficult to predetermine. Clearly, on each of the original *Planets*, with their slightly differing layouts, the fitters adapted the basic principle to suit the completed locomotive. The first foot-pedal arrangement on the replica was unsuccessful as there was insufficient allowance for lateral movement of the crank axle. This was resolved by the insertion of an additional spring on the yoke shaft.

The 1830s drawings show the bifurcation of the main steam-pipe, to serve the two cylinders, within the boiler itself. This seemed to be an unnecessary weakening of the front tube plate, if a single steam-pipe penetration could have

FIG. 5.4
Underside view of the *Planet* replica looking backwards to the firebox.
(Michael R. Bailey)

been achieved. However, all attempts by the project team to divide the steam-pipe within the smoke-box, without throttling the steam flow, failed, and it was apparent that the original design was the only suitable arrangement. This is an example of the way that the project team came to understand the work of the original Stephenson works team.

Valve setting

The shortness of the connecting rods in the confined space between the smokebox back-plate and the firebox front-plate causes significant angularity of the connecting rods to the horizontal. This obliquity causes an effective shortening of the connecting rods, with the pistons being biased towards the rear of the cylinders. The team had to develop a valve setting procedure to counteract this effect, and was achieved by equalising the piston-stroke at the point of cut-off for both forward and backward strokes. In common with the fittings shown on the 1830s illustrations, the valve rods have threaded ends, onto which 'bobbins', driven by reciprocating levers on the rocking shafts, are precisely located, and allow ready adjustments to accommodate the bias. As well as equalising the work done, it reduced the possibility of the locomotive stalling.

The eccentric 'dogs' were originally fitted to achieve a 75 per cent valve cut-off, but were subsequently reset for a 90 per cent cut-off for the second series of performance trials. This had the added advantage of making driving easier, with less incidence of stalling. Once established, the dogs were located, tightened to the axle and positions confirmed by punch marks on the dog rings and axle. There has been no occurrence of the dog rings slipping under operational forces.

Driving

The replica was precisely fitted with the driving apparatus of the original *Planets* in order that the driving skills could be relearned and fully understood (Fig. 5.5).[21] Normally, on restarting the locomotive in the same direction as previously, the slip eccentric remains engaged with the appropriate dog and the valves will continue to be driven when the regulator is opened. However, to reverse the locomotive or, indeed, if it has come to rest in a stall position, the valves have to be operated manually.

To achieve a manual valve change, the eccentric-rods are disengaged by lifting mechanisms from the footplate. The valves are then free to be operated by the hand-levers to suit the new steam cycle. When the regulator is opened, the valve-levers are pulled and pushed for the requisite valve movements, requiring both understanding and dexterity by the driver, and movement commences. At the same time, the foot-pedal is depressed by the driver in readiness for the eccentric sheave to engage with the dog, as the crank-shaft rotates through approximately half a turn.

When the dog is engaged, the eccentric rods are re-engaged to drive the valves, the driver's valve operating levers thereby being driven forwards and backwards. Drivers quickly learned of their threatening presence and the value of the forward-curved valve handles of the later *Planets*, rather than the straight handles of the first locomotives, is immediately apparent.

Alternatively, the locomotive can be and, with developed skill, has been, driven by hand movements of the valves for some distance. This provides valve opening over the full power stroke, providing a very effective form of 100 per cent cut-off. Initially, considerable strength was required because of the high steam chest pressure, but the valves became worn in a 'mirror' finish developed on the faces, making hand operation considerably easier.

The dexterity required in restarting a train 'on the handles' soon encouraged the practice of reversing the valves before coming to a halt. On the last few revolutions of the driving wheel, and with the regulator closed, the eccentric cluster is disengaged and re-engaged by the alternative dog. The foot-pedal spring pressure keeps the eccentric sheave rubbing on the new dog until it coincides with its slot. We can be certain that the footplate crews of the 1830s were also soon encouraged to adopt this practice.

Extending the slot in the eccentric cheek, as may now be seen on the original *Rocket* locomotive, would probably allow for engagement at higher speeds, although this has to be balanced against the potential to shear off the dog.[22] The opening of the regulator after engagement of the dog allows steam into the

FIG. 5.5
Footplate view of the *Planet* replica showing the driving controls.
(D. Boydell)

cylinders, which acts successfully as a counter-pressure brake should this be required. The necessary driving techniques took several weeks of part-time practice for the best drivers to perfect these skills.

Maintenance

The *Planets* required considerable maintenance and modification to keep them operational in the 1830s. The inadequacy of the first materials was well recorded,[23] crank axles, driving wheels, fireboxes and boiler tubes in particular being vulnerable to breakage and wear. This was often serious, leading at best to failures, and at worse, to accidents. The later *Planets* were much more robust, and heavier, than the first locomotives. The use of reliable materials and construction practices for the replica has largely avoided these problems. However, other long-term problems have arisen that were not recorded in the 1830s, and now provide us with a better understanding of certain design and component features.

In particular, the piston forces applied alternately to each crank cause appreciable deflection of the whole wheel-set. This could be reduced by adjusting the bearings on the inner frames to give minimal clearances, but at the risk of binding through misalignment with the outside bearings. The selection of appropriate clearances was one of the many judgements needed in fitting the locomotive.

Mineral grease was first used for lubricating the main bearings, instead of the animal grease (tallow) used on the original locomotives. However, grit entered the bearings during re-greasing and upset the thin film lubrication, which led to overheating of the axle boxes and seizure during the higher speed trials. This demonstrated the advisability of the mid-1830 design of reservoir-fed rape oil and tallow feeding to upper half-bearings within the axle boxes. The replica was accordingly refitted with such bearings fed by mineral oil reservoirs.

After just one season's operation, *Planet's* crews were reporting disengagement of the driving dogs whilst going round sharp curves. This was due to excessive lateral movement by the crank axle, which was being made possible by the progressive 'spreading' of the horn assemblies. The replica, unlike the original *Planets*, has spent half its time going backwards, with the driving wheel-set leading and guiding the whole chassis. The resulting axial force arose from the severity of the track curvature transitions on the Museum's track. After returning the horns to their correct alignment, they were monitored during the second season's operations, and the progressive spread recorded.

Planet drawings dating from 1833/34 show collars on the crank axle ends, which it had been thought were for locating the main bearings within the journals. These had been omitted on the replica, the axle boxes for which were based on a later design for which more detailed drawings were available.[24] The experience with the replica appears to have followed that of the original locomotives, suggesting that these collars had been developed as a restraint to the horn 'spread'. Accordingly, the replica was retrofitted with similar end-collars, held by screws to the crank axle ends, which now successfully restrain the horns, retaining their correct alignment.

Fatigue failure of *Planet's* axles after a little more than three season's operation was revealed following a derailment at the museum in Manchester in 1996. This sheared the carrying axle and cracked the crank axle. The characteristics of the axles, in which the diameter was stepped down for insertion into the wheel-hubs, and further stepped down for the outside bearing journals, followed the practice of the early 1831 *Planet* drawings. In accordance with latter-day design practice, however, a groove had been machined as a stress-relieving measure in the transition between the main axle diameter and the wheel seat.

The breakages, at the shoulder of the axles up to which the wheels had been fitted, revealed stress fractures triggered along the line of the grooves. These were diagnosed to have been stimulated by a 'jacking' action between the wheel hub and the axle shoulder – a further consequence of the sharp axial thrusts on the wheel-sets.[25] The drawings of the original *Planets* strongly suggest that a similar problem was experienced in the first operations of 1831. Subsequent axle drawings show that, by 1834, they were designed with an enlarged diameter for the wheel-fitting (Fig. 5.6).[26]

The experience with the replica now serves to explain that the problems with the crank axles on the original *Planets* were due to the axial forces, rather than always by the piston forces as had previously been supposed. These problems were recorded in Robert Stephenson's specification for the patent of the three-axle *Patentee* locomotive of 1833, the driving wheel-sets of which were flangeless to allow lateral freedom on the sharp curves of the early railways.[27] He wrote:

> My said improvement in such kind of locomotive engines will tend to obviate or diminish the said inconveniences, for by using plain tyres without flanges for the main impelled wheels … the cranked axles of those wheels will not be subjected to any strain endways, in the direction of its length, when the engine enters into sidings, turnings, and crossings of the rails, or passes along curvatures in the line, for that is probably the most severe strain to which the cranked axles have been hitherto subjected, and is thought to be that whereby they have commonly strained and broken … .

Replacement axles, adopting the lessons learned from the failure, were subsequently fitted to the replica, their design being much closer to that of the later-built original *Planets*.

The initial roughness of the valve-faces caused considerable 'juddering' when operating under full boiler pressure. This vibration was communicated to the valve motion, which progressively suffered. It was soon found that the keyways and keys on the rocking shafts were over-stressed and slackness developed. After just a few hundred miles operation the 'sloppy' keyways had reduced the valve travel by about half an inch, causing irregular cut-off. Larger diameter rocking shafts and larger keys reduced the contact stresses, rectifying this problem.

There is no evidence from contemporary authors and drawings regarding the lubrication of the valve chests and cylinders on the *Planets*. The replica was fitted with displacement lubricators filled with steam oil with a molybdenum additive, but the 'juddering' of the valves occurred in spite of this, until they were worn in. Although the original locomotives were operated at a nominal 50lb/in^2, the valve wear, and probably the piston wear, may have been quite severe with inadequate lubrication.

The trial operations undertaken with the *Planet* replica gave opportunity to experience her stability at service speeds, generally between 15 and 30 miles per hour. Whilst the short wheelbase remained the same as in the 1830s, the wheel profiles were machined to the British Rail P3 tyre profile in anticipation of operation on main-line routes. Her ride was remarkably stable, even at 30 miles per hour, with no sensation of the 'serpentine action' which was a feature of the original examples. There is a small degree of pitching as the leading end runs into each rail joint. The use of a modern flange profile on well laid track would suggest that the early locomotives suffered from hunting because of the immature tyre profile on poor track, laid with both lateral and longitudinal imprecision, and accentuated by the short wheelbase.

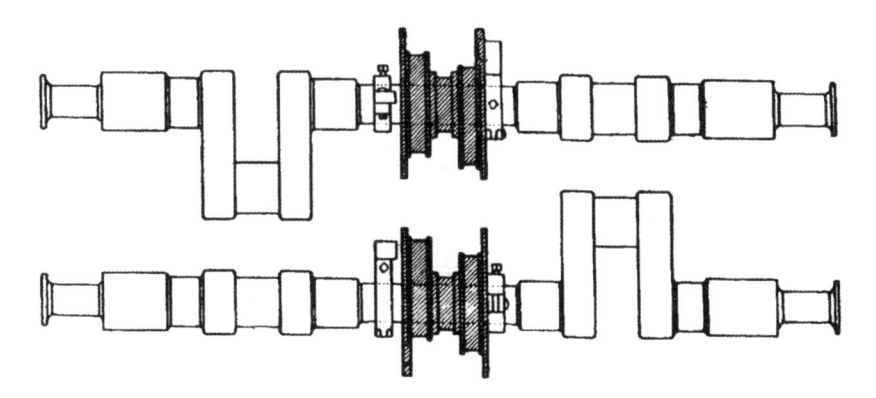

FIG. 5.6

Crank axle fitted to the later *Planets* with larger diameter for fitting into the wheel seats, and fitted with end collars.

(Traced from drawing in De Pambour, Chev F. M. G. (1836): *A Practical Treatise on Locomotive Engines Upon Railways &c.* (London), figs 11 & 12)

Performance

The documented performance trials on the Liverpool and Manchester Railway with the *Planets* and their four-coupled variant, the *Samsons*, undertaken by Robert Stephenson in 1830/31, Dionysius Lardner in 1832 and John Dixon in 1833, were followed up by Chev F. M. G. De Pambour's comprehensive trials in 1834 and 1836.[28] These, together with M. Navier's interpretations in 1835 and Nicholas Wood's of 1838,[29] were written to express a first understanding of railway dynamics and the novel form of motive power, and have provided us with a reasonable appreciation of their performance capabilities. Our subsequent understanding of locomotive performance and efficiency, however, now prompts further research, through experimentation which is well suited to replica operation. Trials were therefore carried out with the replica, not just to repeat the work of De Pambour and the other engineers, nor just to confirm the replica's characteristics, but to develop a more comprehensive understanding of the capabilities and efficiency of the *Planet* design.

The replica undertook two series of performance trials on the heritage rail routes of the Great Central Railway in Loughborough, Leicestershire, in February 1993, and the East Lancashire Railway at Bury, Greater Manchester, in November 1995 (Fig. 5.7). Eight useful trial 'runs' were undertaken on the Great Central, although some were hampered by mechanical failure, largely due to the original grease-lubricated axle bearings. Twenty-one successful runs were undertaken on the East Lancashire line between Bury (Bolton Street) and Ramsbottom. Four further runs were undertaken on the line's extension from Ramsbottom to Rawtenstall.

With the resources available, the data collected during the trials was obtained by spot measurements of indirect parameters, rather than by continuous measurements of performance parameters, such as may be obtained with dynamometer apparatus. Each run was undertaken with a different load and at a different target speed to provide a spread of data for the several gradients on each route. Members of the Museum Friends' project team recorded, at predetermined places, such as quarter-mile posts, the precise passing times, train speeds, boiler pressures and steam chest pressures. Coal consumption and water consumption were recorded for each run, whilst wind speed and direction were recorded on some of the runs.

The purpose of the trials was to provide data from which to determine tractive effort, indicated horsepower, equivalent drawbar horsepower, steaming rates over a range of speeds, steam generated per pound of coal consumed, coal and water consumption (per indicated horsepower per hour), and overall efficiency values. The findings of the trials are fully set out in the original publication of this paper. It was found that, whilst the method of obtaining the data from the trials was rather less satisfactory than would have been obtained from dynamometer equipment, and whilst certain extrapolations have been necessary, particularly arising from the low average steaming rates that were achieved, the efficiency of around 2 per cent was impressive for a class which was the progenitor of main-line railway motive power. At the end of steam use, drawbar/fuel efficiencies remained largely at less than ten per cent, with all the advantages of variable cut-off, superheating and superior draughting arrangements.

Conclusion

The development of the *Planet* locomotives may be regarded as one of the classic achievements of engineering history. Driven by the need to achieve reliable locomotive power within three years, to confirm George Stephenson's tenacious claims to the Directors of the Liverpool and Manchester Railway,[30] the research and development effort by Robert Stephenson and his team set new standards in engineering progress. It is, therefore, fitting to increase our knowledge of this effort through replicating its principal product. Whilst some of the materials of the original locomotives were inadequate, the project has demonstrated the soundness of the *Planet* design. Indeed, we may observe that the majority of the problems experienced when the replica first ran arose from our use of twentieth-century innovations rather than the replication of nineteenth-century features.

The experience gained from the *Planet* project confirms the core thesis of this paper, which is to demonstrate that replication can be used as an important tool in our understanding of the history of engineering and technology. The

FIG. 5.7

Replicas of *Planet* and two second-class Liverpool and Manchester Railway carriages at Loughborough station during the performance trials on the Great Central Railway in 1993.

(F. Beard)

benefits that can be obtained through the processes of designing, building, operating, maintaining and experimentation offer a third resource which can significantly add to our knowledge gained from primary written sources or from surviving structures and artefacts. The plea is not for yet more replication, but that the knowledge gained from present and past projects should be written up and presented to learned audiences, so that, first we may all be the beneficiaries, and, second, to ensure that the knowledge gained by the project teams is not, once more, lost to future generations.

The success of Robert Stephenson and his team was not just an important constituent of the Industrial Revolution, it was the progenitor of the popular transport revolution. Man's latent desire to travel was released in the first months of operations on the Liverpool and Manchester Railway, made possible by the *Planet* locomotives, a desire which has continued to grow all around our 'planet' ever since.

This account is a re-presentation of Dr Michael Bailey's Presidential Address to the Newcomen Society (The International Society for the History of Engineering and Technology) published in the Society's *Transactions*, vol. 68 (1996–97) pp.109–36. It is reproduced here in modified form with the kind permission of the Society.

Notes

1. De Pambour (1836); De Pambour (1840); Lardner (1836); Navier (1836).
2. The work of Manchester University's Historic Structures Group is undertaken by four group members: Tom Swailes of the Civil and Structural Engineering Department, Joe Marsh of the History of Science and Technology Group, Ron Fitzgerald from Structural Perspectives Consultancy, and Stuart Millns from A. A. Associates Consultancy.
3. Hills (1971–72). The model is now on display in the Museum of Science and Industry in Manchester, where it is occasionally worked under steam.
4. A most rewarding and successful exception to this was the steaming of the locomotive *John Bull* by the Smithsonian Institution, Washington DC in 1980 and 1981 for the 150th anniversary of its construction by Robert Stephenson & Co. of Newcastle Upon Tyne and delivery to the Camden & Amboy Railroad, New Jersey, USA. See White (1981).
5. The possibility of a full-scale replica of Davidson's vehicle was at one time under consideration, in anticipation of which Ronald Jarvis (Member of the Newcomen Society) carried out much research. See *The Newcomen Bulletin*, 164, April 1996, pp. 13–14.
6. The author is pleased to acknowledge the views of Anthony Hall-Patch on the definitions of 'replicas' and 'reproductions'.
7. van Heslinga (1996). The author is grateful to Miss van Heslinga for making her paper available.
8. *De Witt Clinton* is no longer worked under steam, and is displayed in the Henry Ford Museum, Dearborn, Michigan, USA.
9. This replica Newcomen engine project was initiated by John S. Allen (past President of the Newcomen Society). The project was undertaken by a group based on the Black Country Museum, under its director, Ian Walden, the group including Mr Allen and fellow members of the Society.
10. Heyerdahl (1950).
11. Severin (1978): *The Brendan Voyage* (London: McGraw-Hill).
12. The Vickers *Vimy*, registration G-EAOU, is

13 van Heslinga (1996).
14 The *Hetton* locomotive is on display in the North of England Open Air Museum, Beamish, County Durham.
15 *Novelty* is on long-term loan to the Museum of Science & Industry in Manchester from the National Museum of Science & Industry. There is also another, operable replica of *Novelty* in the Sveriges Jarnvagsmuseum, Gavle, Sweden.
16 Memorandum on 'The Replica of the *Rocket*, Constructed by Robert Stephenson & Co., Limited, Darlington, Eng., 1929. For Henry Ford, Esq.' Three further replicas were subsequently made from the same specification and drawings, for the Museum of Peaceful Arts in New York, the Museum of Science & Industry in Chicago, Illinois, and a sectioned version for the Science Museum, London. See also Satow, (1979).
17 Replication of ships has been addressed on a previous occasion, namely Crumlin-Pederson and Vinner (eds) (1984), quoted by van Heslinga (1996).
18 Warren (1923).
19 Bailey (1978–79). Also Bailey (1984).
20 De Pambour (1836); De Pambour (1840). Also Anon. (1835).
21 The only known contemporary description of how to drive a *Planet*-type locomotive is contained in a note 'Locomotive by Stevenson [*sic*]' written by a clerk (probably Harris Dickinson) on behalf of Robert Stephenson & Co. to the Mohawk & Hudson Railroad Company accompanying the *Samson*-type locomotive *Robert Fulton*, dated 5 July 1831, Jervis MSS archive, Jervis Library, Rome, New York.
22 *Rocket* is on display in the London Science Museum. For a discussion on the reversing techniques for *Rocket*, see Davidson (1993): 'Reversing Rocket', *The Newcomen Bulletin*, 157, December 1993, pp.14–16
23 Liverpool and Manchester Railway, Minutes of the Board of Directors, National Archives, Kew, PRO Rail 371/2 and Rail 371/3, and Minutes of the Committee of Management, Rail 371/8 and Rail 371/10, *passim*. Also Woods (1838).
24 Marshall (1838).
25 Consultancy advice on the cause of the failure was provided by J. S. Allen and E. S. Burdon, formerly of British Rail, Derby.
26 De Pambour (1836); De Pambour (1840).
27 Patent No. 6484 of 1833, 'Stephenson's Improvement in Locomotive Engines', p. 3.
28 De Pambour (1836); De Pambour (1840). Also Lardner (1836).
29 Navier (1836); Wood(1831 & 1838).
30 Bailey (1980–81).

Bibliography

Anon. (1835): *Locomotives Stephenson circulant en Angleterre et en France* (Brussels: Champon).

Bailey, M. R. (1978–79): 'Robert Stephenson & Co. 1823–1829', *Transactions of the Newcomen Society*, 50, pp. 109–38.

Bailey, M. R. (1980–81): 'George Stephenson – Locomotive Advocate: The Background to the Rainhill Trials', paper read at Commemorative Symposium for the 200th Anniversary of George Stephenson's Birth (Sept. 1981), *Transactions of the Newcomen Society*, vol. 52, pp.171–79.

Bailey, M. R. (1984): Robert Stephenson & Co. 1823–1836, unpublished MA thesis, University of Newcastle upon Tyne.

Crumlin-Pederson, O and M. Vinner (eds) (1984): 'Sailing into the Past', paper read at the International Ship Replica Seminar, Roskilde, Denmark, 1984., quoted by van Heslinga (1996).

De Pambour, Chevalier F. M. G. (1836): *A Practical Treatise on Locomotive Engines Upon Railways &c.* (London).

De Pambour, Comte F. M. G. (1840): *A Practical Treatise on Locomotive Engines Upon Railways &c.*, 2nd edition (London).

Davidson, I. (1993): 'Reversing Rocket', *The Newcomen Bulletin*, 157, December 1993, pp.14–16.

Heyerdahl, T. (1950): *The Kon Tiki Expedition* (London, 1950), trans. from Norwegian book of 1948.

Hills, R. L. (1971–72): 'A One-Third Scale Working Model of the Newcomen Engine of 1712', in *Transactions of the Newcomen Society*, XLIV, (1971–72), pp. 63–77.

Lardner, Rev. Dr Dionysius (1836): *The Steam Engine Familiarly Explained and Illustrated &c.*, 6th edition (London) and subsequent editions.

Liverpool and Manchester Railway, Minutes of the Board of Directors, National Archives, Kew, PRO Rail 371/2 and Rail 371/3, and Minutes of the Committee of Management, Rail 371/8 and Rail 371/10, *passim*.

'Locomotive by Stevenson [*sic*]' written by a clerk (probably Harris Dickinson) on behalf of Robert Stephenson & Co. to the Mohawk & Hudson

Railroad Company accompanying the '*Samson*'-type locomotive '*Robert Fulton*', dated 5 July 1831, Jervis MSS archive, Jervis Library, Rome, New York.

Marshall, W. P.(1838): 'Description of the Patent Locomotive Steam Engine of Messrs Robert Stephenson and Co.', first published in Tredgold, T. (1838); and subsequently published as a separate publication (London).

National Geographic Magazine, 187 (5), May 1995, pp. 2–43.

Navier, M. (1836): *On the Means of Comparing The Respective Advantages of Different Lines of Railway and on the Use of Locomotive Engines*, trans. from the French by John Macneill (London).

The Newcomen Bulletin, 164, April 1996, pp. 13–14.

'The Replica of the "Rocket", Constructed by Robert Stephenson & Co., Limited, Darlington, Eng., 1929. For Henry Ford, Esq.', memorandum.

Severin, T. (1978): *The Brendan Voyage* (London: McGraw-Hill).

Satow, Michael G. (1979): '"Rocket" Reborn', *Railway Magazine*, 125 (942), October 1979, pp. 472–74.

Tredgold, T. (1838): *The Steam Engine: Its Invention and Progressive Improvement* &c., new edition revised and edited by W. S. B. Woodhouse (London).

van Heslinga, Els van Eyck (1996): 'The Use of Replicas, Risks and Opportunities', paper read at a conference of the International Commission of Maritime Museums, London (Sept. 1996).

Warren, J. G. H. (1923): *A Century of Locomotive Building by Robert Stephenson & Co. 1823–1923* (Newcastle-on-Tyne), pp. 252–91.

White, John H. Jr (1981): *The John Bull, 150 Years a Locomotive* (Washington DC).

Wood, N. A. (1831 & 1838): *A Practical Treatise on Rail-Roads &c.*, 2nd edition (London 1831), and 3rd edition (London 1838).

Woods, E. (1838): 'On the Relative Advantages and Disadvantages of Four and Six Wheels for Locomotives', paper read to the Institution of Civil Engineers, January 30th 1838, abstract contained in *The Civil Engineers and Architects Journal*, vol. 1, 1838, p.139.

DORON D. SWADE

Reconstructions as experimental history

HISTORIC COMPUTING MACHINES

Introduction

DURING THE LAST three decades there has been a number of major projects to reconstruct historic computing machines. These projects raise a raft of questions: museological, historiographical, and social. Large reconstruction projects invariably involve high levels of motivation and commitment. None is without heroic tales of setbacks, ingenuity, obstinacy, determination, expertise and sacrifice over sustained periods of time. We applaud, and so we should. But how do we justify these huge undertakings? To what category of object do these new-old machines belong – fictitious antiques, new primary sources, sculpted monuments commemorating episodes in an illustrious past? What is the social and historical value of such projects to their makers, to the scholarly community, and to visitors who view them? What, if anything, is learned that could not otherwise be learned? And what kind of history is it we do when we undertake these enterprises?

The principal preoccupation in this chapter is the construction, completed in 2002, of Difference Engine No. 2, a mechanical calculating engine designed in the late 1840s by the English mathematician and polymath, Charles Babbage (Fig. 6.1). This is one of seven projects used here as case studies. While the Babbage engine project is described in more detail than the others, all will be drawn upon to address issues agitated by such activity. The suite of projects includes:

FIG. 6.1
Babbage's Difference Engine No. 2, designed 1847–49 and completed in 2002. It consists of 8000 parts and weighs 5 tonnes.
(Science Museum/SSPL)

RECONSTRUCTIONS

1. Thomas Fowler's Ternary Calculating Machine (*c.*1840) – digital mechanical calculator using three-state logic devised by an impoverished untutored fell-maker (Fig. 6.2).
2. Babbage's Difference Engine No. 2 (1847–49) – mechanical calculating engine for automatically tabulating (calculating and printing) mathematical functions (Fig. 6.1 and Fig. 6 in art section).
3. Babbage's Calculating Engines (1832, 1849) – Meccano version of principal sections of Charles Babbage's Victorian designs for his Difference Engine and general-purpose computing engine, the Analytical Engine (Fig. 6.3 and Fig. 6b in at section).
4. Konrad Zuse's Z3 (1941) – a general-purpose programmable computing machine developed in Germany during the Second World War using electromechanical components (Figs 6.4, 6.5).

FIG. 6.2 (right)
Thomas Fowler's Ternary Calculating Machine (*c.*1840). Reconstruction with Mark Glusker.
(Bill Leung).

FIG. 6.3 (below)
Difference Engine No. 2 (*c.*1849) in Meccano (2008).
(Doron Swade)

HISTORIC COMPUTING MACHINES

FIG. 6.4 (left)
Konrad Zuse's Z3 (1941).
Physical reconstruction (1967).
(© Deutsches Museum).

FIG. 6.5 (below)
Konrad Zuse's Z3.
Reconstruction (part) (2001).
(Raul Rojas)

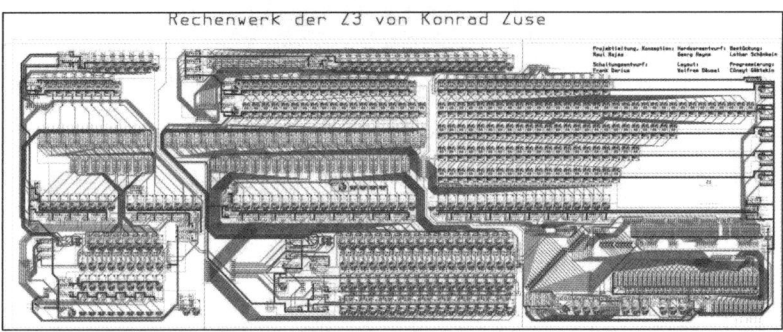

5 Colossus (1943) – a code-breaking computer built at Bletchley Park during the Second World War using electronic vacuum-tube logic and punched paper tape (Figs 6.6, 6.7).
6 Manchester 'Baby' (1948) (a.k.a. SSEM – Small Scale Electronic Machine) – the first computer to run an electronic internal stored program, a feature regarded as defining of the modern digital computer. An experimental machine using vacuum-tube logic and cathode-ray-tube memory (Figs 6.8).
7 Bletchley 'Bombe' (1940) – code-breaking machine built at Bletchley Park during the Second World War. A highly parallel machine using electrically driven mechanical logic (Fig. 6.9).

The original machines, span nearly two centuries and a range of technologies, in turn mechanical, electromechanical and, latterly, electronic. Each was revolutionary, or pioneered new practices, devices, or function in some significant way.

None of the machines survived and in each case there is no original to act as a physical datum on which to base a reconstruction. Some were deliberately destroyed, others accidently so. One was simply lost, and another did not exist

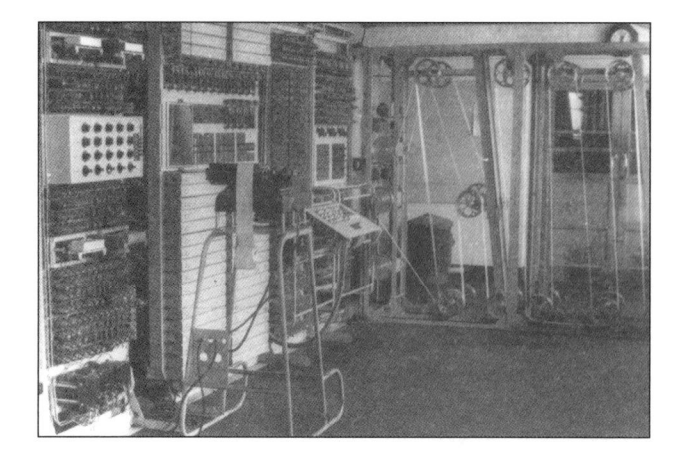

FIG. 6.6
Colossus (1943).
(Science Museum/ SSPL)

FIG. 6.7
Colossus Reconstruction, with Anthony Sale.
(Tony Sale)

FIG. 6.8 (left)
Manchester 'Baby' (1948).
(School of Computer Science, Manchester University)

in the first place. Specifically, after After the Second World War the original Colossus code-breaking computers were dismantled or destroyed. The Manchester 'Baby' was a developmental machine and cannibalised for spares sometime after its landmark run in June 1948.[1] Zuse's Z3 was destroyed in an air raid in 1944,[2] and Thomas Fowler's three-state-logic calculator, made in wood, was lost without trace. In the case of Babbage's Difference Engine No. 2, there was no contemporary original to begin with. Babbage made no attempt to build the machine and it survived only as a set of design drawings until 1986 when the first metal was cut. The physical machine, completed in 2002, is strictly not a reconstruction, but an original completed 160 years after its design.

The distinction between construction and reconstruction prompts reflection on whether there are meaningful distinctions between 'restoration', 'replication', 'reconstruction', 'reproduction' and 'simulation'. By restoration we convey the sense of returning an object to its original state. In the case of devices or machines, restoration may or may not include returning the object to working order. The livery, upholstery, and paintwork of a vintage car, for example, may be restored, but with the engine left non-operational. The essential point here is that restoration invariably presupposes the existence of the artefact to begin with. Replicating a machine usually means creating a copy using an existing contemporary object as a reference datum. The etymology of the word (*replicare* – to fold back) implies a sense of 'taking an impression'. When a replica-maker minimises the physical differences between the original and the copy, the replica is 'exact'. But replicas can be deliberately inexact. Replicas can be scaled copies either larger or smaller than the original, and

FIG. 6.9
'Bombe' Reconstruction by John Harper (2009).
(John Jackson)

there are non-working replicas made without any intention operational intactness. Here again, the process of replication usually implies the existence of an original as the model.

The word 'reproduction' gets us into murky waters and the categories begin to break down. 'Reproduction' usually refers to the act of making a copy, sometimes exact, but usually with the implication of imitation, especially in the context of furniture making. In the way we commonly use the word, the original of a reproduction may or may not exist. Finally we come to 'reconstruction', which can usefully be reserved for the process of recreating an object of which there are no surviving intact examples. In cases where part of the original survives, as in the remaking of the Viking warship, for example, described in chapter 4, the newly-made boat is a hybrid of replication (of the parts or sections that survive) and reconstruction (of those that did not). These distinctions should perhaps not be pursued too vigorously. Casaubon's great enterprise in George Eliot's *Middlemarch* warns us against attempts at comprehensive taxonomies.

So far we have assumed that the act of duplication, in whatever form, involves creating something physical. However, there is a process involving duplication that is not physical – simulation. In the context of computing, simulation is the duplication, in software, of a machine's performance, behaviour or function, on a different hardware platform. Simulation can be seen as 'logical replication': a non-physical form of duplication in which all essential logical predicates are preserved. In cases where there is no physical original, simulation can be seen as logical 'reconstruction'.[3]

Not all reconstructions attempt physical fidelity to the original. In the case of Zuse's Z3, Raul Rojas argues that the purpose and value of the reconstruction is educational – to convey the logical structure to later generations. Rojas makes no attempt to replicate either the contemporary hardware or physical appearance (*see* Fig. 6.5).[4] Another example is Tim Robinson's Meccano implementations of Babbage's Difference and Analytical Engines. Here again there is the deliberate severing of any expectation of physical fidelity or period appearance (*see* Fig 6.3). The Meccano machines, using standardised parts, are working realisations of Babbage's designs.[5] However, the medium of Meccano is far from the manufacturing practices and operational conventions of Babbage's time when parts were individually machined or crafted to order. A third example of the abandonment of physical realism is in the simulation of the Ferranti *Pegasus* computer (Figs 6.10, 6.11a and 6.11b). The illustrations show that little attempt is made to create a visual facsimile of the console controls beyond position in the console layout. But each operator control and function is scrupulously represented. The issue here is not one of visual fidelity, but of the simulation's operational and behavioural exactness. The virtual

FIG. 6.10,
Ferranti *Pegasus* (1959).
(Science Museum/SSPL)

FIG. 6.11a
Ferranti *Pegasus* (1959),
operator's console
(Doron Swade).

FIG. 6.11b
Ferranti *Pegasus* (1959), simulation of operator's console (1990).
(Doron Swade)

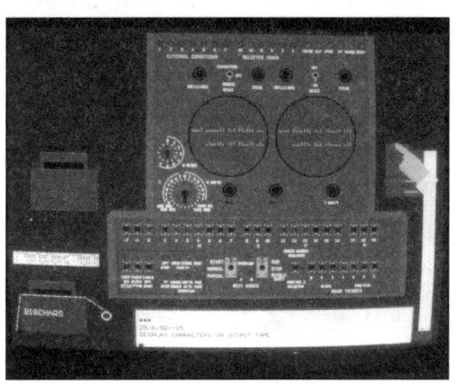

machine can be operated (programs loaded from paper tape, executed, and results punched or printed) by clicking on the controls depicted on screen, and the responses are exact even to the extent of the signal waveforms on the oscilloscope screens. The concern here is operational rather than visual realism.

In each of the seven cases, documentation for the original is incomplete or non-existent. Documentation for Colossus was protected under the Official Secrets Act and has never been fully released. The case of Thomas Fowler's calculator is extreme. Not only did no original object survive, but there are no plans or drawings. If plans ever did exist, they were never released by Fowler who was paranoid about his invention being exploited, and with good reason.[6] There are a few written descriptions of Fowler's calculator, one by Augustus de Morgan, who saw the device demonstrated in the 1840s and described the principle of its arithmetic operation.

In the case of Colossus and the Manchester 'Baby', some fragmentary notes and circuit sketches survived in the private papers of engineers and technicians involved in its design and construction. Physical scale was confirmed from contemporary photographs scanned into a computer, the perspective corrected, and dimensions deduced. In the case of Colossus, a project shrouded in wartime secrecy, a few original components survived in private hands. These were pilfered, or should we say specially cherished, by a few technical staff and kept.[7] These magpie instincts for mementos provided occasional bridges to an otherwise lost artefactual past. In the case of Babbage's Engine, a full set of drawings survived intact. The set is complete, but the drawings are insufficiently detailed to serve as a specification for the manufacture of parts. The drawings needed to be interpreted and a vast amount of supplementary detail supplied before anything could be made. In the case of Fowler's calculator, one source was an image in a stained glass window dating from around 1870 in a church in Devon,[8] and the reconstruction is to some extent necessarily speculative but one that is informed by, and in complete agreement with, the fragmentary sources that survive. In two cases an unexpected source of information was recovered memory. Exposure of several original pioneers to the reconstruction while in progress unlocked technical detail thought to be forgotten. Debriefing the original makers had historical value in itself.

The absence of an original artefact, coupled with incomplete or non-existent documentation, raises inevitable issues of authenticity. This is explored more fully below in the description of construction of the Babbage Engine.

With the exception of Fowler's calculator, all the machines reconstructed are large. Colossus uses some 1500 vacuum tubes and fills a large room. Babbage's Difference Engine weighs five tonnes, has 8000 parts, and measures 3.4 metres long and 2.1 metres high. With one rare exception all the projects called for substantial resources – materials, time, funding and expertise. Many of the reconstructions were led and staffed by specialist volunteers, some retired. Most of the projects would have been unaffordable at commercial rates and would not have been undertaken or realised without volunteer effort. All the projects involved high levels of contemporary and current technical expertise and research. All took several years to complete. The Babbage Engine, for example, was built over a period of 17 years and took about £1 million of direct costs excluding research time and staff salaries.

Babbage's Difference Engine No. 2: Background

Charles Babbage (1791–1871) is widely recognised as the great ancestral figure in the history of computing (Fig. 6.12). The designs for his vast mechanical calculating engines incorporate practically all the essential logical features of the electronic digital computer and rank as one of the startling achievements of the nineteenth century. He is often described as the first computer pioneer and is increasingly celebrated for his ground-breaking work on computing machines.

HISTORIC COMPUTING MACHINES

FIG. 6.12 (far left) Charles Babbage (1860). (Science Museum/SSPL)

FIG. 6.13 (left) Allan Bromley, London (1990). (Doron Swade)

Babbage designed Difference Engine No. 2 between 1847 and 1849. He offered the plans to the British government in 1852 partly in fulfilment of an earlier obligation to build a calculating machine that would automatically calculate and print mathematical tables on which astronomy, navigation, commerce, science and engineering increasingly relied.[9] This was his second design for a Difference Engine. The first engine project, conceived in 1821, was bankrolled by government and abandoned unfinished in 1833 after a decade of design and development, and vast cost. The government rebuffed Babbage's offer of the new designs. He made no further attempt to build the Engine in whole or part and it remained unbuilt in his lifetime. With the exception of small test or demonstration assemblies, Babbage failed to build any of his vast calculating engines and succeeded only in completing a few small test or demonstration assemblies. Historical speculation about the reasons for this was part of the challenge of attempting to do what Babbage himself had famously failed to do. The drawings for Difference Engine No. 2 remained undisturbed in their specially-made wooden storage-and-display case until 1979 when the Australian computer scientist Allan Bromley began to study them in detail (Fig. 6.13).

Description

The main elevation of the Engine is shown in Fig. 6.14 and is the drawing that is most evocative of its overall shape. The drawing depicts a machine 3.4 metres long, 2.1 metres high, and 1.2 metres deep at its deepest, and consisting of 8000 parts. Number values are represented by gear wheels called 'figure wheels' engraved with the numbers 0 to 9 (Fig. 6.15). The figure wheels are

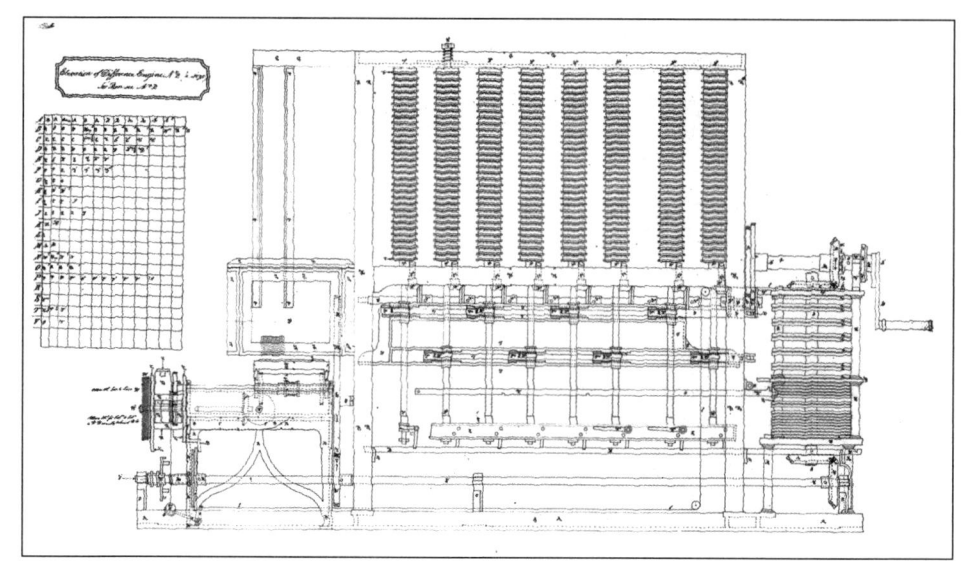

FIG. 6.14
Elevation of Difference Engine No. 2 (c.1849).
(Science Museum/SSPL)

arranged in eight columns, each with 31 wheels (Fig. 6.16). Each column represents one 31-digit decimal number with units on the lowermost wheel, then tens, hundreds, and so on, up the column. The machine adds the 31-digit number on the right-most column to the column immediately alongside and this is repeated for each column in turn.[10] Each cycle of the machine repeats the sequence of seven additions to produce each next result.

The mathematical principle of the Engine is called the 'method of finite differences' from which the machine derives its name. The method, well known in Babbage's day and used by human calculators in the manual calculation of tables, has the convenience of using repeated addition to do what otherwise would require division, multiplication, general addition and subtraction. Mechanisms for addition are significantly simpler than those required for multiplication and division, and so the principle of differences offers both simplicity and engineering economies.

The machine is operated by cranking a handle shown on the right of the drawing. The hand crank drives a stack of 28 cams in a stack immediately alongside, which orchestrate the internal lifting and turning motions

FIG. 6.15
Figure wheel for Difference Engine No. 2.
(Science Museum/SSPL)

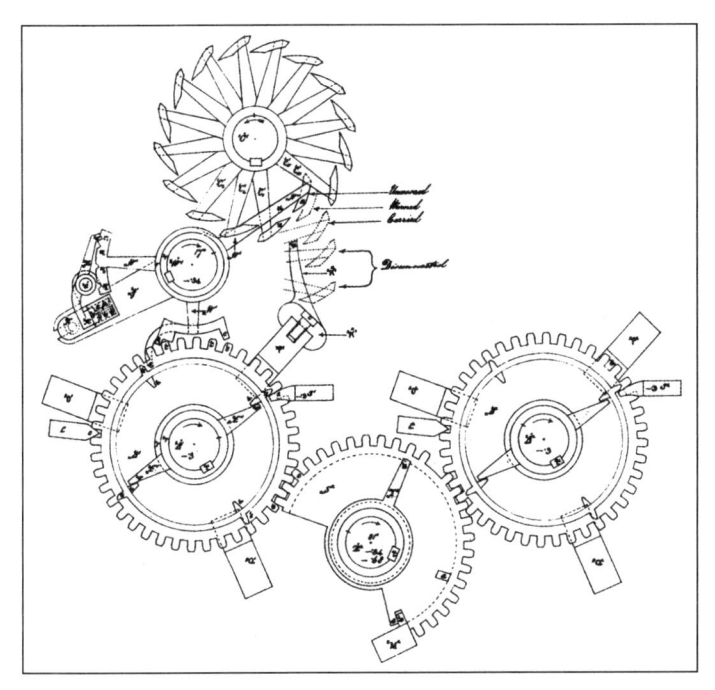

FIG. 6.16
Addition and carriage mechanism for Difference Engine No. 2 (c.1849). (Science Museum/SSPL)

required for repeated addition by the method of differences. Starting values are pre-calculated and entered on the figure wheels by hand. Thereafter each cycle of the Engine (one turn of the handle in the original design) produces each next result in the table. Each tabular result appears in turn on the left-most column and is automatically transferred to the output apparatus: the assembly on the left of the drawing. The Engine calculates and tabulates any seventh order polynomial and automatically prints and stereotypes results to 30 decimal places.

The output apparatus prints an inked hardcopy of the results as a record and at the same time impresses the result on soft material, wet plaster probably, held in trays to produce a stereotype plate used to cast printing plates for use in a conventional printing press. The output apparatus, which prints and stereotypes the results, incorporates the first complete design for the automatic typesetting of numbers. The stereotyping apparatus is unexpectedly versatile in that the format of the results on the page is programmable. By selecting from a suite of pattern wheels, a number of features can be altered – line height, margin width, the number of lines of a page and gaps between groups of lines, for example. Another set of pattern wheels allows the results to appear in one, two, three or four columns, and there is the option for printing line-to-line with automatic column wrap to the start of the next column, or column-to-column with automatic line wrap at the end of a line. Results are stereotyped in two font sizes simultaneously and the line-heights for the two fonts are adjusted automatically. The device and its controls are entirely mechanical.

RECONSTRUCTIONS

Sources

The drawing in Fig. 6.14 is one of a set of 20 left by Babbage supplemented by a few derivative tracings. From Babbage's papers and internal references, the set of drawings appears to be complete: there is no evidence that any are missing and none has suffered the hazards of damage or dirt by exposure to a workshop environment. Once completed, the drawings appear not to have suffered subsequent improvement. This was unusual: Babbage habitually tinkered and endlessly modified. While the drawings depict the parts, their relationship to each other, and constitute a near-complete technical description of the mechanisms, they are insufficiently detailed to serve as a specification for the manufacture of parts in a workshop. They contain no information on what materials parts are to be made from, method of manufacture, precision, finish and typography for the engraved numbers. There is also no information on the order of assembly, setting up, lubrication or testing.

There are other deficiencies. There are dimensioning inconsistencies where the same part is shown differently sized on different drawings. There are instances of omitted mechanisms and at least one case of a redundant mechanism. The inking rollers, for example, have no visible means of support or motive power, and there is an expensive mechanism in the calculating section for the carriage of tens for which no circumstance can arise in which it would be needed. There are layout anomalies. The spacing between the second and third right-most columns in Fig. 6.14, for example, is larger than the spacing for the other columns. This variation in pitch is inconsistent with correct working and conflicts with layouts elsewhere.

By far the most damaging deficiency of the drawings is an error so simple that it precipitated a crisis of confidence in our understanding. As drawn, the direction of rotation of the wheels is inconsistent with correct working. The essential logical function of the machine is the arithmetical addition of one decimal number to another and the carriage of any tens that results. The mechanism for addition and carry, shown in Fig. 6.16, is repeated some 210 times in the machine and is the basis of the design. When the number on the left-most figure wheel shown exceeds ten, the cluster of axes and components connected to it increments by one the wheel immediately above. To operate correctly, the figure wheel must rotate anticlockwise. But the direction of rotation shown in the notational arrows, and from an understanding of the mechanism, drives the wheel clockwise and the machine would jam or the curved finger (shown partially dotted) would deform or break.

Babbage makes no concessions to anyone seeking to understand the logic, design or working of his machines. With one trifling exception, he left no textual description of how the mechanisms work, his intentions, or explana-

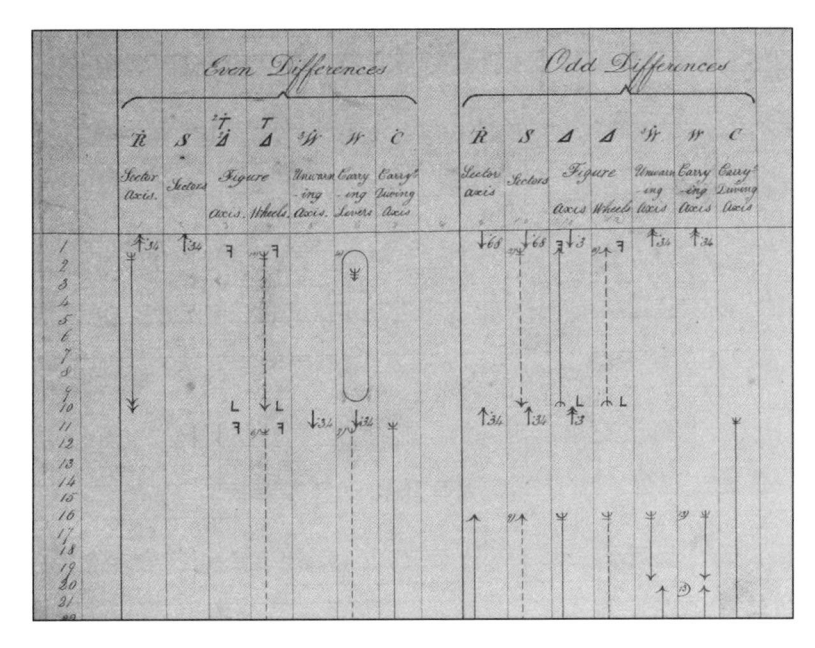

FIG. 6.17
Mechanical Notation, timing diagram form.
(Doron Swade)

tion for the design. The exception is a short description of the procedure to set up the initial values before the start of a calculation. The description is 147 words long and contains a significant though easily remediable oversight. This aside, the function of the mechanisms and the operation of the machine had to be decoded from the drawings and the additional information contained in Babbage's Mechanical Notation.

The Mechanical Notation is a symbolic language of Babbage's own devising that he invented to assist in the design and description of his machines. It is an elaborate system of signs and symbols that describes the connection and motional relationship (fixed, rotational, reciprocating, continuous, intermittent, derivative, prime, etc.) of any part to the rest. The Notation uses several alphabets with a range of superscripts and subscripts and there are three distinct forms. There is a tabular form that features alongside drawings of mechanisms (as in Fig 6.14, for example). There are timing diagrams that indicate the sequence and phasing of motions in relation to each other (Fig. 6.17) and there is a flow-diagram version that shows the overall progression of action from inputs to outputs. An example of the flow-diagram form is shown in Fig. 6.18a and the apparatus that it describes (the output apparatus) is shown in Fig. 6.18b.

The Mechanical Notation is idiosyncratic and thoroughly Baroque in its intricacy. As with the drawings, there is no textual description, explanation or key to the notational scheme. Babbage was inordinately proud of the Notation, regarding it as one of his finest inventions, and he used it extensively to

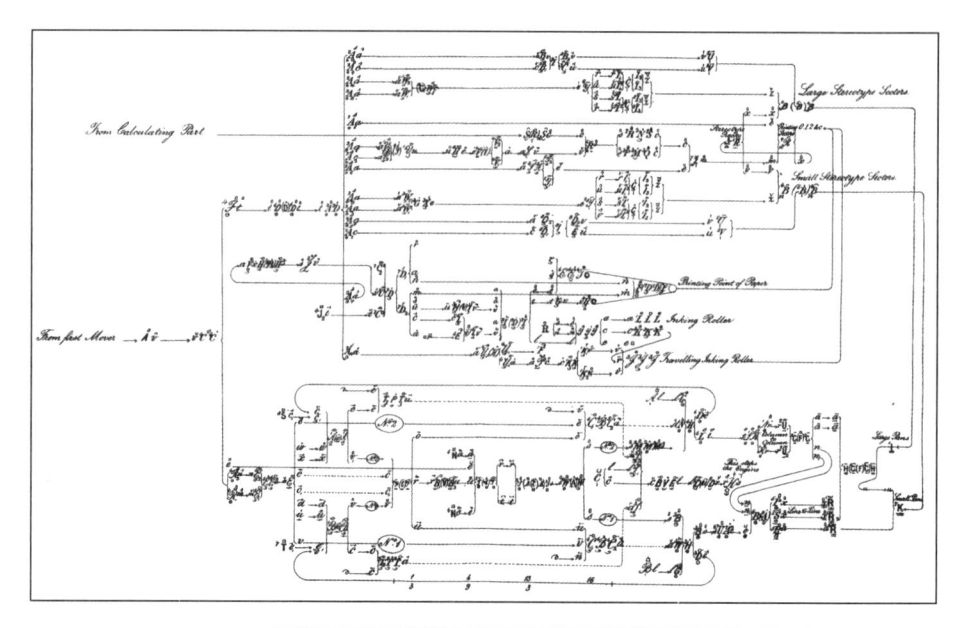

FIG. 6.18a (above)
Mechanical Notation, flow diagram form.
(Doron Swade)

FIG. 6.18b (right)
Output apparatus, Difference Engine No. 2 (2002).
(Doron Swade)

optimise the designs and keep track of the complex trains of actions. However, it has enjoyed spectacular obscurity ever since. We used it only occasionally – to resolve issues that resisted solution using the drawings alone. But these occasions were few, and for the most part our understanding of the machine was based on the drawings alone and these remained the principal source from which to reconstruct his intentions.

Motivation and historical thesis

Babbage failed to build any of his engines in their entirety. Almost all historical accounts give as the reasons for Babbage's failure the limitations of nineteenth-century engineering. It is usually either directly stated or overtly implied that parts could not be made with sufficient precision to meet the exacting needs of the design. The circumstances surrounding the collapse in 1833 of Babbage's first attempt to build his engine are complex: a walk-out by Joseph Clement, Babbage's chief engineer, runaway costs, funding controversies, little credible progress after eleven years, personal vendettas with men of influence, and political instability in Whitehall which funded the project. Experts in England, France and Sweden denied that there was any real need for the engines and George Biddell Airy, Astronomer Royal and *de facto* scientific advisor to government, damned Babbage's engines as 'useless' in his official advice to government.[11] What is curious is that nowhere in these complex considerations does limitation of technology feature as a reason for the abandonment. There were three Royal Society reports (1823, 1829 and 1830) attesting the feasibility, utility and progress of the project. Leading engineers, Donkin, Nasmyth, Whitworth, Clement, and Fairburn, knew of the ambitions for the engines or were directly involved in the project. None warned that the engine could not be built. Babbage's own knowledge of manufacturing techniques and machine tools was encyclopaedic and he had conducted a comprehensive tour of workshops in England and on the Continent. Nowadays those who with intimate knowledge of nineteenth-century engineering, craft and machine tools, dismiss out of hand that achievable precision was a limiting factor.

There is clearly a set of historical theses to explore and challenge. Was the established account, that contemporary technology was the obstacle to completion, correct? Could Babbage have built the engine, or did the circumstances surrounding the failure of his own attempts mask the technical or logical impossibility of his schemes? Had he built the Engine, would it have worked? And there was a related question, that of Babbage's reputation. Ever since his failure to complete an engine, there has been an ambiguity in history's assessment: was he an impractical dreamer or a designer of the highest calibre, gifted with inspired originality? These are some of the questions that began to nag and tease, and this was part of the historical motivation for the reconstruction – to resolve issues of historical dispute through experimental history.

Authenticity

With the issue of achievable precision at the centre of the debate, it was clear that making parts more precisely than Babbage could have would provide little direct insight into whether the engine would have worked had Babbage built it. There was no standardisation at the time and the use of a manufacturing 'tolerance' – the margin within which the dimension of a part will fall – and the interchangeability of parts that this allowed, did not affect practice until the mid-1850s, well after the time-critical window of Babbage's own efforts. Babbage would have instructed his men to 'make to fit' and parts would have been fettled and adjusted by hand until the mechanism worked. Precision for Babbage was not quantified through tolerancing. For us, precision had to be quantified both for manufacture in a modern machine shop as well as to address the historical issue of the role (if any) of precision in Babbage's failures. Measurements had been made by Michael Wright, then curator of mechanical engineering at the Science Museum in London, and Allan Bromley, on parts for his first Difference Engine left by Babbage, to determine how precisely repeated parts could be made in Babbage's time. By taking micro-measurements of contemporary duplicated parts, they established that Babbage, or rather Babbage's engineer, Joseph Clement, could produce repeated parts to at least 1.5 thousandth of an inch. This became the absolute maximum precision throughout and no part was made more precisely.

Without manufacturing techniques with inherent repeatability, die-casting or stamping, for example, Babbage had to make parts one at a time. This was time-consuming and expensive and Babbage lost credibility having only a modest partial assembly to show after eleven years of design and manufacture. So as not to repeat history, we took an early decision not to use period manufacturing techniques or machinery. Though we had wishful notions of building the machine in public view using contemporary machines manned by workshop staff in period costume, we resisted. Instead, we unashamedly used modern methods – numerically controlled machines – for example, but in no instance made parts more precisely than we knew from measurement Babbage could have achieved, but possibly by different means. We welded where Babbage would have forged, and dressed the weld for period look. Such measures nowhere compromised the essential historical agenda.

To provide detailed manufacturing specifications for all 8000 parts, we needed to resolve inconsistencies in the drawings, the design issues relating to omitted and redundant assemblies, and provide information otherwise missing, not least of which was to specify material for each part. Modifying Babbage's description in any way posed a central curatorial dilemma. If the original designs were altered in any way, in what sense could a working engine vindicate Babbage's work? The final object, if achieved, would be an intriguing piece of sculpture, a fine monument to Babbage, but one that would say little about the overall viability of Babbage's schemes. Yet if the drawings were not modified, there was no expectation that the machine would work.

The way out of the impasse was to free ourselves from the constraints of viewing the original drawings as a sacrosanct record of unimpeachable perfection, but instead to view the enterprise as the resumption of an engineering project arrested in 1849 in an incomplete state of development. Had Babbage proceeded from design to manufacture, the deficiencies in the drawings would have become evident. There would have been no way of avoiding them. Our task then was to seek solutions informed by the best knowledge of contemporary practice, a detailed knowledge Babbage's style of working and of the surviving physical relics of his own efforts. Each time we were faced with a question, we asked what Babbage would have done, and sought historically sensitive solutions using the best advice available. Throughout we worked in defence against the hypothetical charge, 'you built the Engine but Babbage could not have'.

For each design issue we first sought solutions that Babbage had used elsewhere and there were few instances in which prior solutions could not be found. The second principle was that of reversibility, that is, to allow the machine to be returned to the state depicted in the original drawings by ensuring any added assembly could be demounted or any modification reversed. Some of the modifications were precautionary, others remedial. One precautionary measure was the addition of counterweights to relieve the deadweight of a set of 30 vertical racks in the output apparatus. This was to prevent possible run-away, i.e. to pre-empt the possibility of the racks free-falling under their own weight in a small unguarded part of the printing cycle. Another was the addition of strong support brackets and legs to relieve possibly excessive stress on the mounting bolts fixing the output apparatus to the main frame. Yet another was the addition of a four-to-one reduction gear added to the main crank in case it transpired that a human, however determined, could not exert sufficient force to drive the machine. These, and other additions, are demountable. There were modifications and additions, the need for which was not foreseen and which were made during the build. No one in Babbage's time had built a machine of such complexity, and what was not foreseen by Babbage in terms of assembly, fault-finding and maintenance, is revealing about the state of knowledge at the time, and a stark reminder of what we now take for granted in our management of complex systems.

The choice of materials was resolved through expert advice from Michael Wright, whose knowledge of nineteenth-century workshop practice is both intimate and extensive, Allan Bromley who had the most detailed knowledge of Babbage's designs of this and his other engines, and a gifted and methodical mechanical engineer, Reg Crick, who was contracted by the Science Museum to prepare detailed piece-part drawings. Gunmetal, cast iron and steel were the principal materials, and our advisory panel selected which of these each of the

8000 parts were to be made of. Gunmetal, a form of bronze, was in common use, but contemporary composition varied. To ensure a close match, we did composition tests using X-ray microanalysis on a small part left by Babbage and then sourced a close modern equivalent.

The possible function of the redundant mechanisms in the drawings was examined carefully, and once all interpretations were exhausted without outcome these mechanisms were dispensed with. In one instance the purpose of a device was unclear. It was nonetheless scrupulously retained. Its purpose became evident when built and its function was a revelation of ingenuity. Omitted mechanisms were designed using a minimum of parts with motions derived from neighbouring parts in keeping with Babbage's style and contemporary practice. The reason for the omissions was not always clear. It is possible that absence was oversight. It is also possible that provision of an omitted feature was regarded as relatively trivial and left to Babbage's engineer Joseph Clement, to supply on his own initiative. The absence of any drive mechanism for the inking rollers is an example of an obvious omission and one that needed solution ahead of the build. There was at least one omission that was unforeseen that became evident in operation. There is a small window in the timing cycle during which the column nearest the crank handle is left unsecured and the numbers might be subject to derangement. An additional device was added to lock the column during the short period of vulnerability with the driving forces derived from parts alongside. There are other examples of added mechanisms. All of these were resolved within the protocols described.

The flaw in the carriage mechanism drawing, where the direction of rotation of the wheels is inconsistent with correct operation, needed to be resolved ahead of any practical attempt at construction. The error is a basic one and difficult to account for. It is possible that it is a simple mistake by the draftsman, who could well have had an imperfect understanding of the device, and which passed unnoticed by Babbage. We also speculated that the error was deliberate and incorporated as a protection against industrial espionage. Whatever the reason, we had satisfied ourselves that it was not an artefact of our own lack of understanding. The mechanism is repeated 210 times so the stakes were high. Of three main solutions considered, we chose the one that preserves the logic and the layout of the arrangement, but mirror-images the mechanism for alternate axes: if one ignores the arrow notations on the drawing, the mechanism will work as shown for the odd-numbered axes but needs to be mirrored (oppositely handed) for the even axes.

A trial piece was built to test the new arrangement and verify our understanding (Fig. 6.19). The device adds two-digit numbers to produce a three-digit sum and the eleven separate steps in the process are activated by hand by

FIG. 6.19
Trial piece for Difference Engine No. 2 (1989).
(Science Museum/SSPL)

lifting and turning the knobs and sliders above the top plate. The trial piece worked as intended and also served as a staging post of credibility for the team, as well as for those we sought to convince that undertaking to build a five-tonne machine was not an act of irrational exuberance. It also had an unexpected benefit – as an aid to visualising the operation of the mechanism during the preparation of the full working drawings. Time and again we returned to the piece to assist understanding.

Little is known about the working relationship between Babbage and his engineers and draftsmen, and it has been a source of speculation as to how much Joseph Clement contributed to the ingenuity and design of many of the mechanisms, how closely the mechanisms were specified by Babbage, and the nature of any supervisory role Babbage might have had. Seeking an explanation for the anomalies in the drawings gave suggestive insights into these relationships. A possible explanation for the flaw in the carriage mechanism layout is that Babbage gave his draftsman a sketch of the design for one digit and instructed that this was to be repeated for each of the axes. Without a detailed understanding of the design, the draftsman would not have known that the mechanism was not needed on the first axis, and that the mechanism, handled only one way, would work on only half the axes. It is also evident from the smaller errors that were easy to correct but were not, that supervision by Babbage was not scrupulously close.

The Build

The first suggestion to construct the Difference Engine was made in 1985 by Allan Bromley who proposed to the Director of the Science Museum that this machine be built in time for the bicentenary, in 1991, of Babbage's birth. I was curator of computing at the time of Bromley's proposal and I took on the project as a personal and professional mission. I assembled a small team of two mechanical engineers who prepared new piece-part drawings and assembled the machine.

The original 20 drawings were expanded to some 160 drawings specifying in manufacturing detail each of the 8000 parts. Conventional drafting techniques were used: pencil, paper and drawing boards. Manufacturing the parts was beyond the production capacity of the Science Museum's in-house workshops, and parts manufacture was contracted out to 46 separate subcontractors. Metal was first cut in 1986 and actual assembly, fitting, modifying and testing of the engine, took a total of four years. This was in two main stints of two years each, the first ending in 1991 with the completion of the calculating section, the second ending in 2002 with the completion of the output apparatus. The rest of the 17 years was spent on research, design, drawings preparation, parts manufacture, fund-raising and inevitable institutional politics.

The Engine was assembled in public view at the Science Museum in London, and the public was free to engage and interact with the engineers (Fig. 6.20). The machine works as Babbage intended. It remains on public display at the Science Museum and is demonstrated from time to time.

Lessons

The machine was built to tolerances achievable by Babbage, using materials available to him, and introducing reversible modifications to the design using solutions used by Babbage elsewhere. Though the number of small modifications was many, and the litany of deficiencies in the drawings long, no flaws were found that compromised the essential logic or principles of the design. Given that the machine works, and was built with conscientious attention to issues of authenticity, the project provided definitive answers to some of the earlier historical questions about Babbage's standing. We can conclude that achievable precision was not a limiting factor and is not in itself the reason Babbage failed. Further, had Babbage built the machine, it would have worked. Finally, the fact of its working vindicates Babbage as a designer of extraordinary practical inventiveness. We can now demonstrably say that Babbage was not an impractical dreamer, but a designer of the highest calibre. These assertions could have been made in the past, but not with the conviction of practical evidence.

There were lessons learned that in principle could have been learned without the construction, but were not. A device for locking of the figure wheels

FIG. 6.20
Assembly of Difference Engine No. 2 in public view
(1990) with Reg Crick (right) and Barrie Holloway (left)
(Science Museum/SSPL)

during the carriage cycle was not fully understood until built and operated, and this revealed a level subtlety and sublime foresight that awed us all. Another contingent finding was the maximum speed of operation and the limiting factor determining this. We found, after the mechanisms had eased with use, that the maximum speed of the machine was ten results per minute. Also that the limiting consideration was a gear that gets thrown out of mesh, something remediable by a stronger spring.

Having a physical machine to operate focused attention on issues of operational procedure and gave insights into the implications for contemporary table making. How many people would be needed to operate the machine, what level of mathematical expertise was required to prepare the initial values, how long was it practical for an operator to crank the handle without fatigue, how difficult was it to ensure a continuous supply of soft plaster of just the right consistency, how reliable was the machine, how long would it need to operate without error to produce a single volume of tables, how sensitive to wear was its reliability, and so on. Simply put, having a machine allowed us to revisit the question of whether or not automatic tabulation by machine was a practical proposition in the nineteenth century. Experience operating the machine allows us to make new assessments of contemporary arguments about the utility of these machines, an issue of heated contention in Babbage's day.

Some of the insights were unexpected. One such is the issue of digit precision. The Engine tabulates to 30 decimal places. Instruments of the day were limited to three or four figures of accuracy. There has been no definitive answer as to why Babbage worked to so many digits of precision. Many have offered reasons. None are conclusive. The operational set-up of the machine offered a new insight. We discovered that the Engine allows the calculation to be split horizontally, allowing two functions to be tabulated at once: the precision can be distributed between problems. This feature would allow one part of the Engine to simply increment the argument for each engine cycle and this would be printed automatically on the same line as the result, so eliminating the need for a separate calculating run. We were not looking for this finding. Hands-on experience setting up the machine for a calculation suggested this unforeseen feature. Distributing the full digit capacity of the machine between the argument and result, or between any two functions, contributes to an ongoing debate about Babbage's apparently extravagant digit precision.

In principle, one could have analysed, inferred and foreseen answers to some or all of these questions. But nobody did. Like playing chess, there is a limit to the number of moves one can foresee. A few counterfactuals in a train of reasoning rapidly escalates uncertainty and undermines confidence in any conclusions speculatively drawn. Some of what we learned was irreducibly empirical and praxis gave conviction to that knowledge that nothing else had.

Conclusion

Reconstructions draw one into a level of intimate detail with the object that rarely occurs through other means. All the projects mentioned here produced contingent findings not foreseen and in some case unforeseeable through analysis or theory. The process of researching, producing and using the objects, offered insights into contemporary ideas, design and operational practice, and allowed new historical assessments of their practical effectiveness, and with this, the implications for contemporary use.[12] Operating the devices provided tacit knowledge. We learn what it might have been like. There is the benefit of documentary completeness: the process of specifying the machine and researching its detail involves the collation and co-location of known material. In several instances, new material emerged, attracted by the public visibility of the project. There were also instances of recovered memory where the reconstructed object unlocked seemingly forgotten detail in the memories of the original pioneers.

There were the benefits of 'social capital' in several forms. The projects gave meaningful work to veteran experts upon whom many of the projects relied, and they recreated professional arenas for pioneers and practitioners. They felt valued and took renewed pride in their expertise and in the practice of past crafts. Reconstructions of lost or incomplete artefacts enhance visitor experience both as spectacle and educationally. Machines as objects of display have pedagogic value both explicit and tacit. They act as a nucleus of visitor attention, define the boundaries of historical discourse, and stimulate new questions and new lines of thought.

Reconstructions are cultural objects in their own right. They memorialise important or extraordinary episodes in history, act as monuments to and placeholders for significant pasts, and as generational bridges.

The Babbage Difference Engine is a sumptuous piece of engineering sculpture. In operation it is a piece of mechanical choreography of arresting beauty that no Victorian ever saw. Witnessing it work prompts us to experience the prospects of a future already past.[13] Such wonder protects us from the 'enormous condescension of posterity' in our attitudes to superseded arts and crafts.[14] Through reconstruction, we rediscover lost arts and renew our respect for those that went before.

Notes

1 Burton (2005), in *IEEE Annals of the History of Computing*.
2 See Zuse (1993).
3 For an expansion of this argument and the implications for artefactual identity, see Swade (2003a) in *Interdisciplinary Science Reviews*.
4 Rojas, Darius, Goktekin and Heyne (2005) in *IEEE Annals of the History of Computing*.
5 For details of Robinson's Meccano constructions, see www.meccano.us/.
6 For background to Fowler's refusal to release drawings, see Swade (2000). For an account of Fowler's machine and its reconstruction, see Glusker, Hogan and Vass (2005) in *IEEE Annals of the History of Computing*.
7 Sale (2005) in *IEEE Annals of the History of Computing*.
8 Glusker, Hogan and Vass (2005), p. 19.
9 For the history of Babbage's efforts to construct his calculating machines and an account of the construction, see Swade (2000).
10 The situation is a little more complicated than suggested. The numbers are not added from one column to the next in a linear sequence. Rather the odd columns are added to the even columns during the first half-cycle and the evens to the odds in the second half cycle. The technique is quite subtle. The end result is the same, but the phased process is more time efficient. For details, see Swade (2005) in *IEEE Annals of the History of Computing*. For further details of this and the working of the whole Engine, see Swade (1996).
11 For an accessible account, see Swade (2000), ch. 7. For an analysis of Airy's role, see Swade (2003).
12 David Link argues compellingly that working for objects the specifics of purpose quickly become lost and that recreating the object and re-enacting its operation is crucial to the recovery and preservation of know-how. He advocates archaeology as a more appropriate model for the study and preservation of technological artefacts even of recent vintage and

uses an early Second World War Bombe as a model. He compares objects not so treated to Kafka's 'Odradek', a paradoxical object in the puzzling no man's land between the function-specific purpose of objects and the apparent functional purposelessness of humans. For reference, see Link (2010), p. 233, fn. 49.
13 This feature of the engine was articulated by Spufford (1996).
14 The phrase is E. P. Thompson's. See Thomson (1991).

Bibliography

Burton, C. P. (2005): 'Replicating the Manchester Baby: Motives, Methods, and Messages from the Past', in *IEEE Annals of the History of Computing*, 27.3, pp. 44–60.

Glusker, M., D. M.Hogan and P. Vass (2005): 'The Ternary Calculating Machine of Thomas Fowler', in *IEEE Annals of the History of Computing*, 27.3, pp. 4–22.

Link, D. (2010): 'Scrambling T-R-U-T-H. Rotating Letters as a Material Form of Thought', in *Variantology 4: On Deep Time Relations of Arts, Sciences and Technologies* (Siegfried Zielinski and Eckhard Fuerlus, eds.) (Cologne: Koenig). pp. 215–66.

Rojas, R., F. Darius, C. Goktekin and G. Heyne (2005): 'The Reconstruction of Konrad Zuze's Z3', in *IEEE Annals of the History of Computing*, 27.3, pp. 23–32.

Sale, A. E. (2005): 'The Rebuilding of Colossus at Bletchley Park', in *IEEE Annals of the History of Computing*, 27.3, pp. 61–69.

Spufford, F. (1996):'The Difference Engine and the Difference Engine,' in *Cultural Babbage: Technology, Time and Invention*, (Francis Spufford and Jenny Uglow, eds) (London: Faber & Faber).

Swade, D. (1996): 'Charles Babbage's Difference Engine No.2', Technical Description: Science Museum Papers in the History of Technology.

Swade, D. (2000): *The Cogwheel Brain: Charles Babbage and the Quest to Build the First Computer* (London: Little, Brown), pp. 311–12.

Swade, D. (2003): 'Calculation and Tabulation in the 19th Century: George Biddell Airy Versus Charles Babbage', PhD, University College London.

Swade, D. (2003a): 'Virtual Objects: The End of the Real?', in *Interdisciplinary Science Reviews*, 28.4, pp. 273–79.

Swade, D. (2005): 'The Construction of Charles Babbage's Difference Engine No. 2', in *IEEE Annals of the History of Computing*, 27.3, pp. 70–88.

Thomson, E. P. (1991): *The Making of the English Working Class* (London: Penguin), p.12.

Zuse, K. (1993): *The Computer – My Life*, (P. McKenna and J. A. Ross, trans.) (Berlin: Springer-Verlag), p. 64.

ELIZABETH CAVICCHI

The spiral conductor of Charles Grafton Page

RECONSTUCTING EXPERIENCE WITH THE BODY, MORE OPTIONS, AND AMBIGUITY

Introduction

PEOPLE IN THE past noticed surprising and intriguing effects in nature, often helped by apparatus that they made or improvised. We can become extended observers in what they encountered by repeating some of their undertakings. Doing this has the potential for us to put ourselves, our bodies, materials, experience, and understandings into relation with those of others at another time and place. What those relations can be, and what we will learn, sense, come to wonder and consider, we cannot infer beforehand. For our research to engage with that potential depends on a widely observant and open curiosity from us. While we have experiences, analyses, tools and background that have arisen subsequent to the historical work, we enhance our overall openness by holding these extra-historical resources provisionally – as much available for us to question, try, doubt and reinterpret as the historical materials.

In this study of a nineteenth-century electrical experiment, exploratory qualities of the original investigation arose within the reconstruction, as present-day materials, improvisations and observations met up in analogue and response to past artefacts and accounts. The exploratory responses of my project that mirror the historical case involved widening the options for configuring and testing the experimental apparatus while working in the midst of ambiguity about its behaviour. This resonance in exploratory qualities came about not by following the original protocol step-by-step, but in the course of many iterations of: my trying out of something in the lab; the experiment not happening as I expected; my revisiting of the historical work together with my own efforts, continued through further experimenting.

I first encountered Charles Grafton Page's experiment with a spiral conductor (1837a) as part of my dissertation project (1999, ch. 20) of constructing an induction coil along with reading nineteenth-century accounts and examining original artifacts (Fig. 7.1a). Page's device intrigued me as transitional; through it electricity was detected in paths that were not identical with where

FIG. 7.1a
Right: Charles Grafton Page
(Robert C. Post Collection)

FIG. 7.1b
Below, left: Top-view diagram of Page's spiral where one conductor serves for both battery current (which can be applied across any pair of connectors) and for human shock (which can be taken across any pair of connectors)

Below, right: Side view and cross-section view of induction coil having two separate wire conductors; one for battery current, the other for shock
(Elizabeth Cavicchi)

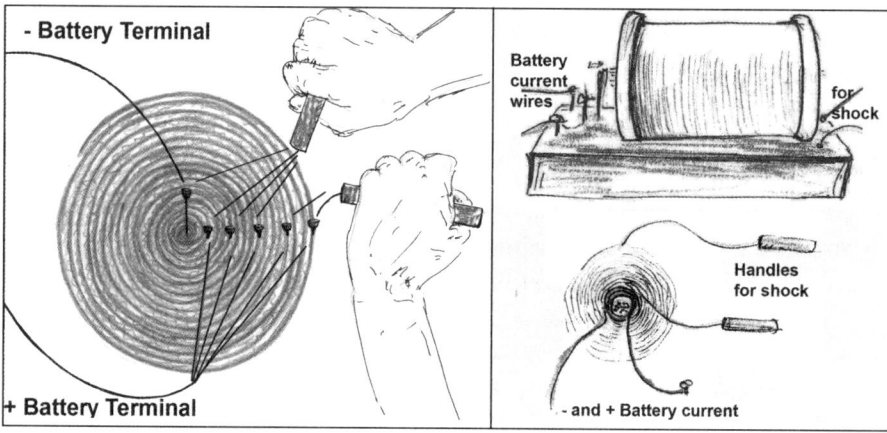

direct battery current went. Whereas the eventual induction coils have two separate wires – one for direct battery current – the other for induced current, in Page's spiral there is only one conductor (Fig. 7.1b). Battery current and induced current occupy overlapping – and at the same time distinct – portions of the spiral's one continuous path. In his brief four-page report about this experiment, Page described behaviors that surprised him. He probed these further, amplifying those effects while providing no explanation.

When, after making my own induction coil, I began this project in response to what Page described doing with a single conductor spiral, I wondered how the spiral's enigmatic electrical effects became a prelude to the seemingly different two-wire instrument. I started by trying to observe Page's findings as voltages induced in a spiraled foil. With my hand-wound induction coils, I routinely used a storage oscilloscope to check for breakdowns and study the high voltage signals. Thus it was a natural extension of my lab practice and study to apply this test equipment in exploring signals induced in the spiraled foil. However, in contrast to the repeatable voltage spikes output by my wire coils, the signals of my foils are variable and inconsistent. Being caught up by that ambiguity, I improvised an experiment, responding both to my observations and Page's report.

My experimental iterations always uncover more for me to notice, rethink, and go on to try. Confusion and ambiguity emerged so recurrently as to be repeatable across my extended efforts. I gradually realised that this *experience* offered matter to work with and research within wider contexts of experimenting, history and learning.

Exploring science and history through reconstructions and teaching

My own curiosity for our lived experience with experimenting, history and learning precedes my investigations of electromagnetic instruments. In college I studied physics and made sculptures; in continuing further with science and visual art concurrently, I experienced alternations between active and reflective pursuits as a mutually supportive exchange. History, science, and our engagement with evidences, stories and materials, became an ever-revising pattern of research and play while I was the researcher for the public science television series The Ring of Truth with Philip Morrison and book (Morrison, 1987). Interweaving across the six films are historical figures – such as Galileo, Cassini, Andrew Ellicott and Cecilia Payne-Gaposchkin – with reconstructions of physics experiments such as the Franck-Hertz experiment, accompanied by science and educational demonstrations such as a jelly doughnut bonfire in illustration of a Tour de France athlete's daily caloric input.

While working at the insides and intersections among these stories and materials of both historical and everyday sources, one passage connected with my personal artistry: Galileo's sepia watercolour sketches of the moon as newly observed through his telescope. Scholars then asserted that Galileo could not have produced these sketches while at his telescope; they regarded the act of producing graduated wash tones and bare white areas surrounded by colour as a studio technique incompatible with observing (Gingerich, 1975: 87–88; Whitaker, 1978: 156). As my response, I looked at the moon through a small refractor with paint brush in hand, sketching it in ink and watercolours, night after night across several years. My watercolour renderings included: dark circles bordered in white rings within black washes; arcs and rays of darker tone overlaid on lighter tones; intense pigment bleeding into dilute regions; small white dots standing out against black (Fig. 7.2). Painting with ink and colour fluidly extended my observing, both by the act of watching and recording changes of light and shadow, and by sharing the excitement of

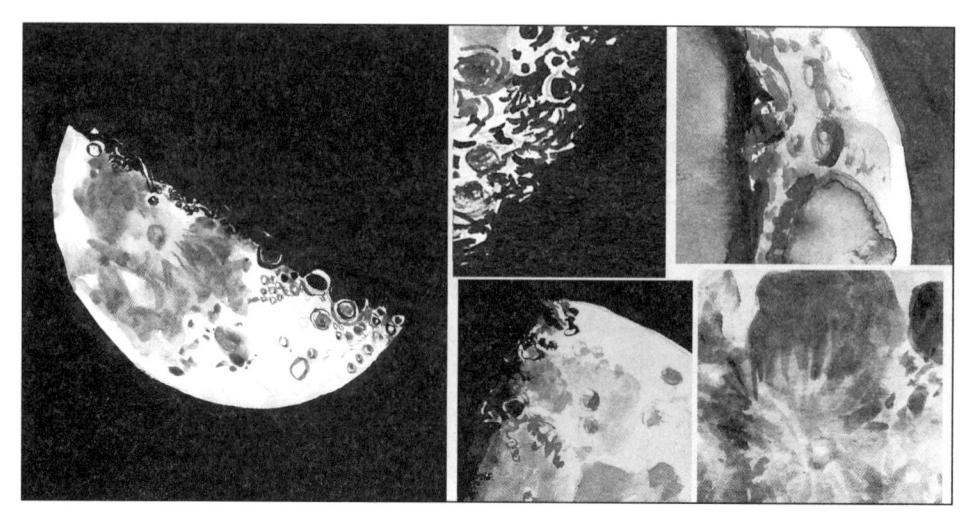

FIG. 7.2
My sketches of the moon, as viewed through a small telescope, painted in watercolour and India ink.
Left: 15 October 1988.

Middle: Details showing ring craters and points of light in black, 9 September 1989 (top) and 20 August 1988 (bottom).
Right: Details showing bleeding of color and layers of tones in lunar rays, 25 October 1987 (top) and 27 September 1988 (bottom).
(Elizabeth Cavicchi)

Galileo's discoveries (1991). In subsequent reassessment, on looking more closely at how five of Galileo's seven lunar sketches are arranged on one side of a watercolour paper, the scholars reconstructed how the sheet was turned for each next sketch. Evidences from the page, of its composition sequence, enabled them to retrace how the 'sheet makes sense as an original record' of direct telescopic observing (Gingerich and Van Helden, 2003: 256).

Later, as a physics teacher, my sense that lecturing did not elicit physical understanding for my students moved me to seek more lively, interacting and inquisitive participation by students with phenomena. My initial attempts at teaching by having students experiment without being told what outcomes to find, evoked such creativity in their science that I went on to create explorative experiences and research the educational developments occurring within them. In sessions where I brought a few students together with evocative materials, I began learning to practice the pedagogy of 'critical exploration' that Eleanor Duckworth (1987, 1991, 2005) developed for the classroom from the historical work of Jean Piaget (1926), Bärbel Inhelder (Inhelder, Sinclair and Bovet, 1974) and the 'Elementary Science' series (1970). In uncovering properties of magnets (1997), batteries and bulbs (1999), light and shadow (Cavicchi, Hughes-McDonnell and Lucht, 2001), or water (Cavicchi, 2005a), these learners became: invested in their own inquiries; observant and surprised by what happened; generative of new experiments; and reflective on what and how they learned (Figs 7.3a–e). Not only were these developments unlike conventional instructional paths in the corresponding contents, but confusion and uncertainty – usually treated as something to overcome in a classroom – recurrently emerged as instigators of learners' new productive work. As the teacher following and seeking to extend exploration, I wondered if history might offer analogies and provocative insights for what we were doing. As with the explorative teaching and learning, that the students and I had to develop interactively, the history that I sought would not be written down somewhere already. To learn the evolving experiences of historical investigators responding to physical phenomena which were also unknown for them, I would have to investigate their work, in turn.

The investigative method that I undertake with history is iterative, reconstructive and reflective, as noted above. With my watercolours of the telescopically observed moon, personal artistry became a means to re-express and reopen historical experience. Similarly, with my studies of nineteenth-century electromagnetism (Cavicchi, 1997, 1999, 2003, 2005b, 2006a, 2008b) a personal experience – being a teacher of explorative science – brings about awareness, questions, and possibilities that open the reconstructive experience and its historical heritage (Fig. 7.3f–g). Reconstructing past experiments, like teaching, involves looking into another's experience and our own, following

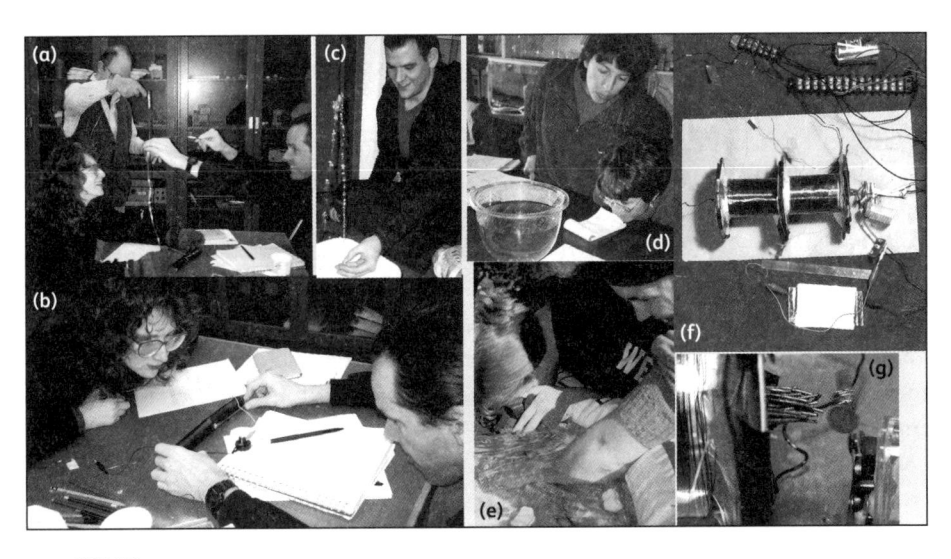

FIG. 7.3
- (a) My students participating in critical explorations with conductive wire;
- (b) batteries and bulbs;
- (c) A homemade fountain;
- (d) Looking at water;
- (e) Looking underwater;
- (f) My hand-wound, two section induction coil;
- (g) Detail of its contact breaker.

(Elizabeth Cavicchi; Coil photos Joe Peidle).

the sense being formed on its own grounds, and partaking in the con-fusions or ambiguities along with their genuine productivity for continuing on.

The theme of ambiguity – a touchstone for me as a teacher – recurs in other historians' studies of experimenting where science understandings were in flux. Friedrich Steinle (1997, 2002; Ribe and Steinle, 2002) documented explorative creativity on the part of both Michael Faraday and André-Marie Ampère in their initial responses to Hans Christian Oersted's 1820 announcement about conducting wires' magnetism. David Gooding (1990: 46–47, 118) discerned that subsequent to this initial exploratory phase, Ampère abandoned his openness and focused on bolstering his theoretical commitments. By contrast, Faraday persisted in puzzling over what he did *not* understand: the magnetism's circularity.[1] By staying with that physical ambiguity – exploring it further – Faraday brought about experiences foundational to his invention of the first motor, a device that uses electromagnetism's circular action to revolve a conductor around a fixed magnet. An example from early twentieth-century biology researched by Evelyn Fox Keller (2002: 123–47) illustrates a mode of development inverse to Faraday's, where ambiguity in the means of thought supported the investigators in recognising and tolerating ambiguity in what they observed. The amorphous term 'gene action' gave biologists a way to talk about hereditary transmission and work with evidence of it before they had access to explanatory mechanisms, such as DNA. By sustaining generative relationships with ambiguity, Faraday and the biologists extended their experimental process without settling on a premature result having definite but artificially constrained design.

Reconstructing Page's experiment gave me first-hand experience with this kind of physical ambiguity, and with working through my own resources and limitations in conceiving new options for tests and pursuing these experimentally. I faced variability, ambiguity and confusion that held my interest during more than 90 laboratory sessions over six years. The challenges of my reconstruction put me into a role, like Page, of dealing with confusion, although he and I might describe our confusion differently and approach it with differing tools and expertise. One such area, where he depended on an expertise and tradition which is now long out of practice, is that of taking bodily shocks and comparing their strength to evaluate an electrical device.

Background practices of putting the body into the circuit

Human bodies were integral components of the eighteenth-century circuits that first manifested many properties of electricity and its conduction. In April 1730, British pensioner Stephen Gray (Gray, 1731–32: 39–42) suspended an eight-year old boy from the ceiling on clothesline, so the child rested either prone down or up. When Gray placed a glass rod, electrified by rubbing, near the boy's feet, a brass leaf indicator, set up near his head, deflected attractively toward the face. Subsequent technology amplified the electrical effect. Hand-cranked friction machines rubbed glass against leather at high rotation rates; the Leyden jar stored this electricity for later uses (Figs 7.4a and b). The body's responsiveness to the electrification and shocks delivered by these devices provided entertainment in public science lectures, and cheap medical

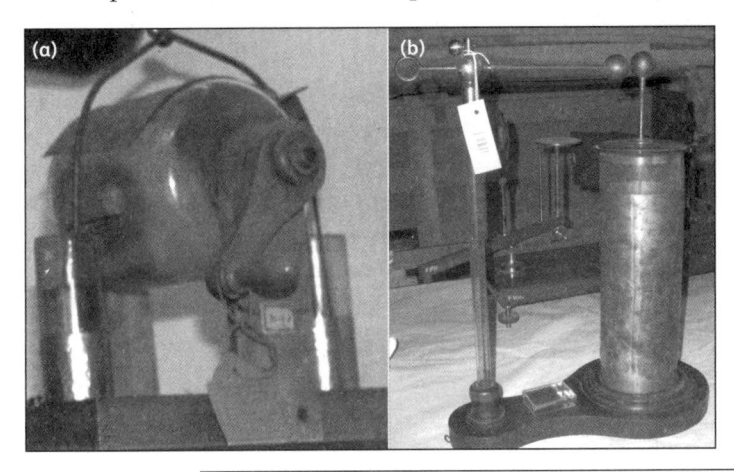

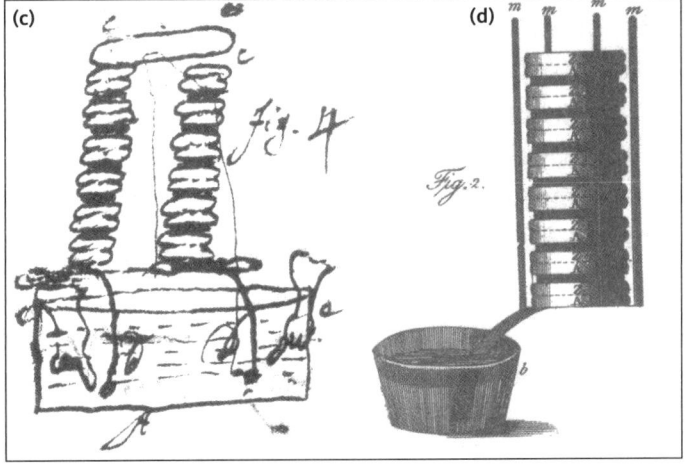

FIG. 7.4

(a) Cranked friction machine made from a glass bottle; (b) Leyden jar and discharger at the Norsk Teknisk Museum.

(Elizabeth Cavicchi)

(c) Volta's sketch showing how his hands made contact with his alternating pile of zinc (topmost disc in each stack and repeat set of three discs), silver (middle disc in each set of three) and moist cardboard (black, smaller diameter disc in each set of three).

(Volta, 1918, vol. 1, pl. XXII).

(d) Volta's published diagram where the pile links to a saltwater basin where he placed one hand, while putting the other on the top of the pile.

(Volta, 1800).

therapy accessible to the poor (Rowbottom and Susskind, 1984; Bertucci, 2001, 2006; Hochadel, 2001).

Body parts were components in the trial assemblies of dissimilar metals and moist substances by which late eighteenth-century Italian investigators produced electricity by non-frictional means. While everything touched in a circle of contact, these body parts reacted unmistakably, exhibiting a new-found electricity. To Luigi Galvani, a frog leg's twitch indicated an electricity originating in life processes. Convinced otherwise, Alessandro Volta substituted a sensitive instrument for the frog and still detected electricity. But Volta soon realised that this instrument's internal materials produced some of the electricity it detected. By stacking metals and liquids in analogy to the electric fish's anatomy, Volta eventually constructed a chemical battery whose enhanced potency he demonstrated by using only himself to close its circle (Pancaldi, 2003: 183; Figs 7.4c and d). The body was back in Volta's circuit, but he viewed its function as only to manifest shock, not to generate it.

As voltaic sources of electricity became available around 1800, some physicians substituted these for friction electrical machines in electrical therapies (Rowbottom and Susskind, 1984; Bresadola, 2001). In doing so, they had to adapt to the distinction between the high tension (voltage) and low quantity (current) of the friction machines, and the higher quantity at low tension of voltaic sources. The body's resistance to voltaic electricity introduced a barrier that was not present before. Just to get electricity past the skin's high resistance, practitioners imposed wounds into their patients' bodies to receive electrodes. Avoiding this need to wound, British surgeon Charles Wilkinson (1804, vol. 2: 444) introduced the technique of placing broad metal discs (attached to electrodes) in close contact with moist skin. At the same time, care had to be taken to regulate and limit the voltaic battery's currents within a safe range.

The electric circuit intruded further into patients' bodies through the 'electropuncture' technique innovated by French physicians who reintroduced the Chinese method of acupuncture to Western medical practice around 1820.[2] Demonstrating that the needles affect an electrometer, these physicians inferred that acupuncture's effect involves electricity. To augment it, they attached a voltaic battery's electrodes to a pair of needles inserted in a patient's body, bringing on 'more pungent pain' (Morand 1825: 36). Members of the American medical community took interest in 'the growing importance of the remedy' (Morand, 1825: 3) as it was introduced by Benjamin Franklin's great-grandson, physician Franklin Bache, in his translation of a French volume on acupuncture. Not long after Page's experiment, a Kentucky physician delivered a presentation acquainting his colleagues with French techniques for applying galvanism through acupuncture needles (Peters, 1836 [ASP, 2010: 18]). Entering this community as a Harvard medical student who was authoring the dissertation 'On the Ear' (Anon., 1836), Page attended to these new therapies and extended them.

The body's reaction to voltaic electricity interested experimenters as well as physicians. Their own bodies, not a patient's, provided a convenient detector of electricity. Sometimes this detection was inadvertent – and a harbinger of new electrical behaviour. In 1834, British amateur Mr William Jenkins got shocked upon disconnecting a battery from a coiled helix whose ends he grasped in either hand. He had not expected this: experimenters working with direct current from one or two cells ordinarily felt no shock. Jenkins told Michael Faraday.[3] Faraday's ensuing investigation set off the network of experimenting which Page furthered with his spiral. Faraday elaborated that the body's reception of Jenkins' shock depended on good contact between the body and the electrical conductors:

> On holding the two copper handles tightly in the hands, previously moistened with brine, and then alternately making and breaking the contact of the ends of the helix with the electro-motor [*battery*], there was a considerable electric shock felt … . (1834: 351)

When Page put his body into the circuit of a spiralled conductor, he applied these experimental and medical practices in new ways. Like Faraday and Jenkins, Page took the shock hand-to-hand directly through his body's core. In some configurations of his test circuit, Page scarcely felt the shocks, so he amplified his sensitivity by piercing his fingertips with acupuncture needles available from a Boston medical supply shop (White, 1828).[4] Without either the therapeutic intent or the direct battery current which characterised medical 'electropuncture' techniques, Page's use of these needles as assists in detecting electricity was innovative. No other detector than his body would as compellingly report the marginally observable electricity induced in parts of the spiral remote from the battery current's direct path.

While Page used his body as a detector in research, he regarded the shocks he took as having potential in electrical therapy. Active in the local medical community as a teacher and student,[5] he sent a one-paragraph notice about his research on 'Medical Application of Galvanism' to the *Boston Medical and Surgical Journal* (1836a). While disclosing no details about his apparatus, Page promoted its suitability for a French electropuncture technique where needles burned flesh between them, or transmitted medicines.[6] In doing so, he demonstrated conversancy in novel treatments that were outside conventional medical instruction (Eve, 1836). Page's experience with shocks applied in treatments is evidenced by his remark that shocks from his device were 'quite unlike and less disagreeable than' those of a conventional galvanic source (Page, 1836a). A medical journal based in Atlanta, Georgia published a frustrated inquiry about Page's device, offering a fifty-dollar premium. Although the Boston journal republished the southern medical society's query (Page, 1836c), Page never responded in print.

Bodies and circuits combined in fluid relation throughout the investigations by which voltaic electricity was originally observed and explored. The participants' understanding of that relation shifted: from Galvani's assumption that the frogs' bodies produced the electricity, to Volta's exploitation of bodily shock to demonstrate his inanimate pile, to the French physicians' inference that acupuncture needles tapped into bodily electricity. With voltaic electricity's expanding use, experimenters like Faraday routinely took shocks to check their circuit and healers applied it in therapies. As Page drew on both these experimental and therapeutic practices, he participated in a larger trend toward directly involving the body in its medical treatment that Michel Foucault has identified (1963). While eighteenth-century doctors diagnosed without touching patients, anatomist Xavier Bichat broke from this tradition by establishing diagnosis criteria that related pathologies inside patients' bodies to a disease's usual progression, as charted through autopsies. Devices like Page's spiral intervened further by sending electricity into the body (Page, 1836c: 183).

Page's experiment

Page's parents' home in Salem, MA, housed his spiral conductor experiment, as it had his many electrical adventures from childhood on (Fig. 7.5, left). At nine, he climbed onto its roof, three stories up, to catch lightning with a shovel during a storm. The next year, he converted his mother's lamp glass into an electrical friction machine (like that in Fig. 7.4a). Page augmented these

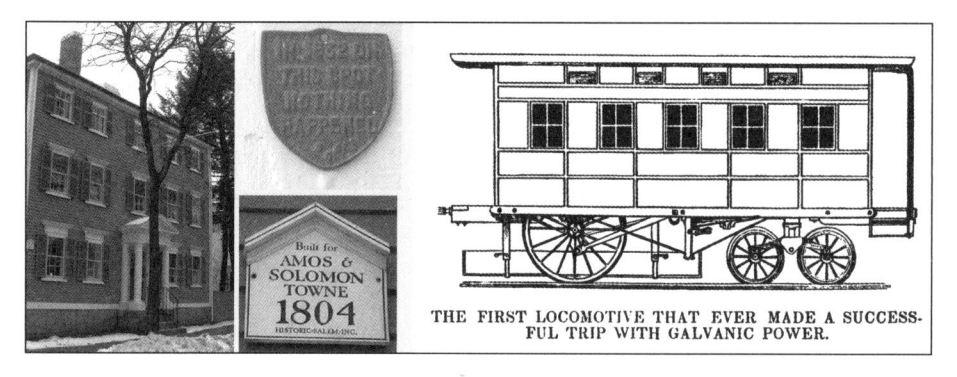

THE FIRST LOCOMOTIVE THAT EVER MADE A SUCCESSFUL TRIP WITH GALVANIC POWER.

FIG. 7.5

Left: The Page home in Salem MA today, with its historical plaques.
(Elizabeth Cavicchi)

Right: Page's electromagnetic locomotive.
(Greenenough, 1854; 257)

THE SPIRAL CONDUCTOR OF CHARLES GRAFTON PAGE

electrostatic pursuits by applying his chemistry studies at Harvard (class of 1832) in constructing voltaic batteries, organising a college chemistry club, and giving public science lectures in Salem. Page's parents' home hosted his brief medical practice where in 1836–37 his long-time mentor witnessed Page's 'miniature magnetic engine' speeding laps on a scaled-down railway track! (Lane, 1869: 2–3) This miniature train was a forerunner of the electromagnetic locomotive which Page constructed in 1851, being pre-dated only by the Scottish inventor Robert Davidson. Running off huge zinc-platinum cells and funded by a Senate allocation, Page's full-sized train limped back to Washington DC from its truncated test run (Fig. 7.5, right). It won him great notoriety then, but little notice by historians (Post, 1972, 1976a).

Just as my spiral reconstruction was a response to Page's terse report, Page's original experiment was in itself an effort to replicate one that he read about in a short notice by Princeton professor Joseph Henry (Bache, 1835; Henry, 1835).[7] In turn, Henry's experiment was done in a haste incited by Michael Faraday's latest work (1834, 1835) that opened into an area where Henry had a prior observation (1832). Adding yet further to the chain, Faraday was researching the strong shock that, as described above, he first learned about from Jenkins.

The circuits constructed by Faraday and Henry were composed of loops. In Faraday's circuits, one loop consisted of a wire helix connected across the plates of a voltaic cell; a second loop circled from that cell, through his body by

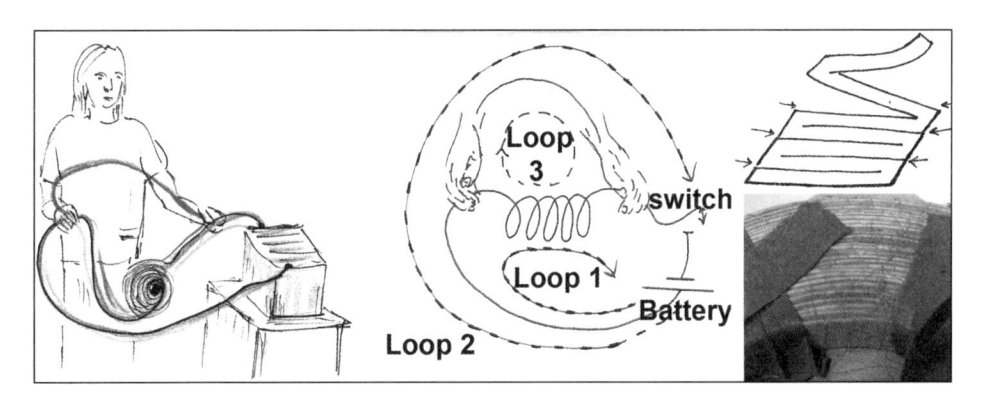

FIG. 7.6

Left: A person holding both ends of a coil feels shock when the coil breaks its connection to the battery.

Middle: Current traverses Loop **1**, from the trough battery, through the spiral or coil, and back. The person adds a second Loop **2**, running from battery and then through their body. When the switch opens, the shock takes Loop **3**, running between the person and the coil.

Right, above: My diagram of the Page's method of slitting a copper sheet from opposite sides (arrows) so that it would open as a zig-zag strip.

Right, below: Fabric wrapping around the copper ribbon of a spiral used by Joseph Henry. (Cat. no. 181,540, National Museum of American History.)

(Elizabeth Cavicchi)

137

way of his hands (Fig. 7.6, left, middle). When a break in battery connection stopped current in these loops, electricity arose in a third loop whose circle joined the helix with the body (Faraday, 1834). Henry's circuits were similar. In place of Faraday's helix he substituted conductors having other configurations, obtaining the most intense electrical effects with a spiralled copper ribbon (Bache, 1835; Henry, 1835, 1837).

Both Faraday and Henry gauged the effect's intensity by two means. One was the brightness of a spark that appeared where the battery was disconnecting from the circuit. The other was the severity of shock felt only during battery disconnection, and while both hands spanned the long conductor. Whenever battery current was maintained steadily, no perceptible current passed through the body's high resistance and the experimenters felt nothing. Thus Faraday distinguished the felt electricity when the battery stopped from that of the unfelt direct battery output.

In Faraday's view, the shocks were due to an electricity brought about, or *induced*, by the stopping of battery current. This induced electricity had an intensity (voltage) heightened above that of the battery current. Faraday realised it related to his seminal 1831 finding that a changing current induces currents in nearby separate conductors. Yet something new and different was going on: the changing current acts on *itself* and induces another current in that *same* wire which exhibits *differing* electrical properties.

Page had read only Henry's notice and not Faraday's. Tantalised by Henry's claim that the maximum shock of a spiral was 'not yet determined' (Henry, 1835: 328), Page constructed a spiral more than twice the length of Henry's. Having no continuous copper ribbon, he cut flat copper sheets in zig-zag strips, soldered these lengthwise together, and wound up the length with fabric (Fig. 7.6, right). He assembled this apparatus by hand and covered it in a box.

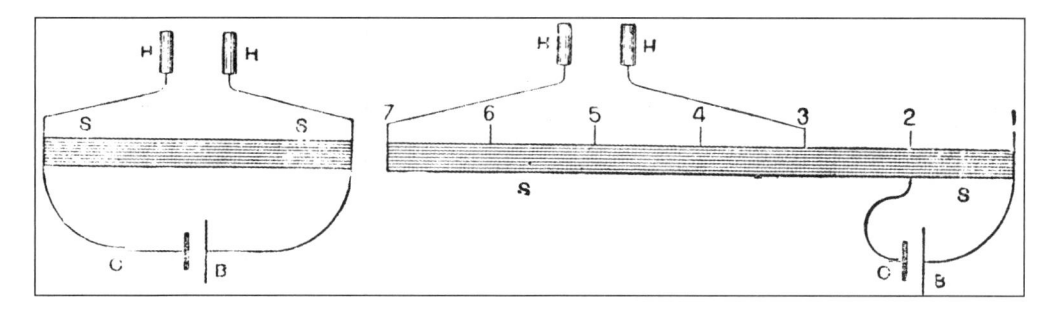

FIG. 7.7

Left: Henry's spiral unwound; the shock is taken across the handles **HH**, while the battery is applied across the same span.

Right: Page's spiral unwound; the shock may be taken across parts of the spiral that may differ from the segment carrying the battery current. (Fleming, 1892, vol. 2: 6, figures 1 and 2)

THE SPIRAL CONDUCTOR OF CHARLES GRAFTON PAGE

The homemade construction of Page's spiral belied an experimental flexibility more sophisticated than Professor Henry's. Instead of sensing shock only while battery current went through the entire conductor as Henry and Faraday had, Page set up connector cups at six positions along the length of the spiral (Fig. 7.7). Each cup contained mercury; on dipping a bare wire into a cup, a good electrical connection was quickly formed that could be easily undone just by its removal. With these cups, spaced at different distances apart, Page could direct current through part of the conductor, and take shocks across that same part — or any other part. But there was more. The cups made the spiral into a research tool whose options he recognised and explored over time.

Extending Henry's finding that a longer conductor gave greater effect, Page lengthened the conductive path successively within his one spiral, instead of making separate, longer ones. He did this by putting one battery wire into the central cup 1, and the other wire into cup 2, and then observing the spark when either wire was removed from its cup (Fig. 7.8a). Leaving the first wire in cup 1, he then placed the other in cups 3, 4, and so on, observing the spark produced upon each wire's removal (Cavicchi, 2008b; fig. 1, left). These sparks were brightest and loudest when the break was made from cup 3, and declined as the battery current was sent through more of the spiral. Page suggested that if mercury cups were soldered to every spire, the exact location of the turnaround in spark brightness could be determined.

Shock intensity worked different from sparks. While an assistant operated the battery connections across the same successively widening span, Page took the shock by way of handgrips running to the same pairs of mercury cups. The longer the span traversed by both battery current and bodily connection – up to the whole length – the greater the shock. Layering water over mercury in the cup amplified these shocks, puzzling Page: 'the rationale I am unable to give' (Page, 1837: 139). Page then perceived other options for configuring the experiment. The battery and the body could be inserted across *different* intervals of the spiral. Page's tests of these options yielded results that he found 'curious … difficult to explain' (Page, 1837: 139).

First, he kept the battery's connectors placed across the spiral's inner turns (cups 1 and 2). One hand grip remained always at the inner cup (cup 1); the

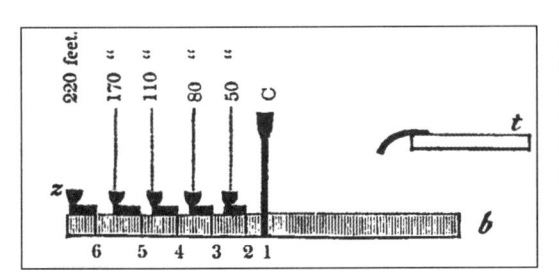

FIG. 7.8a

Side view of Page's spiral showing numbered connector cups spaced across its length The handgrip is below letter **t**.

(Page, 1837a, p. 137)

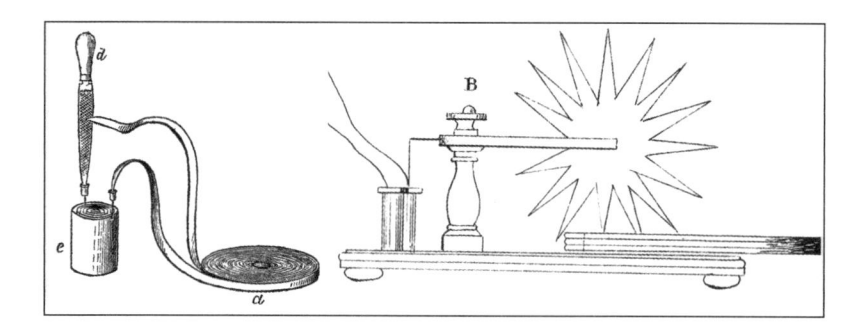

FIG. 7.8b
Left: The rasp interrupter, invented by Page.
(Henry, 1839, fig. 1: p. 304)

Right: The spur wheel interrupter, introduced by Page.
(Clarke, 1837)

other was placed at each of the other cups in turn. The loop defined by the battery connections remained fixed; the loop passing through his body traversed more of the spiral. When his assistant broke that battery connection, Page reported a greater shock than if his hands spanned just the cups that took the battery current. This shock increased in severity as his hands encompassed more of the spiral. The instrument delivered its greatest shock of all when the battery current traversed about half the winds (1 and 4), and his hands spanned the entire spiral (1 and 6; Cavicchi, 2008b). That the shock was not as strong when the current went through the entire spiral, suggested that turns extending beyond the current's path were electrically operative, by some means which Page termed 'lateral cooperation' (Page, 1837: 139).

Page expanded the experimental options to put battery and body across non-coincident spiral intervals, and met with astonishment. 'Contrary to expectation', on disrupting battery current from traversing the inner turns (cups 1 and 3), he felt a weak shock while his hands spanned *only* the outer ones (5 and 6). Page amplified his sensitivity by piercing acupuncture needles into his thumb and finger. Now the shock felt 'extremely painful' (Page, 1837: 139). The needles also enabled Page to greatly reduce the scale of the battery activating the spiral, from a large 'calorimotor' such as Henry had used – a cell with large plates that put out high currents at low tension – to a 'single pair of plates of only four inches' (on a side) (Page, 1837: 141).

Something was happening even where direct current had not passed, which could be sensed throughout the spiral. Page realised that this sensual detection distinguished whatever it was that he felt, from the battery's direct – and insensible – current. Page checked that this was so, by putting his body *in series* with the battery's direct path. In this case, nothing could be felt, even when he again heightened his sensitivity by inserting 'fine needles deep into the thumb and fore finger' (Page, 1837: 140). By contrast, the sudden stopping of battery current within the spiral gave rise to a momentary electricity of high enough intensity in that same conductor, to overcome skin resistance and shock a human body in parallel with it. Spiralling the copper magnified this, and the body functioned as an acute detector.

Spark and shock occurred only on breaking battery connections. Realising the technical import of this finding, Page innovated switches that operated repetitively. The first contact breakers were Page's steel rasp that when scraped by another conductor emitted sparks accompanied by intolerable shocks (Fig. 7.8b, left) – and a spurred wheel that rotated its conducting tines in and out of mercury (Fig. 7.8b, right).[8] Sparks shone where each tine exited mercury, making beautiful stroboscopic effects in the dark (Page, 1837b: 141).[9] Page made this wheeled switch self-actuated by positioning a magnet so its gap was crosswise to the conducting tine, thereby rotating the wheel as a motor.

The spiral, as accessorised by its spaced cups, acupuncture needles, single-cell source and switches, provided electricity across a graduated range of outputs. More forthrightly than in his preliminary notice to the medical journal, in presenting the spiral to the scientific community Page ascribed its suitability for 'medical galvanism' to this instrumentally manipulable feature: 'shocks of all grades can be obtained' (Page, 1837: 141).[10] Yet while Page identified a therapeutic value for the electricity newly accessed by the spiral, he did not go on to pioneer therapies in this new area of medical galvanism. It was the interaction between electricity and his instruments that held his curiosity for further research.

Starting with a circuit which was the forefront research of Faraday and Henry, Page took it further. His tools – cups, needles, rasp and wheel breakers – opened up possibilities. His body functioned as conductor, detector, and potential beneficiary, yet throughout *he* was the agent of change. What Page learned and felt kept the experiment going. As his means of detection made evident electricity where no one expected it to be, he flexibly reconfigured and extended his instrumentation. In observing behaviours that violated 'received theories of electromotion', Page put forward no explanation, yet followed those behaviours productively, amplifying the effects and widening contexts of detection (Page, 1837: 139).

In the lab with spiraled tape and its confusing signals

Where a science experiment or technology is redone as part of a historical study, there are many possible expressions for the relationship between the historical work and materials, and those of the researcher. While some studies emphasise close following, reproduction or reuse of original artifacts and accounts (Withuhn, 1981; Heering, 1994; Weber and Frercks, 2005), others incorporate practices and instruments of disparate contexts, historical and otherwise (Finn, 1966; Settle, 1996; Tweney, 2006; Chang, 2007, 2008). Similarly, the role of the researcher ranges from studies which highlight experimental results while sidelining human involvement (Mills, 2002, Usselman, et al., 2005), to others where the historians' actions, interactions and personal reflections are a source for the interpretations brought to light by the experimental project (Gooding, 1989; Tarver, 1995; Tweney, 2005; Heering, 2008). Because my study's concern was to understand the experience and process of an exploratory experiment, my starting point was my personal involvement in observing phenomena, including responding by a variety of means, including with instruments that were not available for the original investigation. Rather than replicating an instrument or a particular experimental path from (necessarily) incomplete historical resources, I look to understand the range of possibilities that can emerge within the experiment. For example, using modern test equipment to observe similar phenomena is a way of uncovering complexities that were not mentioned historically, yet invisibly influenced what was historically observed. By freshly entering the historical inquiry along any path, experiences and improvisations undertaken along the way become means for expressing and opening understandings, where actions and phenomena of experimenting in the past interrelate with our current experimenting (Settle, 1996; Heering, 2008; Cavicchi, 2008a).

FIG. 7.9

Left: My copper tape spiral.
(Omari Stephens)

Below: Diagram of my circuit with battery applied across part of the spiral, and oscilloscope probes connected across a wider span.
(Elizabeth Cavicchi)

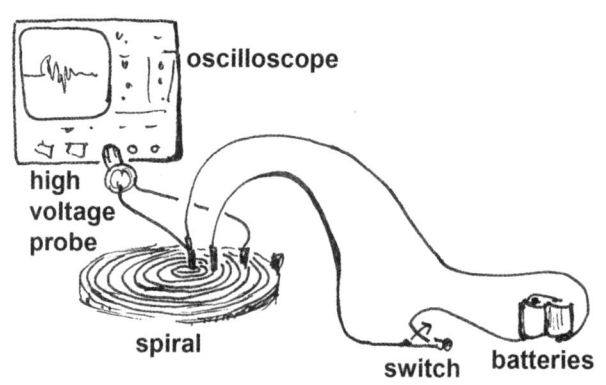

The path by which I entered a relation with Page's spiral passed through materials readily at hand in the electronics student lab where I developed this project and in crafts that I pursue as an artist. For the spiralled conductor, I repurposed copper foil tape from stained glass art (Fig. 7.9, left). The conductive foil spirals outward in an unbroken path, while its paper backing insulates successive turns from direct contact. At first, I used this foil just as it came on a spool; later I rewound foil off spools, into tight spirals winding out from a centre (2005). At intervals along these spirals, I inserted copper strips to function like Page's cup supports; on revising the spiral, I soldered these strips onto the spiral foil. I joined batteries and other apparatus to these strips using alligator clipleads, in place of wire dipped in mercury cups. Having discerned from my previous reconstruction projects that D cell batteries suffice for demonstrating many nineteenth-century electromagnetic effects (1999, 2003), I used two of these, or a 3-volt power supply, as a source. Initially, I broke the circuit with a mechanical leaf switch that I had found to act consistently during my prior experimenting; eventually I tested and constructed a range of switching mechanisms.

Being smaller in scale and less robust than Page's copper sheet spiralled in fabric, the foil spirals of my improvisations would not stand up to the amperes of direct current that Page's 'calorimotor' may have delivered – nor would such high currents be necessary to induce electricity in my foil's tight windings. Those high currents were among many features of Page's practice that are now understood to pose health and safety risks. Others include: battery acids and unvented fumes; contact with mercury and its vapours; taking unknown electric shocks bodily; piercing the skin without sterile needles and medical cause (Butterfield, 1975). For each of these hazards, I substituted safer means through modern technology. Where Page relied on liquid mercury for making and breaking electrical contacts, I employed a range of techniques including alligator clipleads, mechanical switches, a frequency generator, and my own analogue to Page's spur wheel where the liquid metal alloy galinstan substituted for mercury. In place of taking bodily shocks, I viewed voltage traces on a storage oscilloscope, having a high voltage probe to protect the instrument from the high voltages (Fig. 7.9, below; 7.10, left).[11] While the shocks afforded by my instrument would be much reduced in scale from those Page took, they were present nonetheless, as I experienced when handling it carelessly with wet hands.

While the spiral's behaviours surprised Page, initially I did not expect to be surprised. I had the outcome of his experiment before me, along with the subsequent interpretation of its phenomena of 'self-induction', where a disruption of current in a coil brings about a transitory voltage in that same coil. I supposed that by following Page's practice of testing different parts of the spiral, I would recognise distinctive features in the corresponding oscilloscope signals. Where Page had applied battery current across a pair of cups, such as 1 and 3, I hooked battery leads across the first and third tabs of my spiral and put oscilloscope probes across parts corresponding to where Page put his hands (Fig. 7.9, below). Then I switched the circuit on and off. Typically, whenever the current *stopped*, the probe picked up a brief pulse of high voltage, which showed on the oscilloscope screen. A typical trace displays a voltage spike of several hundred volts, followed by lesser peaks declining within a damped envelope whose periodicity lies in the microsecond range (Fig. 7.10, left).

I started by using an analogue storage oscilloscope, having no means of recording such a trace. Finding that I could not remember the traces produced by each switched event, I paused after each to write down the peak value or sketch its appearance. Then I switched the circuit again. I was constantly re-doing how I worked with the circuit and instruments. For example, through noticing the oscilloscope's two channels, I formed the idea of observing two intervals of the circuit at once. I put one probe across the part of the circuit where battery current went, and the second across a different, longer segment, like that of Page's body. Over and over, I repeated the cycle of switching the current, observing a trace, and repositioning the connectors. Sometimes the signal taken across more of the spiral was more pronounced than that taken across a lesser interval; often it was hard to tell. Every event seemed so different. The probe from the second channel seemed to perturb the overall signal. I took it out.

The circuit resonated after each switching. The peaks of those resonances varied so that I could not tell what was going on. In discussing this with others in the lab, drawing on interpretations of electricity from our training, we noted

FIG. 7.10

Left: A typical oscilloscope screen image showing voltage (vertical 200V/div.; horizontal 2 μs/div.) induced in the spiral when the switch opens. Middle: The human body model (such as resistor) is connected in parallel with the oscilloscope probe.

Right: Photograph of the test equipment, including digital oscilloscope and monitor (top) with spur wheel and spiral in the foreground. (Elizabeth Cavicchi)

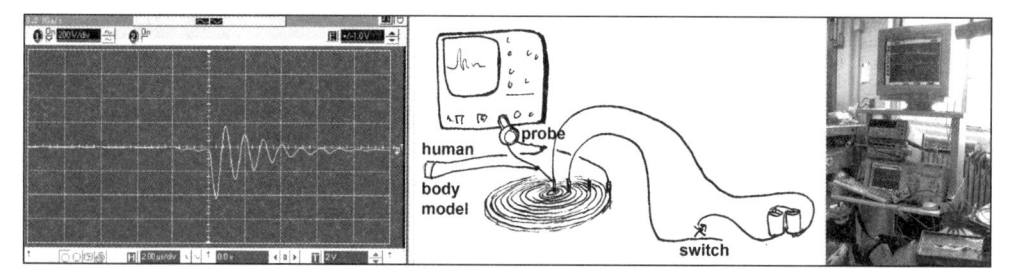

that a source of electrical resistance absent in my circuit, but present in Page's, was his body. Maybe Page's *body* contributed to the electrical behaviors he described!

Having this idea was an intriguing moment in my study. It expanded my awareness of roles in the historical circuit beyond my initial interpretation. I realised that I had regarded Page's body only as a detector, not as a circuit component. Perhaps Page also had this view. Using an oscilloscope as a detector was a different case; presenting a very high resistance (MΩ) to the circuit, under most conditions an oscilloscope can be regarded as a passive detector. The experiments where Page detected the spiral's electricity by observing only a visual spark without taking shocks, would be more analogous to the configurations I had tried with an oscilloscope probe than those where bodily shock provided Page a means of detection. Like me, Page may have considered these two modes of detection interchangeable.

Acting on this idea, I sought to add into my circuit something that would function as an analogue to the human body. Through discussion with other experimenters and readings, I considered many methods for providing an electrical substitute for the body. Most interpretations represent the body as a pure resistance whose value decreases with moister skin, or as a resistance combined with inductance and capacitance. I constructed several alternative models consisting of a resistor, or of resistors combined with other elements (Cavicchi, 2005, 2008b).

This question about the electrical role of the body increased the options of what to include in the circuit. It was a lengthy process to test these further options across many spiral configurations by placing each in parallel with the oscilloscope (Fig. 7.10, middle). Would an overall pattern emerge in the voltage traces? It did not. In response to this impasse, I let off doing more with the spiral. Its fleeting signals did not register my interventions and each sequence of trials appeared undifferentiated. My experimental work stalled where my expectations for identifying particular trends met with ambiguous outcomes.

When I later resumed the experiment, I overhauled the apparatus, doubling the spiral's length, improving connections, and substituting a digital oscilloscope for the analogue one (Fig. 7.10, right).[12] This switch to a digital oscilloscope again added many more options for experimental tests and their analysis. Through visual observing and sketching voltage traces, I was not taking in enough of what the oscilloscope detected to make out any pattern in the electrical behaviours. With the digital oscilloscope, the values in voltage and time making up a trace could be saved as a file containing paired numerical values. I subsequently plotted these values in Excel or Matlab (1994–2010). By superposing plots of traces taken under different arrangements of

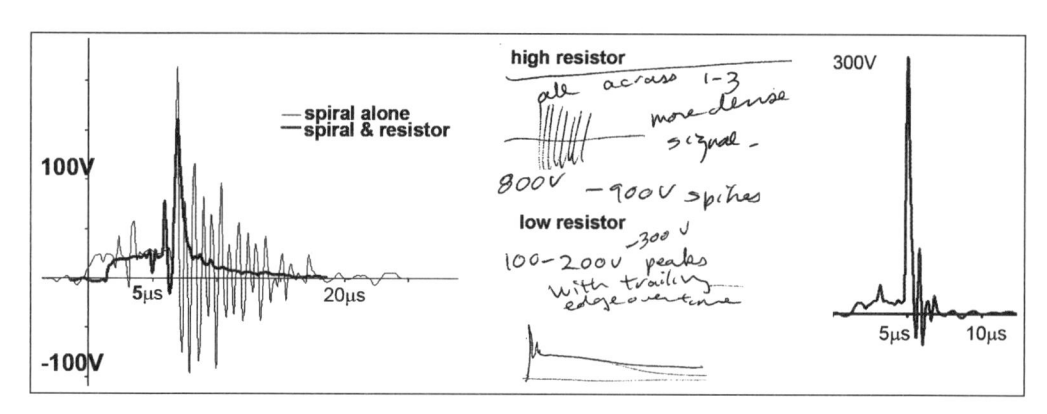

FIG. 7.11

Left: The light gray line is a voltage trace induced across a part of the spiral when the switch opens. The dark line shows a trace induced across the same portion of the spiral when a resistor (1 kΩ) is put in parallel with the probe.

Middle: My handwritten notes showing my first observation of a difference in the trace resulting from using a high resistor (vertical lines representing ringing, top sketch) and a low resistor (single peak, bottom) as models for the body.

Right: The voltage trace produced with a human volunteer connected across my spiral, in place of the resistor. The peak voltage is 300 V.

(Elizabeth Cavicchi)

the circuit, I look for trends that in turn raise questions, setting off further trials. Providing feedback to my interventions, these means of recording and analysis moved me out of the impasse, opening a window on the electrical behaviours and experimental options that was not available before.

An example from my early use with these analytic means involves the effect of inserting a resistor in parallel with the oscilloscope probe, as a stand-in for the body. When the resistor's value was high, the traces resembled those produced with no resistor: an initial high-voltage peak followed by a resonant 'ringing' of many peaks whose value declined successively. When the resistor's value was low, there was a difference in the traces' overall appearance; the voltage of that initial peak was lower voltage, after which the signal declined without ringing (Fig. 7.11, left).[13] My hand-written sketches and notes made while using the analogue oscilloscope contain these same features (Fig. 7.11, middle). At the time I did not appreciate this finding and was expecting instead a distinct trend in peak voltage values that was not manifested. The data analysis plots assisted me in seeing what I had observed but failed to appreciate. A confirmation that the placement of a body (substitute) in the circuit alters the induced signal, yielding a single prominent voltage peak without ringing, came when Professor Pancaldi of Bologna voluntarily put himself in parallel with the probe – feeling no shock (Fig. 7.11, right). This characteristic shape held for traces produced with other substitutes for the body that I tested.[14]

However, while I had characterised the effect on the spiral's circuit of adding a body or substitute, this effect applied to all circuit configurations. It did not seem to correlate with the differing severities of shock that Page described, exhibiting greater intensity when more of the spiral was included in a test.

Whenever I switched the circuit, the peak values of the voltage traces varied. Sometimes the peaks were indeed highest when I probed my spiral in ways analogous to what Page did when he reported strongest shocks. Then my experiment seemed to cohere with Page's findings. In analogy to Page's heightening sense of shock when battery current and his body were put together across successively longer spans of the spiral, my comparable tests demonstrated an increase in voltage. Similarly, with the configuration that had most astonished Page, that of feeling weak (acupuncture-amplified) shock from turns outside the battery current's direct path, I too detected small voltage peaks from a probe placed across intervals entirely outside my battery current's direct path.

However, when my probe spanned the battery current's path plus an additional length, I could not always tell whether the peak voltage had increased along with that addition. The peak voltages were so variable as to render ambiguous any attempt at interpretation. Gradually I noticed that this underlying ambiguity occurred with my spiral, but not when I performed the comparable test on wire coils that I wound onto iron cores. With my iron core coils, putting a probe across more length resulted in consistently greater induced voltage (Fig. 7a in art section, left and middle; Cavicchi, 1999, 2006a).

Although I did not realise it for some time, the variability attending my spiral observations tended to overwhelm whatever characteristics might be due to the different spiral configurations that I tested. I demonstrated this variability by overplotting successive traces taken from the same circuit configuration after I switched the battery off successively, without changing anything else (Fig. 7a in art section, right; Cavicchi, 2005, 2008a).

Did this variability have to do with the switch? As with my inquiry about the role of the body, this question about the switch brought me to reconsider the circuit. I reconstructed the circuit with my own homemade analogues of its historical components. I applied electronic test equipment to stimulate the circuit as I modified it, and to observe its response. I went on to explore effects of switching in the spiral circuit in two ways: with mechanical switches; and by substituting periodic electronic pulses for switched battery current.

My testing of the effects of mechanical and periodic switching brings into play many further options for consideration in each experiment. Mechanical switching is an inherently irregular process; the two surfaces of a switch are jagged at a microscopic scale. When separating, these surfaces disconnect and reconnect, making for erratic momentary voltages in the circuit (Fig. 7.12, left). When applied to a circuit containing the spiral, these switches induce complex signals that are composed of high and low frequencies. As I sought

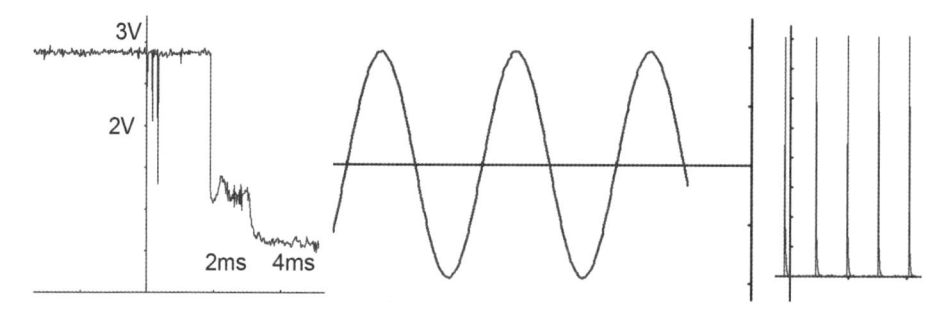

FIG. 7.12

Left: A voltage trace taken when a knife switch opens a circuit consisting of just the switch and two flashlight batteries (3V). The trace shows a complex structure as the switch contacts break away and reconnect erratically.

Middle: A voltage trace showing a periodic waveform produced by a function generator; the frequency of the wave can be adjusted within the instrument's range (Hz to MHz).

Left: A voltage trace showing a periodic pulse, exhibiting a very sharp rise and fall in voltage, produced by a pulse generator. The frequency and shape of the pulse can be adjusted.

(Elizabeth Cavicchi)

ways of examining the circuit's behaviour across the diverse frequency ranges of its mechanical switching, the number of experimental and interpretive options increased. By contrast, pulse and frequency generators output just one well-defined waveform at a time (Fig. 7.12, middle, right).[15] From the complex range of frequencies manifested in a switching event, a wave generator provides only one frequency to the circuit for any given test. To view the circuit's response to different frequencies, I select and test frequencies of differing logarithmic domains, from Hertz to Megahertz. This practice, of sampling across representative frequencies, results in numerous test options for each circuit configuration.

I expanded my experience with mechanical switches by constructing analogues to Page's metal rasp (Fig. 7.13) and rotating spur wheel (Fig. 7b in art section). Through my novice efforts in a machine shop, I made a succession

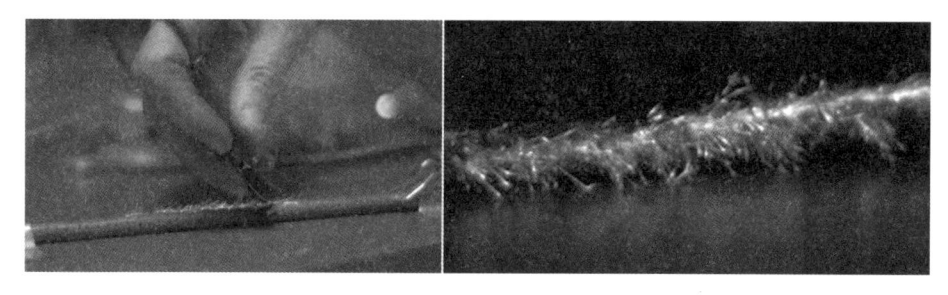

FIG. 7.13

A metal rasp is used as an interrupter for my spiral circuit (see **Fig. 7.8b**, left) by connecting one end to the battery and the spiral. A cliplead coming from the spiral is scraped across the rasp teeth. Sparks arise in the dark.

(Omari Stephens)

FIG. 5 (above) **(BAILEY)**
The *Planet* locomotive replica.
(Courtesy, Museum of Science & Industry in Manchester)

FIG. 6 (below) **(SWADE)**
Difference Engine No. 2, broadside view, 2002.
(Doron Swade)

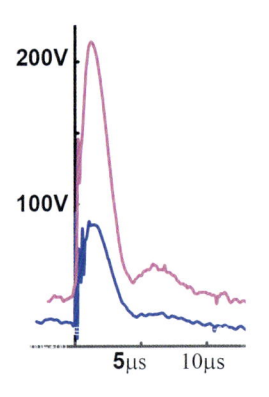

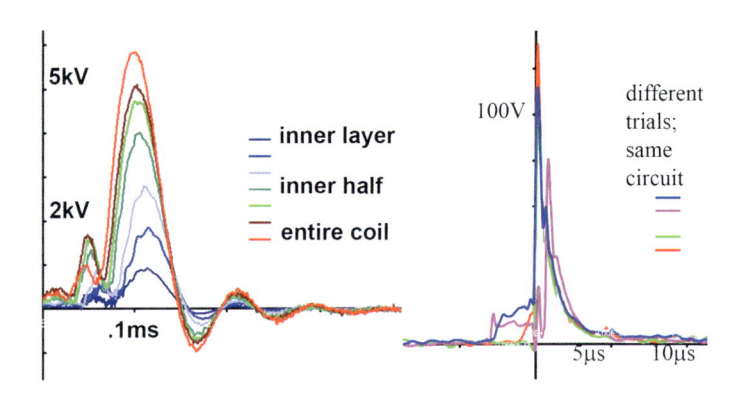

FIG. 7a (above) **(CAVICCHI)**

Left: A superposition plot showing two cases of voltage induced in parts of a coil that I wound from one continuous wire, over a core of thin iron wires. Battery current was applied across only an inner portion of this coil. For the trace indicated by the blue line, the probe was placed across the portion of the coil that carried battery current. For the trace indicated by the pink line, the probe was placed across the entire coil (including the current-bearing segment); the voltage peak was higher in this case, analogous to Page's findings with the spiral conductor. Voltage peaks induced in my wire coils having iron cores were more stable from event to event, than with my spiral.

Middle: In one of my induction coils, voltage traces are taken across successively longer segments of the outer (secondary) wire from the coil's inner layer (dark blue) to its entire length (red), when battery current stopped flowing in the inner (primary) wire.

Right: An overlay plot illustrates the variation in voltage traces exhibited by my spiral. Each of the five traces, designated by different colours (blue, pink, red, lavender, green), was taken across the same interval of spiral, when my wheel interrupter switched off the current.

(Elizabeth Cavicchi)

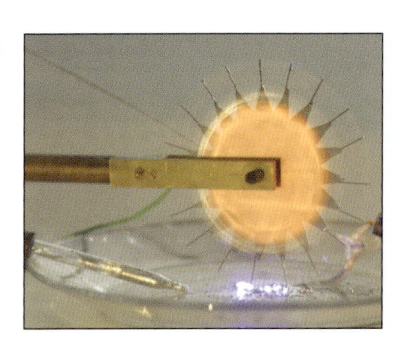

FIG. 7b (CAVICCHI) Three versions of my spur wheel interrupter.

Left: Purple sparks show where the spur's points leave the galinstan pool in this time exposure of my first interrupter, pulled by a string (left) wound around its hub. (Jeff Tinsley)

Below: **(a)** The spur wheel spins with purple sparks in the gap between two magnets that I hold. (Jeff Tinsley).

(b) My second wheel interrupter, connected to batteries and the wire coil for the experiment discussed in **Fig. 7a** (left). This second interrupter with the spiral were used in producing the traces of **Video 3** (see additional information at Cavicchi chapter end).

(c) My third interrupter, photographed with a spiral, spins like a motor in **Videos 1** and **2** (see additional information).

(Elizabeth Cavicchi)

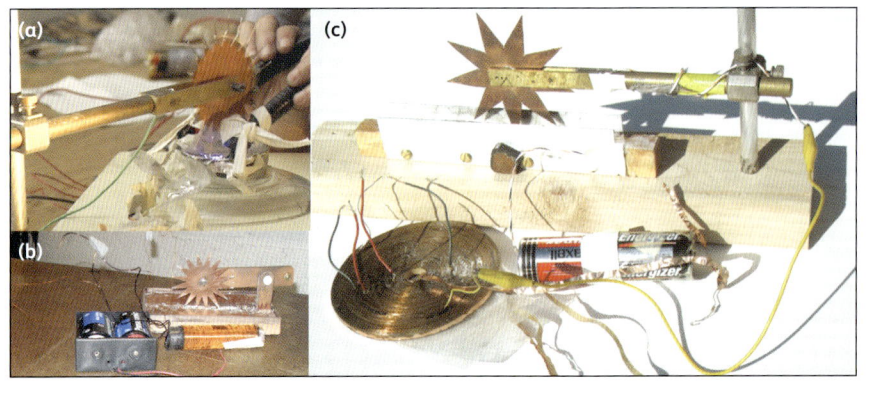

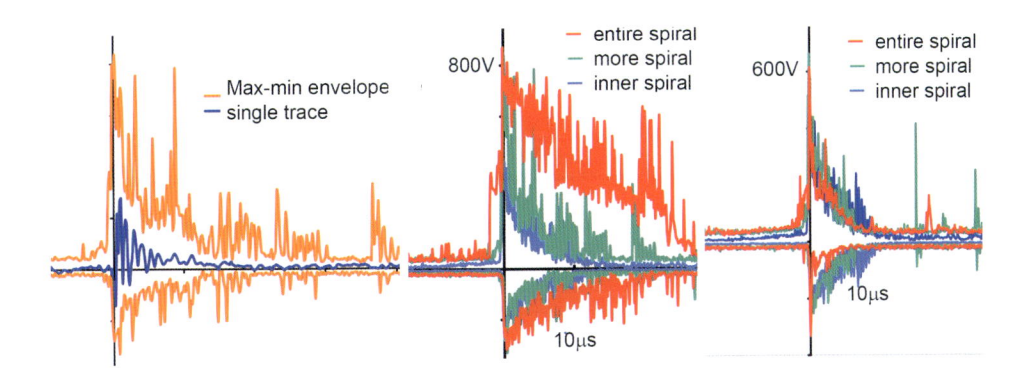

FIG. 7c (CAVICCHI)

An oscilloscope function records extremal values of voltages taken at each time, cumulatively across many (1024) events where the spiral was switched by the spur wheel with no body substitute in the circuit.

Left: An individual trace (maroon) appears within the spiky extremal boundaries (orange).

Middle: Overlay of extremal boundaries that resulted across inner spiral (blue), mid-spiral (green), entire spiral (red). Both battery current and probe were applied together across each region; the boundaries separate and increase in voltage.

Right: Envelope curves resulting where the battery current is applied across the inner spiral, and the probe is placed across longer spans show more overlap and ambiguity among the boundaries.

(Elizabeth Cavicchi)

FIG. 8 (FRERCKS)

Students experimenting with the rebuilt electrical machine.

(Courtesy of Collaborative Research Centre 482, University of Jena)

FIG. 9 (above) **(HEERING)**
Projection with a Dollond solar microscope at the Deutsches Museum in Munich.
(Peter Heering)

FIG. 10 (below) **(STAUBERMANN)**
One of the three Nicol prisms found in the photometer at the Deutsches Museum. The photo clearly shows the cork mounting, the blackened outer surfaces, and the two cemented halves. The prism is less than three centimetres long.
(European Southern Observatory)

FIG. 11 (STEADMAN)

Photograph by Trevor Yorke (top), taken to simulate *The Music Lesson* (bottom).

(Trevor Yorke;
The Royal Collection © 2011, Her Majesty Queen Elizabeth II)

FIG 12a (above) **(OSTERMAN)**

Scully & Osterman Studio. This illustrates the skylight area which features adjustable curtains on all the windows that are controlled by a system of chords and pulleys visible on the far left. Also shown, from left to right; highlight diffuser, posing table, posing chair, head immobiliser (behind chair), tilting cloth background frame and reflector on far right.

FIG 12b (left, both images) **(OSTERMAN)**

Collodion plates have very limited colour sensitivity. This comparison chart of a gelatin silver print made from a collodion negative and a conventional chromogenic colour print illustrate both the blue sensitivity and the insensitivity to warm hues of the spectrum.

(Courtesy, Scully & Osterman)

of spur wheel switches. When I first operated a wheel as a switch for the spiral, what I saw amazed me. Manually turning it through a glob of the liquid metal galinstan[16] in the dark, for the first time, I saw purple sparks and heard snaps (Fig. 7b, left in art section)! Caught up by the beauty, I followed a practice of Page's by affixing bits of metal leaf to the star's tips. When current combusted through them, the sparks were colored corresponding to the metal (Cavicchi, 2005: 131–32, figures 10–11).

Page discerned brightest sparks where battery connection broke from half of the spiral's length. When I tried to compare spark brightness, I experienced ambiguity again. I could not tell whether sparks were brighter when the whole spiral was interrupted, or just the half. Only when I reduced the battery source from two cells to one, did overall sparking diminish to where the midpoint brightness stood out from dimmer glows at other points.

Page had his wheel spinning on its own as a self-actuated switch by placing the gap between a magnet's poles crosswise to where the wheel spur contacts the mercury pool. My attempts to produce this motion have met with setbacks and given rise to my reconstructing several versions of the wheel switch (Cavicchi, 2005). The motor effect is minutely sensitive to the relative positions of magnet and wheel (Fig. 7b in art section, middle, right). That motion is astonishing! (Videos 1 and 2, *see* 'Additional information' on p. 170.)

Whether my wheel turned by hand, or as a motor, the repetitions of its breaking contact revealed more than the individual switching I had done before, where only one event at a time showed on the oscilloscope screen. In contrast, my star wheel interrupter made it possible to quickly spin through many events. Successive voltage traces appear on the oscilloscope screen while I turn the wheel, switching the circuit and observing by the oscilloscope probe connected across it. Voltage peaks dance like an animated movie of fluctuations; nothing is steady (Video 3, *see* 'Additional information on p. 170).

With this method of observing, the variability underlying the mechanical switching became more apparent to me. As I noticed distinctive patterns, I stopped the spinning to select those events to save on disc for plotting. However, high peaks flitted past too quickly for me to capture their traces by this manual selection method.

I experienced an opening of another kind of window on the trace's behaviour when I began to use oscilloscope functions that save data from successive events in real time (instead of single selection) and compute averages of the voltage values taken in these events. Keeping the circuit at one configuration, I employed various of these oscilloscope functions to collect and average many hundred switching events produced by spinning my spur wheel; I then repeated this procedure at another circuit configuration, and so on. Stabilised curves

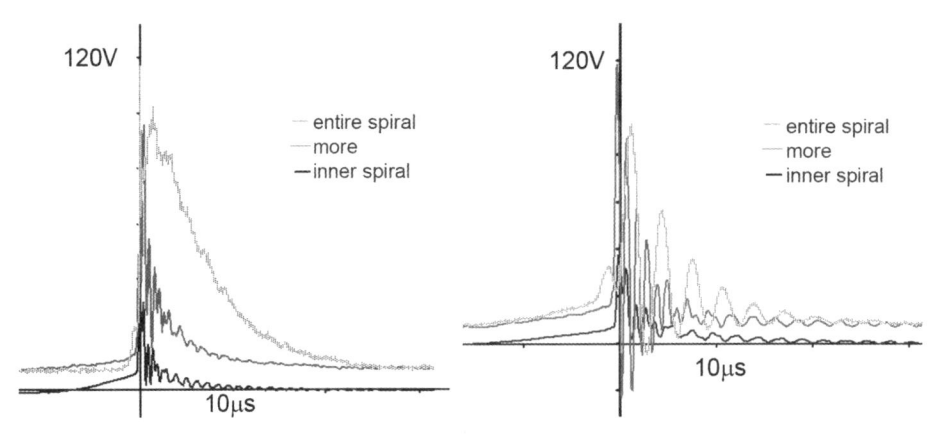

FIG. 7.14

An oscilloscope function records the average value of voltage from all those that occurred at each time slot for 1024 events switched by the spur wheel. The spiral's voltage was observed across its inner segment (black), mid region (gray), and entire length (light gray).

Left: Battery current is applied across each of the three regions along with the probe; the average curves show an increase in voltage and extension in duration as more of the spiral is stimulated.

Right: Battery current is applied only across the inner region; the average curves show a periodicity of greater duration and lesser value in the initial peak as more of the spiral is observed.

(Elizabeth Cavicchi)

resulted for each circuit configuration. By contrast, there was erratic variability among the traces of the individual events making up each set of averages.

For example, one oscilloscope function computes the average – at each discrete point in time – of all the voltage values (from 1000 collected traces) that occurred at that discrete time in the history of the triggered event trace. I applied this function when taking traces across three regions of the spiral: the first taken across the inner portion of the spiral; the second taken across a longer segment of the spiral; and the third taken across the entire spiral. I conducted this three-region study for two cases, in analogy to Page's experiment. For the first case I applied both the battery current and the probe across each region together (Fig. 7.14, left). As more of the spiral is stimulated with battery current, the averaged signal has an increase in both the peak value of voltage, and its duration. For the second case I applied the battery current only across the inner region of the spiral, and placed the probe across that same region, then the mid-region, then the entire spiral (Fig. 7.14, right). The average taken across the spiral's current-bearing region is high in voltage and brief in time; where the entire spiral is observed, the average exhibits a peak value somewhat lower in voltage, while the signal persists longer overall.

Another oscilloscope function automatically records the extremal voltages (high and low) occurring at each time position in the time history of successive triggered events. This function outputs an upper and lower boundary envelope for each set of sampled traces; any individual trace falls between those two boundary limits (Fig. 7c in art section, right). The complexity of voltage spikes in these upper and lower boundary envelopes depicts the varied, ambiguous behaviour that I had encountered before through individual events. I applied this function for the same two cases identified above, and across the same three regions of the spiral. For the first case, where battery current and probe are applied together, the extremal boundaries indicate an increase in overall maximal voltages when the entire spiral is involved (Fig. 7c in art section, middle). For the second case, the extremal boundaries are approximately overlapping, with perhaps lesser high voltage spikes when more of the spiral is observed (Fig. 7c in art section, left).

These two functions of average and extremal boundaries, viewed by superposition constructed from averages or compilations made over many successive switching events, depict an overall voltage effect across the spiral, an analogy of what Page reported. I wonder if these methods of averaging and accumulating sequences of hundreds of traces relate more to a blurring within Page's sensation of shock over many spins of the wheel, than do the separate transient voltage traces.

In addition to, and alongside, these experiments with mechanical switching by my wheel devices, I stimulated the spiral by periodic means. Periodic stimulation removes the variability which figures so prominently in my mechanically switched events; however, each test's finding holds only for one specific frequency. To survey over many frequency domains, I conducted extensive tests of the spiral and its intervals, applying periodic waves and pulses.

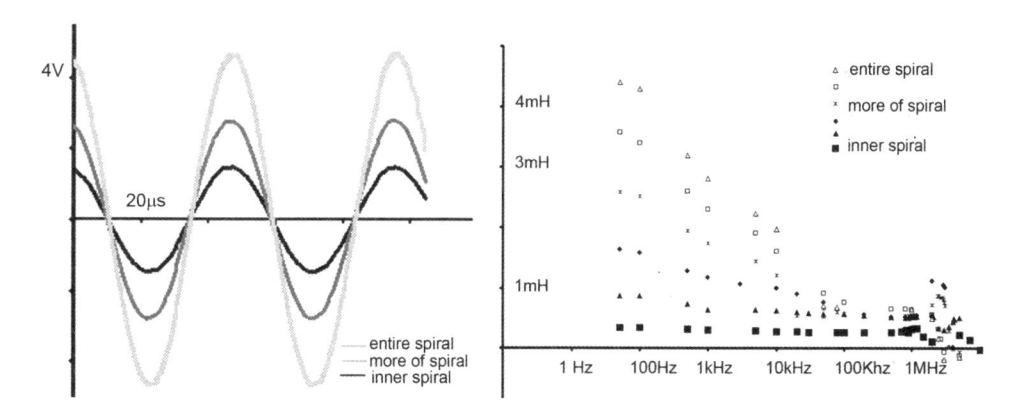

FIG. 7.15

Left: A constant frequency of 20kHz was applied to three intervals of the spiral in succession (inner, middle, outer). The observed voltage is superimposed, showing an increase in peak voltage across the spiral.

Right: Plot of inductance measured at several frequency decades, for different spiral intervals, its value increases as more of the spiral is covered. Overall, these values decline as frequency rises, and an anomaly appears above 1 MHz.

(Elizabeth Cavicchi)

RECONSTRUCTIONS

With a pure 20kHz sine wave, I first recorded a case where voltage distinctly increases as the probe is put across more of the spiral than the current-bearing segment (Fig. 7.15, left; Cavicchi, 2005). This result correlates with Page's report that his sense of shocks increased when he put himself across more of the spiral than where battery current passed. It also is consistent with my spiral having a greater value of electrical inductance – as this property is interpreted today – the more of its length is included in an observation. I checked this implication further by using an inductance meter to directly measure the spiral's inductance across each of the successive intervals where I have positioned connector tabs in analogy to Page's cups. As with the pulse generators, the inductance meter operates at one frequency at a time. At low frequencies, the inductance increased across the spiral (Fig. 7.15, right).[17] At high frequencies, its value declined overall, displaying an anomaly above one megahertz.

This megahertz anomaly intrigued me; I was curious about the spiral's differing electrical response when stimulated by frequencies of different magnitudes (e.g. Hz, kHz, 10kHz, 100kHz, MHz). During any mechanical switching event, the spiral is exposed to a range of frequencies, including high values. Might the spiral's apparent anomaly at very high frequencies contribute to the variability that I observed with mechanical switching?

I realised that a method of further investigating this possibility lay in applying Fourier transform analysis to my data; this analysis takes as input a function having values in time, and outputs a breakdown of the component

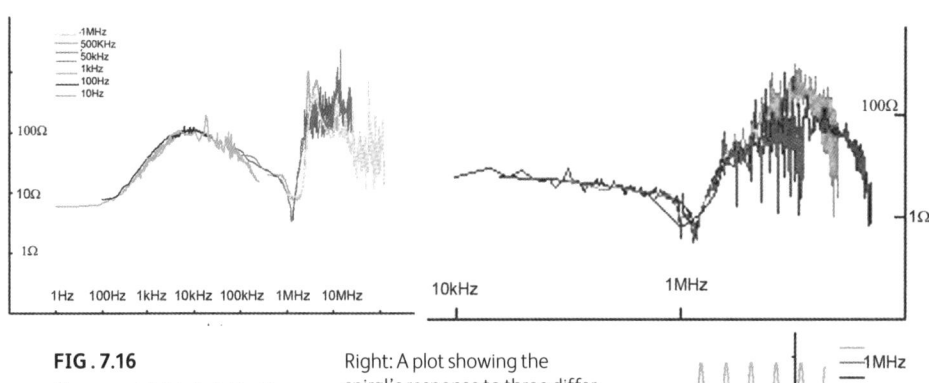

FIG. 7.16

Above and right: A distinctive dip in impedance occurs at about 4MHz in these log-log plots of observations taken across the same spiral interval. The spiral was stimulated by waves of different frequencies in each frequency range.

Right: A plot showing the spiral's response to three different sine waves (black: 500kHz; gray: 1MHz; light gray: 1.7MHz). The amplitude (height) of the sine wave is greater for the MHz waves that are near the spiral's resonant frequency.
(Elizabeth Cavicchi)

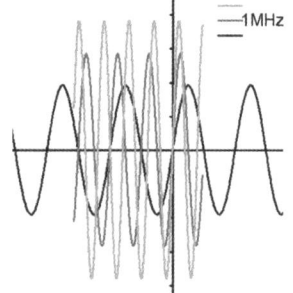

frequencies of which that function is constituted. For example, it converts a pure sine wave, where voltage varies with the same period for all time, into a function having one value – the value of the frequency corresponding to that period. An input function consisting of two superimposed sine waves converts to the two frequency values associated with those waves. A more complicated function, such as my voltage traces, will convert to a multivalued function of frequency, where the value at each frequency represents the relative weighting of a wave of that periodicity in the overall make-up of the original signal. The digital oscilloscope that I used had rudimentary software for computing this analysis; eventually finding it inadequate, I submitted my voltage trace values to the Fast Fourier Transform (FFT) program in Matlab (1994–2010). Applying these programs to my data, I produced log-log plots of the log of impedance against the log of frequency; in these plots, frequencies that are particularly resonant with the spiral appear as dips not peaks.

To study the spiral's responsiveness across different frequencies as computed by the FFT software, I stimulated the spiral with either my mechanical wheel switch, or a square pulse of different durations (frequencies) that was an output option of the signal generator. The experimental options widened yet again, as I alternated among these multiple modes of stimulation, and, as before, among all the intervals of the spiral's length across which these stimuli and the oscilloscope probes could be placed. My initial studies showed a distinctive dip above 1MHz in the spiral's response both to the wheel's mechanical switching of battery current, and to periodic stimuli. This dip shifted slightly in its frequency when the spiral was observed across different intervals and by other varied conditions (Fig. 7.16, above).[18] In some later tests, this Megahertz dip did not recur. As an alternative means of checking the

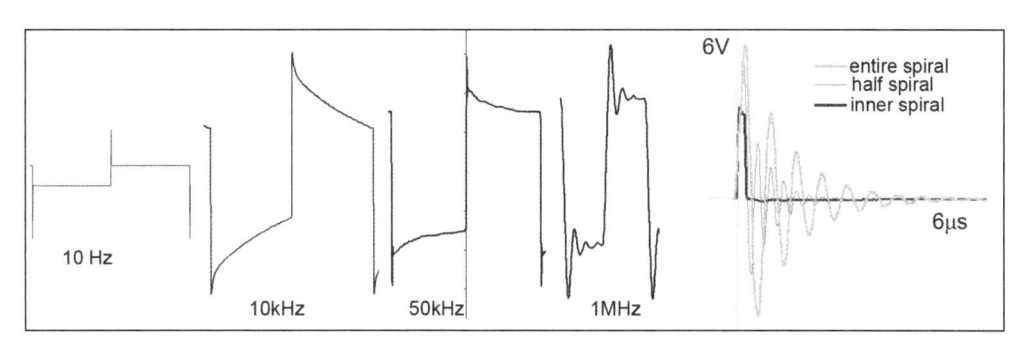

FIG. 7.17

Left: A periodic square wave of different frequencies (10Hz, 10kHz, 50kHz, 1MHz) is applied across half the spiral, and the probe is placed across the entire spiral. The transmission of the wave in the spiral distorts it; this distortion is greatest for the high frequency MHz wave.

Right: The spiral's inner interval is stimulated by a narrow voltage spike (.3 s duration); when viewed across more of the spiral, the observed signal stretches out in time, and may increase in voltage. (Elizabeth Cavicchi)

spiral's resonance frequencies, I stimulated it with a pure sine wave whose frequency I varied by successively dialling the generator through all values from hertz to megahertz. Resonances showed as an increased amplitude in that wave as detected by a probe placed across the spiral (Fig. 7.16, right).

I came across other frequency-related effects. For example, I applied square waves from different frequency domains (10 Hz to 1 MHz) to part of the spiral, and applied the probe across the entire spiral. The detected waveform was not so square (Fig. 7.17, left)! In the spatial extent of the spiral beyond where the square wave is applied, the signal becomes distorted, especially at high frequency. As another illustration of how a distinctive input signal is affected when it is observed across the entire spiral, I applied a narrow voltage spike to the spiral's inner section, and placed the probe across more of it. Further out, the pulse spreads in time and rings with peaks of decreasing height (Fig. 7.17, right).

In looking into the spiral by means not available to Page, I – like him – find electrical behaviours to wonder about. Always, the overall effect is amazing: on putting a spiral into the circuit, the voltages induced exceed my flashlight batteries' 3 volt input by over two orders of magnitude. This heightened voltage, its variations in degree, and its presence in winds outside the battery current's direct path, had intrigued Page and was also unmistakable for me. Having no stable sense of what to expect while exploring apparatus and effects that were new to him, Page improvised the intermediate cups and circuit breakers – that opened original experimental options – that revealed otherwise unseen behaviours. In writing about an experimental development by Michael Faraday that is commensurate with Page's, historian David Gooding characterised it as 'experimenting to *realise* possibilities, not to *decide* between two distinct or incompatible interpretations' (Gooding, 1990: 124). Page and Faraday functioned productively in an environment of ambiguity through generating experimental options or 'possibilities'. Analogous to how the extra cups in Page's spiral expanded the experimental options, so amid the ambiguity that arose in my study, the oscilloscope, test equipment and analysis techniques offered new opportunities. Realising possibilities meant coming up with more to try, widening the world of options beyond what prior explanations might prepare me to consider.

While Page and I both experienced ambiguity in our work with the spiral conductor, this ambiguity was expressed in differing forms. In part, Page developed his experiment by qualitatively comparing the shock or spark exhibited by one circuit configuration with that of another. Usually there was sufficient difference between effects being compared, for him to report which one seemed stronger. However, on discharging a Leyden jar through different regions of the spiral, he could not distinguish one case from another. Characterising these results as 'somewhat equivocal', he suspected that the fabric separating the spiral's turns provided inadequate insulation (Page, 1837a: 140). Page's sense of something 'equivocal' going on in his Leyden jar tests correlates with the ambiguity that is so prevalent for me in comparing voltage traces from the same, or differing, circuit configurations. Leyden jar discharges involve high voltages and frequencies: the same regimes that showed anomalous and resonant effects in the spirals of my studies.[19]

Along with the ambiguity of electrical effects that makes it 'equivocal' to compare and describe them, ambiguity of a second form was involved in construing and interpreting the underlying behaviours. While Page's experiment accessed the inductive phenomena in ways that often made the effects more distinctive than in mine (leaving him perhaps with less awareness of ambiguity regarding those effects than I encountered), he was immersed in the second form of ambiguity, as characterised here. The means available to him for interpreting the new electrical effects were insufficient, plunging him in uncertainty about what was going on. His paper reveals this uncertainty in such remarks as: 'the rationale I am unable to give'; 'still more curious ... difficult to explain'; 'contrary to expectation' (Page, 1837a: 139). Under Page's hands – and through them! – the experiment changed and complex electrical relationships became apparent even while he lacked an explanatory description. By contrast, I had access to electronic tools and analyses by which the self-inductive properties of a conductor can be identified and described. I did not pretend not to have this access, yet I soon found that my expectations for particular experimental outcomes could be both unfulfilled, and limiting. I traversed no direct path to demonstrating Page's findings. Even with modern tools at hand, the spiral conductor experiment retained ambiguity and complexity for me, as it had for Page. Instead of dispelling ambiguity with definitive outcomes and answers, the instrumental resources and analyses allowed me to observe effects that I had not expected and to move beyond the limits of my expectations in flexibly developing experimental options for exploring that ambiguity. Experiencing confusion in this way, I became as much an explorer of the spiral conductor as Page.

Revisiting the spiral

Page's electrical investigations soon took over to the extent that his medical practice fell by the wayside. The most productive period (1837–39) in Page's scientific contributions to electromagnetism spanned either side of his 1838 relocation with his parents from Salem, Massachusetts, to a town outside Washington DC. Page's unique electrical expertise came to public prominence through his roles as US patent examiner, key witness in the 1848 Morse vs

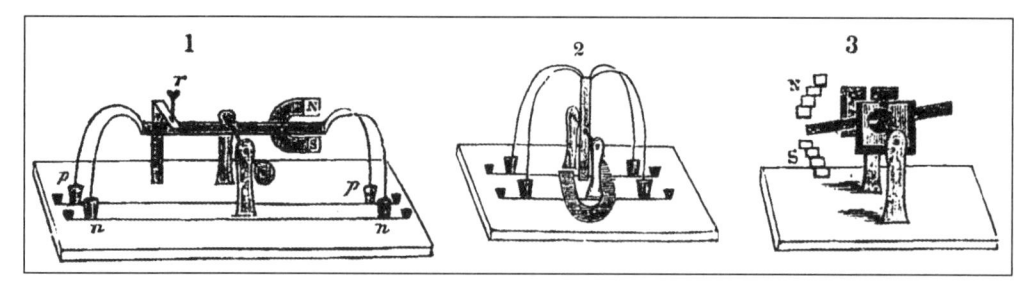

FIG. 7.18

Page's homemade circuit-breakers, invented for use with his lengthened spiral.

Left and middle: small cups (labelled **p** or **n** on left) contain mercury and connect to the battery's + or – terminal. A current-bearing wire bobs up or down between the poles of a magnet; its downward curving wire ends rock into or out of these cups, breaking or making the circuit at a rate which Page described as 'inconceivably rapid'.

Right: Page's revolving interrupter scattered drops of mercury as it turned in the presence of a horseshoe magnet (not included in the diagram).

(Page, 1837b: 356)

O'Reilly lawsuit, and independent inventor (Post, 1976a). A US Senate allocation of $20,000 launched his electromagnetically powered locomotive whose fortunes collapsed even before its beleaguered test run. The foundational yet rudimentary spiral of 1836 was core to Page's death-bed appeal to the US Congress for a retrospective patent on it and his subsequent double coils. The resulting 'Page Patent', sweepingly interpreted to cover circuit breakers and other essential telegraphic apparatus, garnered a fortune for Page's heirs and ill will from the telegraphic community (Editor, 1872; Post, 1972, 1976a, 1976b).

After his first publication on the spiral, Page continued finding more to try and in the process deepened his understandings of electricity. A year later, the spiral was 100 feet longer, with four more mercury cups for making connections (Page, 1837b). Page acknowledged Ampère in naming it the 'Dynamic Multiplier' and in describing its function as 'Electro-dynamic'. Page's exhaustive inquiries into its inductive sparking under series and parallel battery configurations led to original research on the battery, resulting in a more compact, stable cell. Going beyond the acid battery, Page also activated the spiral with a thermo-electric source. To replace the human operator's action in opening the circuit, and to overcome problems he encountered with his spur wheel switch, Page pioneered rocking and spring-loaded forms of the self-actuated switch (Fig. 7.18): 'I have tried a variety of means and succeeded in the contrivance of several beautiful pieces of apparatus' (Page, 1837b: 355; Sherman, 1988).

These innovations heightened the spiral's effects so that it was no longer essential to put the body in the circuit. Page applied the spiral's heightened electricity to standard demonstration tests of the time, including sparking across separated charcoal points and decomposition of water. He alluded to the body only in indirectly mentioning 'acupuncture' shocks showing the thermo-electric source's influence on the outer spiral (Page, 1837a: 358).

But even if the body was superseded, it remained essential to experimental development that something fill its role as detector. For example, Page found the faster the contact breaker went, the more ferociously foamed bubbles of water decomposed by the spiral's induced electricity. Just as differing shock intensities had enabled Page to evaluate successive placements of his hands across the spiral, now he used this bubbling as feedback while improving the interrupter. He achieved greatest rapidity when his rotary wheel operated electromagnetically as a motor powered by a miniature battery separate from that which ran the larger circuit.

In the widening range of experimenting brought about by the spiralled conductor, Page observed yet another new phenomenon, one which contributed to the future of telephony. Instead of resting the spiral horizontally, this new set-up involved vertically mounting a lighter-weight spiral of cotton-covered wire so that it resided edgewise within the horizontally oriented gap between a horseshoe magnet's poles. On each interruption of electrical current through the spiral, the magnet rang with a characteristic musical tone; different sized magnets gave different tones (Page, 1837c, 1838a).[20] Alexander Graham Bell opened his ground-breaking lecture at the American Academy of Arts and Sciences on 10 May 1876 by crediting Page's 1837 discovery of 'galvanic music' with kindling world-wide inquiry on sounds associated with magnetisation and demagnetisation, including his own research (Bell, 1876–77: 1).

But the magnet's singing merited only passing notice in Page's 1869 obituary. Page was long out of the top ranks of American science when he died penniless of sufferings that chemical exposures perhaps had exacerbated (Lane, 1869). Public laurels in telegraphy and telephony came to rest on others, both during Page's abbreviated life and subsequently. In contesting this injustice, Page's biographer Robert Post sheds light on culturally imposed expectations about the conduct befitting a scientist, whose violation by Page resulted in marginalisation during his own day, and in succeeding historical assessments (Post, 1976a).

Extending and interpreting the spiral's effects

The spiral was the first of Page's electrical contributions, and through communicating about it he engaged a broader community of electrical experimenters. Two figures, one foreign, one local, are conspicuous in supporting young Page's development. London-based William Sturgeon reprinted Page's papers in his journal with his own commentary (Post, 1976a: 207–13; Morus, 1998); Boston instrument-maker Daniel Davis Jr collaborated with Page in making new and production apparatus (Post, 1976a; Sherman, 1988; Cavicchi, 2006a). From both, Page's interest in following inductive phenomena deepened, and his instrumental work shifted from the spiral and toward electromagnetic coils. However, the spiral remained to offer yet new inductive behaviours to later researchers.

Sturgeon first heard about Page's shocking device from a traveller from Salem who visited the Adelaide Gallery of Practical Science in London, then a locus for science experimenters and the curious public. Crucial details about the instrument – including that its inventor was allegedly Henry! – were garbled in the informal transmission (Sturgeon, 1837a; Page, 1867: 11, footnote). Sturgeon sought to replicate its effect of heightened tension, but in his version there was no spiral at all! Just as Page responded to Henry's spiral by improvising with copper sheet, and I to his with artist's foil that I had on hand, Sturgeon appropriated two helical coils of wire from a magneto for his reconstruction. Joining these two coils in sequence, he sent battery current through one and took shocks from it alone and both together (Fig. 7.19, left). The second coil failed to enhance shocks, so he dispensed with it and configured a single coil variously for shocks. In the process, Sturgeon rediscovered the shocking effect reported first by Jenkins and reinvented many of the tests by which Faraday extended it, while being unaware of their prior work. Initially ecstatic over 'bringing to light a novel principle' (Sturgeon, 1837a: 75), Sturgeon subsequently conceded Faraday's precedence in doing some experiments (Sturgeon, 1837b) and reprinted Faraday's paper in his journal (Faraday, 1835).

Later when Sturgeon received Page's actual text, he grasped what he had misunderstood before. In republishing Page's paper, Sturgeon remedied Page's lack of interpretation by appending his own to it:

FIG. 7.19

Left: Sturgeon's two linked coils **A** and **B**; he found the shock was not increased by adding coil B
(Sturgeon, 1837a, plate ii, fig. 16).
Middle: Sturgeon's shocking coil where current flows through an inner coil and shock is taken from the handles **rr** of a second coil that is wound over the inner one
(Sturgeon, 1837d, plate xv, fig. 125).
Right: Sturgeon's shocking coil in the London Science Museum (no. 1860–72).
(Elizabeth Cavicchi)

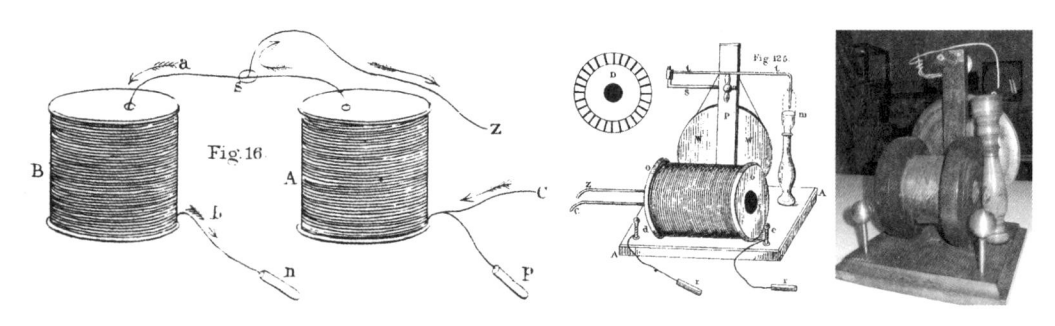

In every instance the phenomena may be traced to the *collapsion* of the electro-magnetic lines. In some instances the phenomena proceeded from a *primitive* current; in others, from a *secondary* current; and in others from both *primitive* and *secondary* (Sturgeon, 1837c: 294).

'Primitive' current came direct from the battery; 'secondary' had another path. To illustrate each of the three cases, Sturgeon cited an example circuit from Page's paper. Sturgeon admitted that he too had not formerly understood that wires bearing these two currents needed to be located 'within the influence of each other'. With his new understanding from reading Page's paper, Sturgeon redid his experimental reconstruction – with wire coils, not with a spiral. Only after Sturgeon over-wound the secondary on his primary coil, was the combined coil's shock greater than that of the primary alone (Fig. 7.19, middle, right). He marketed this instrument with a revolving contact breaker whose sparks combusted in differing colors (depending on its interchangeable metal discs) like the effect Page produced with his spur wheel tipped with metal leaf (Sturgeon, 1837d; Cavicchi, 2006a).

Sturgeon's published commentaries moved Page's thinking. Previously Page had not speculated about the spiral's electricity. As a result, Page wrote about it as an example of electromagnetic induction and made sense of such aspects as the interrupter's role. Subsequent experimenting developed his ideas so far as to reject an earlier, now 'irrational', view that conjoining 'primitive and secondary' currents (as in the spiral) was what produced shocks:

… the sparks and shocks indicating a new and secondary current are directly consequences of the dissolution of the primitive current … due solely to magnetic excitation, and have no connexion with that primitive, except that of cause and effect. (Page, 1838b: 366)

Secondary currents did not offshoot directly from battery current, but instead arose from changes in a magnetic medium surrounding them, as envisioned in what Page called Sturgeon's 'beautiful theory of electro-magnetic lines' (Page, 1838b: 367). Page's acknowledgment of Sturgeon's contribution meant much to the recipient; Sturgeon excerpted it in the last publication of his life, adding, 'I know of no philosopher more capable of close reasoning on electro-magnetics and magnetic-electrical physics than Prof. Page, M.D.' (Sturgeon, 1850: vii).

As spatial relations among coils and magnets became increasingly critical in Page's experimenting, he consulted the 'ingenious' Daniel Davis Jr, first American manufacturer of electromagnetic demonstration instruments. Their collaboration was reciprocal. Davis refined Page's prototype devices and marketed them through his shop, trade catalogues and textbook. Page illus-

FIG. 7.20

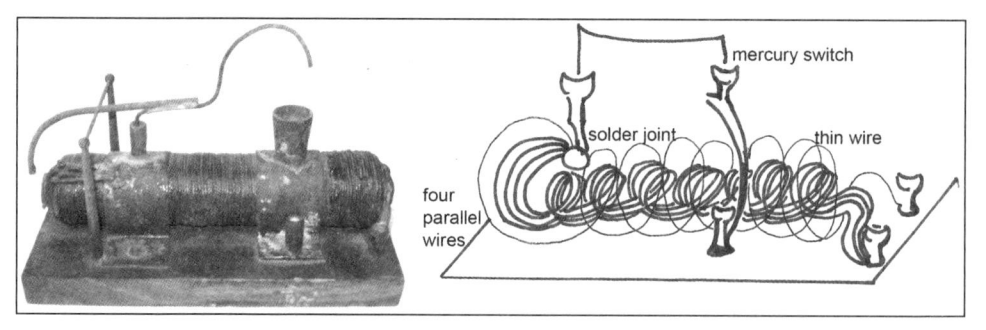

Left: Shocking coil number 2002.1.35088 in the Allen King Collection of Scientific Instruments, Dartmouth College. A solder joint on the Dartmouth shocking coil unites the four thick wires that bear current to the thin 'secondary' wiring from which shock is taken.

Right: My diagram shows the coil's solder joint and wiring.

(Elizabeth Cavicchi)

trated his scientific papers with Davis' distinctive apparatus and acknowledged Davis' contributions to his work. The instruments and understandings that Page and Davis developed together elucidated electromagnetic phenomena elegantly and went into wide instructional use (Davis, 1838, 1842; Sherman, 1988; Greenslade, n.d.).

But before new instruments attained commercial viability, Page improvised them experimentally. Page reacted to Sturgeon's reports about iron core coils by constructing coils of varied dimensions, wiring, cores, and batteries and by testing their magnetic pull, sparks and shocks. In revising his instruments, Page applied what he learned from these tests about the differing characteristics of primitive and secondary currents. For example, he employed thick wire to carry primitive currents, and thin for secondary.

A shocking coil that may represent the early Page-Davis association is now in Dartmouth College's Allen King Collection of Scientific Instruments (Fig. 7.20).[21] Appearing to be a prototype, without Davis' usual high craft, design features link this instrument to Page while its materials correlate with Davis. A similar, presumably subsequent, coil was first offered for $8.00 in Davis' 1838 catalogue and illustrated in Page's 1839 paper and Davis' 1842 textbook (Fig. 7.21) (Davis, 1838, 1842, 1846; Page, 1839; Channing, 1849: 20; Garrett, 1876: 49–50, 53).[22] These publications describe an instrument having two separate, concentric coils: one for battery current, the other for shock. The Dartmouth instrument is wired differently (Pantalony et al., 2005). A solder joint affixes the secondary coil directly to the current-bearing coil. Shock may be taken either across the secondary alone, or across the combination of both coils.

This solder joint preserves the continuity between secondary and primitive paths that Page's spiral first exhibited. It embodies the transitional moment

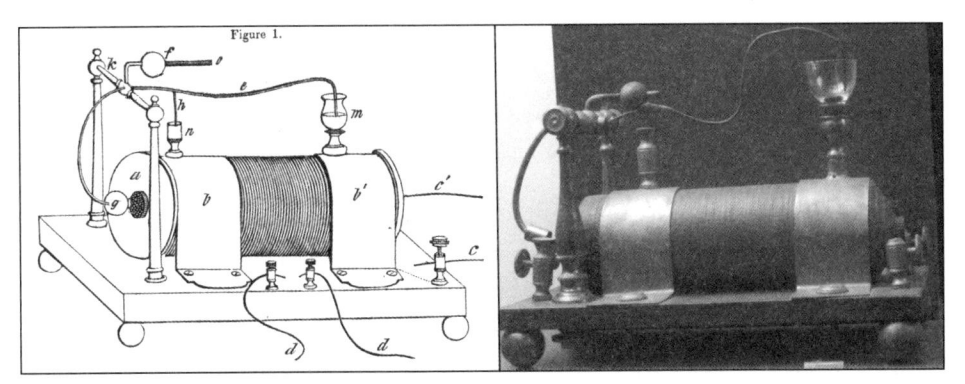

FIG. 7.21
Left: Page's double helix coil, where **cc'** are battery terminals and **dd'** are connectors for taking shock. The rocking wire **e** interrupts the primary circuit when the coil's magnetism attracts iron ball **g**, lifting **e** from mercury cup **m**, giving rise to sparks in cup m and shocks at **dd'** (Page, 1839, Fig. 1: p. 258).
Right: Page's patent model for his double helix coil, displayed in the National Museum of American History, catalogue number 309: p. 254.
(Elizabeth Cavicchi)

before Page rejected as 'irrational' the notion that the elevated, shocking electricity depends on continuity between these two paths (Cavicchi, 2006). While the solder joint reflects ambiguity in understanding electromagnetic behaviours, it also accommodates multiple options regarding which parts of the coil can be used for taking shocks. Those multiple options, originating in the intermediate tabs of Page's spiral, are eliminated in Davis' commercial version of Page's coil and in Page's patent model of it (Fig. 7.21, right), where shock was taken only across the secondary. The spiral's intermediate cups, and the coil's solder joint, were gone. Where these connectors opened up multiple options for electrical paths, the commercial instruments limited current to one fixed path instead. Being constructed to reproduce pre-existing effects, not create new ones, the commercial apparatus closed off experimental space and access to ambiguity.

The spiral had a further experimental history, but not in medicine where electromagnetic coils like Page's soon gained wide currency for electrical therapies.[23] Instead, it figured in landmark experiments with wireless transmission of electricity from place to place. Joseph Henry pulled pranks with his wireless apparatus by mounting a battery-interrupted spiral on one side of a wall and putting a spiral with handles on the other side so that someone grasping its handles received mysterious shocks (Henry, 1839, 1843). A half-century after Page's spiral experiment, Heinrich Hertz experienced a 'surprise' analogous to those Henry and Page encountered. Whenever Hertz electrically stimulated one spiral, sparks appeared in the air gap between ends of a distant spiral. The sparks' high frequencies (100MHz) represented wavelengths long enough to make lab experimenting practical. This pivotal observation launched Hertz's research of the electric waves that Maxwell predicted.[24] Thus the spiral contributed to wireless communication, where the high frequencies amplified through spiral resonances transport electrical imprints of speech without any mediating body.

Conclusions

Page's spiral experiment opened up options for electricity's paths; these paths showed themselves to be more complex than simple flow between two endpoints. Electricity arose interactively inside conductors: Page experienced it as shocks from spiral intervals where he did not expect electricity to be. His body was both a constituent of those new paths, and a reporter on what was going on. It filled in where no measuring apparatus then available could, by sensing momentary pulses induced in the spiral's many turns.

Body and circuit are partners, with each being a locus for inquiry and intervention, in the experimenting of Page and his peers, from Volta through Faraday, Henry and Sturgeon. The instruments bear out this partnership: medical acupuncture needles became electrodes, and coiled conductors become therapeutic aids. The analogy goes further: Page opening up the spiral to probe its interior resembles a physician looking into the body. His thought experiment to put a mercury cup on every spire applies surgical precision to circuital intrusions. Once inside, both body and circuit were baffling; the sensational observations disclosed electrical activity, but left the workings obscure. Page communicated his observations in all their ambiguity, proffering no explanations until a community extended his findings with their own.

Multiple factors confounded in the effects Page used and detected, and these melded together for him. Only through his later extensive experimenting with circuit breakers, batteries, and electromagnets of varying construction and wiring, did he work out such characteristics as induction's enhancement under abrupt stimulus, lengthened coiling, and differentiated thick and thin wires. Change in both time and space matter to electromagnetic induction: Page's spiral exemplified this by the *timing* of its switching and by its spiral extent in *space*. Time, space and magnetic lines changing in that space came to have interactive roles in Faraday's more mature thinking about fields. Yet in the late 1830s, years before Faraday began investigating thoroughly what goes on in a powerful electromagnet's gap (Gooding, 1980, 1981; Cavicchi, 1997), Page, Sturgeon, Henry and others had already engaged directly with electromagnetic field phenomena. Although lacking the field analysis that pervades subsequent science and engineering, they worked productively with the whole web of electromagnetic effects and ambiguities to develop instruments that manifested and amplified inductive behaviours.

Faraday, Henry, Page, Sturgeon and I responded to instruments and reports of others through initiating experiences of our own with apparatus culled from whatever was ready at hand – including hands! None of these reconstructions of self-inductive phenomena literally redid effects of the others. Each engaged with ambiguity by differing forms: from those entangled with the phenomena, such as Page's 'equivocal' Leyden jar tests and my variable voltage traces; to others compounded by such sketchy communications as Henry's hasty notice or the Salem traveller's faulty memory; to the ambiguity of evolving one's sense of what is going on through proposing and doing new experimental work. As diverse as these instruments and experiences were, the experimenting interrelated, to retrace and open new options that kept extending the work. The participants' emerging understandings of electromagnetic induction were overlapping but not identical, enriched by the particular observations and paths of each.

My reconstruction of Page's experiment recovered something of those past experiences of dealing in the unknown. My observations following electrical effects across ever-wider spiral intervals did not readily confirm Page's sensations of heightened shock. If Page's experiment had translated directly into my improvisations and instruments, my project might have concluded as a success in replication while remaining unaware of the experience with ambiguity that was core to the original investigations. In this sense, my experimental journey re-expresses what Page's biographer, Robert Post, described as Page's engagement with 'the baffling complexity of things' (Post, 1976b: 26–27, quoting Beard, 1927: 741–42). That complexity, both in nature and our curious responses, becomes hidden from view by subsequent formalised and purpose-driven packaging that, like Davis' commercial coil, constrains the options for engaging it around a particular favoured outcome. Where such options are limited and ambiguity is masked, it is hard to explore; we suppose we know in advance where any path will go.

A challenge inherent in reconstructing a past experience lies in recovering our access to complexity and ambiguity sufficient that genuine opportunities for investigation emerge from options which under present practices and knowledge might be unnoticed, discounted or unexpected. Those options and inquiries may take on different forms for us – for example, here requirements of health and safety reframed the instrumental context. Our involvement with historical and reconstructive material deepens through coming upon passages in the work that open our vulnerability, such that we can find ourselves in ambiguity and begin to explore.

These reflections spiral back to my aspirations as a teacher seeking to facilitate exploratory experiences among students. Conventional practices in classrooms make daunting the challenge of engaging students in their own genuine, sustained experiences with the 'baffling complexity' of any subject matter. The prevalence of didactic explanations, along with students' expectations for such answers, closes down options, leaving nothing to explore. Historical reconstructions offer evidence of what an alternative pedagogy might encompass: on going into a material seemingly well-known, the learners find there the unknown, not just about what someone else already did, but also within their own understandings. For historical investigators like Page, widening personal experience supported by a community made possible their unique exploratory work; similarly there is a role for educators to bring about environments where each student's curiosity evolves by undertaking its own explorations in relation with a community of other explorers.

Acknowledgements

Robert Post encouraged and deepened my study of Page. I thank the Edgerton Center at MIT for access to lab space and instruments, ongoing experimental discussions with James Bales, Ed Moriarty, Anthony J. Caloggero, Fred Cote and other assistance from the staff. Thomas Cavicchi, Chen Pang Yeang and Markus Zahn discussed my experiment and methods of analysis; Thomas Cavicchi responded daily to my experimental struggles. Grant Suter, Lourenco Pires and Wayne Ryan provided other technical support. My understanding of Page's experiment and replication developed through many thoughtful discussions with Ronald Anderson, Michael Dettelbach, Abigail Lustig, Peter Heering, Giora Hon, Evelyn Fox Keller, Richard Kremer, Frank Manasek, Ben Marsden, Alberto Martinez, Arthur Molella, Philip Morrison, Giuliano Pancaldi, David Pantalony, Robert Post, Martin Reuss, Wolfgang Rueckner, Mike Schiffer, Roger Sherman, Yunli Shi, Christian Sichau, Chris Smeenk, Friedrich Steinle, Klaus Staubermann, Ryan Tweney and Chen Pang Yeang. Comments from James Bales, Alva Couch, Eleanor Duckworth, Kate Gill, Sabine Hoidn, Philip Morrison, Joshua Ryoo, Bill Shorr, Klaus Staubermann and Chris Smeenk improved the paper. The former Dibner Institute for the History of Science and Technology at MIT provided support. Larry Ford furthered my study through an invitation to present it. Eleanor Duckworth and my students helped me realise the possibilities of being in confusion. I thank Klaus Staubermann for encouraging my participation in this historical reconstructions workshop. Alva Couch wrote plotting programs and sustained my spirits throughout many uncertainties. Inspiring our wonder about all the options in a circuit or the complexity in an experience, the teaching of Philip Morrison here continues in his memory, through an experiment we discussed together. This essay honours the memory of Ronald Anderson SJ.

Notes

1. See also Cavicchi (1997, 2006b) for related discussions of Faraday's exploratory work.
2. The introduction of 'electropuncture' is attributed to Jules Cloquet, Jean Baptist Sarlandière and Fabré-Palaprat in 1825 by Gwei-Djen and Needham (2002, pp. 295–302) and by Rowbottom and Susskind (1984). Boston physician William Channing (1849) credited it to M. Berlioz.
3. A recent physics textbook adapts the story of Jenkins and Faraday in its introduction to self-induction and associated exercises (Saslow, 2002, pp. 533–44).
4. *Boston Medical and Surgical Journal* is the predecessor of *New England Journal of Medicine*. All historical articles cited from *Boston Medical and Surgical Journal* and *American Journal of Science* are now available in the Proquest digital resource American Periodical Series (2010).
5. Praise of Page's electromagnetic inventions and teaching of a chemistry course in chemistry appeared in the *Boston Medical and Surgical Journal* (26 April 1837), p. 195 and (22 November 1837), p. 256. Page published notices in that journal (1836a, 1836b).
6. Page referred to 'M. Palabrat's discovery … transmission of remedial substances … .' (1836a). Fabré Palaprat described his electropuncture technique in La Beaume, 1828: 36–61. Channing summarised it (1849), pp. 38–39.
7. Henry presented his work with the spiral on 6 February 1835, but his full paper was not published until 1837. To secure credit for Henry (Faraday was publishing related work at the time), Alexander Bache composed an abstract by Henry, which was immediately published in *Journal of the Franklin Institute* (March 1835), and in *American Journal of Science* (July 1835) with the addition of a brief appendix by Henry (1835).
8. Page's spur wheel was an adaptation of Barlow's wheel (Barlow, 1822).
9. Also in 1836, Charles Tomlinson (1837) produced similar effect with a sparking motor. In 1831, Faraday used a slotted spinning wheel to explore optical deceptions associated with the persistence of vision which rendered Page's wheel apparently stationary (Tweney, 1992).
10. Page did not specify the dimensions of his initial 'calorimotor'. Page followed Henry's preliminary notice, which was vaguely worded in recommending 'one of Dr. Hare's Calorimotors' (Henry, 1835, p. 329). Henry later stated that he employed one pair of large plates having 1.5 square feet of zinc surface area (Henry, 1837, p. 224).
11. I worked with the following storage oscilloscopes in successive phases of my study: HP 54600B; Lecroy 9450A; HP Infinium 54810A. Hewlett Packard's product Infinium is now serviced under Agilent, 2010–2011.
12. HP Infinium 54810A.
13. In the trial illustrated, I varied the resistor's value from a low of 330Ω to a high $560k\Omega$. These values correspond to those tabulated for the human body's resistance to current: dry skin $\sim 500k\Omega$; wet skin $\sim 1K\Omega$; internal body length $\sim 400\Omega$ (Jefferson Lab, n.d.).
14. In addition to the resistors, these substitutes included neon bulbs; a metal-oxide varistor; a resistor in series with a capacitor, and several variations on the Siconolfi model (Siconolfi et al., 1996). This human body model consists of a resistance in series with a capacitance, in parallel with another resistance in series with an inductance. Steven Siconolfi provided data enabling me to construct models. A $1.87k\Omega$ resistor is in series with a 2.2nF capacitor; this is then in parallel with a 510Ω resistor and an inductance (of 27 H in parallel with 56 H). For more extensive empirical and modeling studies of the human body's impedance (Reiley, 1998). Observed impedances range from over a $k\Omega$ at low frequency, to below 500Ω at high frequency.
15. I use the HP33120A function/arbitrary wave generator for sine and square waves up to 15 MHz. For higher voltage square pulses (up to 150 V at periods down to .01ms), I used a Grass S44 Stimulator of Grass Medical Instruments, Quincy MA.
16. I used the liquid metal galinstan, a product of Geratherm Medical Diagnostic Systems (n.d.), a safe substitute for mercury. It will, however, be mistaken for mercury by security detectors (e.g. airports). For more description of the reconstructed spiral and wheeled switch, see Cavicchi (2005, 2008b).
17. A HP 4192A impedance analyzer was used. The spiral's overall inductance was on the order of 4mH at low frequency; its resistance went from 5.6Ω at low frequency, into the $k\Omega$ range at 50 kHz.
18. The electrical property of admittance is the reciprocal of impedance. Impedance (measured in Ω) is the ratio of the complex voltage, V, to

the complex current, I; where both these are real, that ratio is the familiar electrical resistance. Impedance depends on frequency. I compute spectrums of spiral impedance from the ratio of the Fast Fourier Transform (FFT) of a voltage trace to that of a simultaneously observed current trace. As inputs for stimulating these traces, I use square waves generated at selected frequencies, as well as excitations made by dipping the spur wheel into liquid metal. Present results suggest that the megahertz regime where the spiral impedance drops may represent a transition from capacitive to inductive behavior. See Lehar, n.d. for a pictorial depiction of Fourier Transforms, and Brigham, 1988 for a more complete discussion of the FFT.

19 I am curious to apply Leyden jar discharges to conductive spirals or coils. Working with the high voltages might entail further modifications to the spiral experiment and methods of detection.

20 In my unsuccessful attempt to reproduce 'galvanic music', the spiral mounting was so insecure that the current-bearing spiral moved into contact with the horseshoe magnet's pole.

21 The Dartmouth instrument, accession number 2002.1.35088, was listed in an 1870s inventory as 'Page's apparatus for shocks with mercury break' (Pantalony et al., 2005, pp. 157–59; Cavicchi, 2006a, pp. 351–53).

22 Page's 1868 patent model of this instrument is on display in the National Museum of American History in Washington DC, catalogue no. 309 254, accession no. 89 797.

23 J. B. Zabriskie, physician in Flatbush, Long Island, reported on his experimental spirals (1837), but the spiral did not become a standard medical device. The electromagnetic coil's medical context is described in reference to physician Golding Bird in Beard & Rockwell, 1871; Morus, 1998.

24 H. Hertz described his work with 'Reiss or Knochenhauer spirals' (1892/1900, p. 2). The experiment is discussed in Buchwald, 1994, pp. 217–27. Further references to historical experimenting with spirals are given in Gluckman, 1993.

Bibliography

Agilent (2010–11): 'Agilent Technologies' [online]: www.home.agilent.com/agilent/home.jspx?cc=US&lc=eng

Anon. (1836): *Salem Gazette*, 9 September 1836, XIV (73), p. 2; American Periodical Series online, 2010.

APS (2010): American Periodical Series [online], via Proquest: www.proquest.com/en-US/catalogs/databases/detail/aps.shtml

Bache, A. (1835): 'Facts in reference to the Spark, &c. from a long conductor uniting the poles of a Galvanic Battery by Joseph Henry', in *Journal of the Franklin Institute*, March 1835, pp. 169–70; reprinted in *American Journal of Science*, July 1835, 28, pp. 327–29.

Barlow, P. (1822): 'A curious electro-magnetic Experiment', in *Philosophical Magazine*, 59, pp. 241–42.

Beard, C. A. and M. R. Beard (1927): *The Rise of American Civilization*, 1940 edition (New York: MacMillan).

Beard, G. M. and A. D. Rockwell (1871): *Practical Treatise on Medical and Surgical Uses of Electricity* (New York, NY: William Wood & Co.).

Bell, A. G. (1876–77): 'Researches in Telephony', in *Proceedings of the American Academy of Arts and Sciences*, 12, pp. 1–10.

Bertucci, P. (2001): 'The Electrical Body of Knowledge: Medical Electricity and Experimental Philosophy in the Mid-Eighteenth Century', in *Electric Bodies: Episodes in the history of medical electricity* (P. Bertucci and G. Pancaldi, eds) (Bologna: Università di Bologna), pp. 43–68).

Bertucci, P. (2006): 'John Wesley and the religious utility of electrical healing', in *British Journal for the History of Science*, 39(3), pp. 341–62.

Bresaldo, M. (2001): 'Early Galvanism as a Technique and Medical Practice', in *Electric Bodies: Episodes in the history of medical electricity* (P. Bertucci and G. Pancaldi, eds) (Bologna: Università di Bologna), pp. 157–80.

Brigham, O. (1988): *The fast Fourier transform and its applications* (Englewood Cliffs, NJ: Prentice Hall).

Buchwald, J. (1994): *The Creation of Scientific Effects: Heinrich Hertz and Electric Waves* (Chicago, IL: University of Chicago Press).

Butterfield, W. (1975): 'Electric Shock – Safety Factors When Used for the Aversive Conditioning of Humans', in *Behavior Therapy*, 6, pp. 98–110.

Cavicchi, E. (1991): 'Painting the moon', in *Sky and Telescope*, September, 82(3), pp. 313–15.

Cavicchi, E. (1997): 'Experimenting with magnetism: Ways of learning of Joann and Faraday', in *American Journal of Physics*, 65, pp. 867–82.

Cavicchi, E. (1999): 'Experimenting with wires, batteries, bulbs and the induction coil: Narratives of teaching and learning physics in the electrical investigations of Laura, David, Jamie, myself and the nineteenth century experimenters – Our developments and instruments', unpublished doctoral dissertation (Cambridge, MA: Harvard University).

Cavicchi, E., F. Hughes-McDonnell and P. Lucht (2001): 'Playing with Light', in *Educational Action Research*, 9, pp. 25–49.

Cavicchi, E. (2003): 'Experiences with the magnetism of conducting loops: Historical instruments, experimental replications, and productive confusions', in *American Journal of Physics*, 71(2), pp. 156–67.

Cavicchi, E. (2005a): 'Exploring Water: Art and Physics in Teaching and Learning with Water', in *Facilitating Watershed Management: Fostering Awareness and Stewardship* (R. France, ed.) (Lanham, MD: Rowman & Littlefield Publishers), pp. 173–94.

Cavicchi, E. (2005b): Sparks, Shocks and Voltage Traces as Windows into Experience: The Spiraled Conductor and Star Wheel Interrupter of Charles Grafton Page. *Archives des Sciences*, 58, pp. 123–36.

Cavicchi, E. (2006a): 'Nineteenth century developments in coiled instruments and experiences with electromagnetic induction', in *Annals of Science*, 63, pp. 319–61.

Cavicchi, E. (2006b): 'Faraday and Piaget: Experimenting in Relation with the World,' in *Perspectives on Science*, 14 (1), pp. 66–96.

Cavicchi, E. (2008a): 'A Witness Account of Solar Microscope Projections: Collective Acts integrating across Personal and Historical Memory', in *British Journal for the History of Science*, 41(3), pp. 369–83.

Cavicchi, E. (2008b): 'Charles Grafton Page's Experiment with a Spiral Conductor', in *Technology and Culture*, 49, pp. 884–907.

Chang, H. (2007): 'The Myth of the Boiling Point' [online]: www.ucl.ac.uk/sts/staff/chang/boiling/index.htm

Chang, H. (2008): 'The Myth of the Boiling Point', in *Science Process*, 91(3), pp. 219–40.

Channing, W. (1849): *Notes on the Medical Application of Electricity* (Boston, MA: Daniel Davis Jr).

Clarke, E. M. (1837): 'Voltaic Battery and Pole Director', in *Annals of Electricity*, 1, (W. Annan, Lithographer), plate VIII, figure 55.

Davis Jr, D. (1838): *Catalogue of Apparatus* (Boston, MA: Daniel Davis Jr).

Davis Jr, D. (1842): *Manual of Magnetism* (Boston, MA: Daniel Davis Jr).

Duckworth, E. (1987): 'Teaching as research', in E. Duckworth (2006): *'The having of wonderful ideas' and other essays on teaching and learning*, 3rd edition (New York: Teacher's College Press), pp. 173–92.

Duckworth, E. (1991): 'Twenty-four, forty-two and I love you: Keeping it complex', in E. Duckworth (2006): *'The having of wonderful ideas' and other essays on teaching and learning*, 3rd edition (New York: Teacher's College Press), pp. 125–55.

Duckworth, E. (2005): 'Critical exploration in the classroom', in E. Duckworth (2006): *'The having of wonderful ideas' and other essays on teaching and learning*, third edition (New York: Teacher's College Press), pp. 157–72.

Editor (1872): 'The Page Patent – The Attempts to Enforce it to be Resisted', in *Scientific American*, 26 October 1872, 27, p. 256; APS Online.

ESS (Elementary Science Study) (1970): *The ESS Reader* (Newton, MA: Educational Development Center).

Eve, J. (1836): 'Medical Education', in *Southern Medical and Surgical Journal*, 1, pp. 216–23.

Faraday, M. (1831): 'On the Induction of Electric Currents', first series, read 24 November 1831, *Experimental Researches in Electricity*, 3 vols, 1839 reprint (Sante Fe, NM: Green Lion Press) 2000, vol. 1, pp. 27–32.

Faraday, M. (1834): 'On the Magneto-Electric Spark and Shock, and on a Peculiar Condition of Electric and Magneto-Electric Induction', in *Philosophical Magazine*, 5, pp. 349–54.

Faraday, M. (1835): 'On the influence by induction of an Electric Current on itself: and on the inductive action of Electric Currents generally', in *Experimental Researches in Electricity*, 3 vols, 1839, reprint (Santa Fe, NM: Green Lion Press), 2000, vol. 1, pp. 1048–1118; reprinted in *Annals of Electricity*, 1837, 1, pp. 160–62 and pp. 169–86.

Finn, B. S. (1966): 'Alexander Graham Bell's Experiments with the Variable-resistance Transmitter', in *Smithsonian Journal of History*, 1(4), pp. 1–16.

Fleming, J. A. (1892): *The alternate current transformer in theory and practice*, 2 vols. (London: 'The Electrician' printing and publishing company).

Foucault, M. (1963): *The Birth of the Clinic: An Archaeology of Medical Perception* (trans A. M. Sheridan Smith, 1973) (New York, NY: Pantheon Books).

Garratt, A. (1876): *Guide for using Medical Batteries* (Philadelphia, PA: Lindsay & Blakiston).

Geratherm (n.d.): Geratherm Medical AG [online]: www.geratherm.com/en/

Gingerich, O. (1975): 'Dissertatio cum Professore Righini et Sidereo Nuncio', in *Reason, Experiment, and Mysticism in the Scientific Revolution* (M. L. Richini Bonelli and W. R. Shea, eds) (New York, NY: Science History Publications), pp. 77–88.

Gingerich, O. and A. Van Helden (2003): 'From Occhiale to printed page: the making of Galileo's *Sidereus Nuncius*', in *Journal for the History of Astronomy*, 34, pp. 251–67.

Gluckman, A. (1993): *The Invention and Evolution of the Electrotechnology to Transmit Electrical Signals without Wires* (Washington, DC: Washington Academy of Sciences).

Gooding, D. (1980): 'Faraday, Thomson and the concept of the magnetic field', in *British Journal for the History of Science*, 13, pp. 91–120.

Gooding, D. (1981): 'Final steps to the field theory: Faraday's study of magnetic phenomena, 1845–1850', in *Historical Studies in the Physical Sciences*, 11, pp. 231–75.

Gooding, D. (1985): '"In Nature's School": Faraday as an Experimentalist', in *Faraday Rediscovered: Essays on the Life and Work of Michael Faraday 1791–1867* (D. Gooding and F. A. J. L. James, eds) (New York, NY: Macmillan Press), pp. 105–35.

Gooding, D. (1989): 'History in the laboratory: Can we tell what really went on?', in *The development of the laboratory: Essays on the place of experiment in industrial civilization* (F. A. J. L. James, ed.) (New York, NY: American Institute of Physics), pp. 63–82.

Gooding, D. (1990): *Experiment and the Making of Meaning: Human Agency in Scientific Observation and Experiment* (Dordrecht: Kluwer Academic Publishers).

Gray, S. (1731–32): 'A Letter to Cromwell Mortimer … containing several Experiments concerning Electricity', in *Philosophical Transactions*, 37, pp. 18–44.

Greenough, J. J. (1854): 'The first locomotive that ever made a successful trip with Galvanic Power', in *American Polytechnic Journal*, 4, pp. 257–63.

Greenslade, T. (n.d.): 'Daniel Davis, Jr Apparatus', in *Instruments of Natural Philosophy* [online]: www2.kenyon.edu/depts/physics/EarlyApparatus/

Gwei-Djen, L. and J. Needham (1980/2002): *Celestial Lancets: A History and Rationale of Acupuncture and Moxa* (London: Routledge-Curzon).

Heering, P. (1994): 'The Replication of the Torsion Balance Experiment, the Inverse Square Law and its Refutation by early 19th-century German Physicists', in *Restaging Coulomb: Usages, Controverses et Réplications autour de la Balance de Torsion* (C. Blondel and M. Dörries, eds) (Florence: Leo S. Olschki), pp. 47–66.

Heering, P. (2008): 'The enlightened microscope: re-enactment and analysis of projections with eighteenth-century solar microscopes', in *British Journal for the History of Science*, 41(3), pp. 345–68.

Henry, J. (1832): 'On the Production of Currents and Sparks of Electricity from Magnetism', in *American Journal of Science*, 22, 403–8.

Henry, J. (1835): Appendix to the above, in *American Journal of Science*, July 1835, 28, pp. 329–31.

Henry, J. (1837): 'On the Influence of a Spiral Conductor in increasing the Intensity of Electricity from a Galvanic Arrangement of a Single Pair', in *American Philosophical Society Transactions*, pp. 223–31; reprinted in *The Scientific Writings of Joseph Henry*, vol. 1 (Washington, DC: Smithsonian Institute, 1886).

Henry, J. (1839): 'Contributions to Electricity and Magnetism: On Electro-Dynamic Induction No. III', in *Transactions of the American Philosophical Society*, 6, pp. 303–37.

Henry, J. (1843): 'Contributions to Electricity and Magnetism: On Electro-Dynamic Induction No. IV', in *Transactions of the American Philosophical Society*, 8, pp. 1–35.

Hertz, H. (1892): *Electric Waves* (trans. D. E. Jones, 1900) (London, New York: Macmillan and Co.).

Hochadel, O. (2001): '"My Patient told me how to do it": The Practice of Medical Electricity in the German Enlightenment', in *Electric Bodies: Episodes in the history of medical electricity* (P. Bertucci and G. Pancaldi, eds) (Bologna: Università di Bologna), pp. 69–90.

Inhelder, B. H. Sinclair and M. Bovet (1974): *Learning and the development of cognition* (trans. by S. Wedgwood (Cambridge, MA: Harvard University Press).

Jefferson Lab (n.d.): Electrical Safety. Environment, Safety and Health Manual [online] [updated 8 October 2009] at: www.jlab.org/ehs/ehsmanual/index.html

Keller, E. F. (2002): *Making Sense of Life: Explaining Biological Development with Models, Metaphors, and Machines* (Cambridge, MA: Harvard University Press).

La Beaume, M. (1828): *Du galvanisme appliqué à la médicine et de son efficacité dans le traitement … .* (Paris).

(Lane, J.H.) (1869): 'Charles Grafton Page', in *American Journal of Science*, 48, pp. 1–17.

Lehar, S. (n.d.): 'An Intuitive Explanation of Fourier Theory' [online]: http://cns-alumni.bu.edu/~slehar/fourier/fourier.html

Matlab (1994–2010): 'Matlab: The Language of Technical Computing', The MathWorks, Inc., [online]: www.mathworks.com/products/matlab/

Mills, A. A. (2003): 'Early Voltaic Batteries: an Evaluation in Modern Units and Application to the Work of Davy and Faraday', in *Annals of Science*, 60(4), pp. 373–98.

Morand, M. (1825): *Memoir on Acupuncturation* (trans. by F. Bache) (Philadelphia, PA: Robert Desilver; Clark & Raser).

Morrison, P. (1987): *The Ring of Truth with Professor Philip Morrison* [video/film] (Public Broadcastings Associates, Cambridge, MA and Alexandria, VA: PBS video). Six one-hour videos.

Morrison, P. and P. Morrison (1987): *The Ring of Truth: an inquiry into how we know what we know* (New York, NY: Random House).

Morus, I, (1998): *Frankenstein's Children: Electricity, Exhibition, and Experiment in Early-Nineteenth-Century London* (Princeton, NJ: Princeton University Press).

Page, C. G. (1836a): 'Medical Application of Galvanism', in *Boston Medical and Surgical Journal*, dated June 18, issue of June 22, p. 333.

Page, C. G. (1836b): 'Insect Dissections', in *Boston Medical and Surgical Journal*, 13 July, pp. 364–65.

Page, C. G. (1836c): 'Medical Application of Galvanism', in *Southern Medical and Surgical Journal*, 1, pp. 183–85; excerpt reprinted in *Boston Medical and Surgical Journal*, 21 September, p. 113.

Page, C. G. (1837a): 'Method of increasing shocks, and experiments, with Prof. Henry's apparatus for obtaining sparks and shocks from the Calorimotor', in *American Journal of Science*, 31, pp. 137–41; reprinted in *Annals of Electricity* (1837), 1, pp. 290–94.

Page, C. G. (1837b): 'On the use of the Dynamic Multiplier, with a new accompanying apparatus', in *American Journal of Science*, 32, pp. 354–60.

Page, C. G. (1837c): 'The Production of Galvanic Music', in *American Journal of Science*, 32, pp. 396–97.

Page, C. G. (1838a): 'Experiments in Electromagnetism', in *American Journal of Science*, 33, pp. 118–20.

Page, C. G. (1838b): 'Researches in Magnetic Electricity and new Magnetic Electrical Instruments', in *American Journal of Science*, 34, pp. 364–73.

Page, C. G. (1839): Magneto-Electric and Electro-Magnetic Apparatus and Experiments', in *American Journal of Science*, 35, pp. 252–68.

(Page, C. G.) (1867): *The American Claim to the Induction Coil and its electrostatic developments* (Washington, DC: Intelligencer Printing House).

Pancaldi, G. (2003): *Volta: Science and Culture in the Age of Enlightenment* (Princeton, NJ: Princeton University Press).

Pantalony, D., R. L. Kremer and F. J. Manasek (2005): *Study, measure, experiment: Dartmouth's Allen King Collection of Scientific Instruments* (Norwich, VT: Terra Nova Press).

Peters, R. (1836): 'On the application of Galvanic Electricity to medicine', in *Transylvania Journal of Medicine and the Associate Sciences (1828–1836)*, Oct–Dec 1836, 9 (4), p. 641. In APS (2010).

Piaget, J. (1926): *The child's conception of the world* (trans. J. and A. Tomlinson,1960) (Totowa, NJ: Littlefield, Adams).

Post, R. C. (1972): 'The Page Locomotive: Federal Sponsorship of Invention in Mid-19th-Century America', in *Technology and Culture*, 13, pp. 140–69.

Post, R. C. (1976a): *Physics, Patents & Politics: A Biography of Charles Grafton Page* (New York, NY: Science History Publications).

Post, R. C. (1976b): 'Stray Sparks from the Induction Coil: The Volta Prize and the Page Patent', in *Proceedings of the IEEE*, 64, pp. 1279–86.

Reiley, J. P. (1998): *Applied Bioelectricity: From Electrical Stimulation to Electropathology* (New York, NY: Springer).

Ribe, N. and F. Steinle (2002): 'Exploratory Experimentation: Goethe, Land and Color Theory', in *Physics Today*, July, pp. 43–49.

Rowbottom, M. and C. Susskind (1984): *Electricity and Medicine: History of Their Interaction* (San Francisco, CA: San Francisco Press).

Saslow, W. (2002): *Electricity, Magnetism and Light* (Amsterdam, Boston: Academic Press).

Settle, T. (1996): *Galileo's Experimental Research* (Berlin: Max Planck Institute for the History of Science).

Siconolfi, S., M. Nusynowitz, S. Suire, A. Moore Jr and J. Leig (1996): 'Determining blood and plasma volumes using bioelectrical response spectro-scopy', in *Medicine & Science in Sports & Exercise*, 28, pp. 1510–16.

Sherman, R. (1988): 'Charles Page, Daniel Davis, and their electromagnetic apparatus', in *Rittenhouse*, 2, pp. 34–47.

Steinle, F. (1997): 'Entering New Fields: Exploratory Uses of Experimentation', in *Philosophy of Science*, 64 (Proceedings), S65–S74.

Steinle, F. (2002): 'Challenging Established Concepts: Ampère and Exploratory Experimentation', in *Theoria, Segunda Época*, 17(2), pp. 291–316.

Sturgeon, W. (1837a): 'On the Electric Shock from a single Pair of Voltaic Plates, by Professor Henry, of Yale College, United States: Repeated, and new Experiments (28 Sep 1836)', in *Annals of Electricity*, 1, pp. 67–75, reprinted in W. Sturgeon (1850): *Scientific researches, experimental and theoretical in electricity, magnetism, galvanism, electro-magnetism and electrochemistry* (Bury), pp. 282–289.

Sturgeon, W. (1837b): 'On the production of Electric Shocks from a single Voltaic pair', in *Annals of Electricity*, 1, pp. 159–60.

Sturgeon, W. (1837c): Explanation of the Phenomena, &c., in *Annals of Electricity*, 1, pp. 294–95.

Sturgeon, W. (1837d): An experimental investigation of the influence of Electric Currents on Soft Iron … , in *Annals of Electricity*, 1, pp. 470–84.

Tarver, W. T. S. (1995): 'The Traction Trebuchet: A Reconstruction of an Early Medieval Siege Engine', in *Technology and Culture*, 36(1), pp. 136–67.

Tomlinson, C. (1837): 'On an Optical Illusion observed during the action of Professor Ritchie's horizontal artificial voltaic magnet', in *Annals of Electricity*, pp. 108–11.

Tweney, R. (1992): 'Stopping Time: Faraday and the Scientific Creation of Perceptual Order', in *Physis*, 29, pp.149–64.

Tweney, R. (2005): 'On replicating Faraday: Experiencing historical procedures in science', *Archives des Sciences*, 58(2), pp.137–48.

Tweney, R. (2006): 'Discovering Discovery: How Faraday Found the first Metallic Colloid', *Perspectives on Science*, 14 (1), pp. 97–121.

Usselman, M. C., A. J. Rocke, C. Reinhart, and K. Foulser (2005): 'Restaging Liebig: A Study in the Replication of Experiments', in *Annals of Science*, 62(1), pp. 1–55.

Volta, A. (1800): 'On the Electricity excited by the mere Contact of conducting Substances', in *Philosophical Transactions*, part 2, pp. 289–311.

Volta, A. (1918): *Le Opere di Alessandro Volta* (Milan: Hoepli).

Weber, H. and J. Frercks (2005): 'Replication of Replicability: Schmidt's Electrical Machine', in *Archives des Sciences*, 58(2), pp. 113–122;

Whitaker, E., (1978): 'Galileo's Lunar Observations and the Dating of the Composition of "Sidereus Nuncius"', in *Journal for the History of Astronomy*, 9, pp.155–69.

White, C. (1828): 'Ad for acupuncture needles', in *Boston Medical and Surgical Journal*, 1 April, p. 112.

Wilkinson, C. H. (1804): *Elements of Galvanism in Theory and Practice*, 2 vols (London: John Murray).

Withuhn, W. L. (1981): 'Testing the John Bull, 1980', in J. White: *The John Bull, 150 years a locomotive* (Washington DC: Smithsonian Institution), pp. 107–21.

Zabriskie, J. B. (1837): 'Experiments upon the Induction of Metallic Coils', in *American Journal of Science*, 32, pp. 308–13.

Additional information

The author has uploaded three short videos to the Internet.

Video 1 can be found at:
    www.youtube.com/watch?v=99Ar-mzNLV8
Video 2 can be found at:
    www.youtube.com/watch?v=7V1bBZ41T1o
Video 3 can be found at:
    www.youtube.com/watch?v=mVvX_SQMGRM

JAN FRERCKS

Experience and self-reflection
AN ELECTRICAL-HISTORIOGRAPHIC-DIDACTIC EXPERIMENT

Introduction

IN THIS ARTICLE I will try to make sense of a history-of-science lab-course, which was based on the use of replicated instruments. I use the lab-course to tackle the question of whether the replication of past experiments is practised and experienced (rather than intended to be) as a didactic or as a historiographic method. In order to do this, I will turn the lab-course into an electrical-historiographic-didactic experiment. This is only made possible by a marvellous report produced by the participants which contains both the minutes of the experiments and the meta-minutes of their retrospective self-reflection on their own experiences during the lab-course.

Readers who are mainly interested in replication used for historiography of science may profit from this article since it puts replication itself to an experimental test. It will reveal to what degree the use of rebuilt scientific instruments is tied to historiographic research. Readers who are mainly interested in a historical approach to teaching science will find presented here a case of such an approach applied in practice. In contrast to most earlier studies, however, the learners themselves get a chance to speak, indirectly first in my analysis of their report, and then directly at the end of this article.

Both my interest in the nature of historical replication and the method deployed to tackle will be explained in section 2. In section 3, I will examine the students' minutes of their experiments, in section 4, I will examine the students' meta-minutes of their own experimental practice and the status of the lab-course, and in section 5, I will state the need for self-reflection on my part which will result in allowing for a comment by the participants in section 6. I will start, however, by giving some basic information about the lab-course.

RECONSTRUCTIONS

Section 1: A history of science lab-course

The lab-course was organised by Olaf Breidbach, Heiko Weber and myself. It took place in September 2004 at the University of Jena (Germany) and comprised a week of full-time work. The participants were four advanced students of history of science at the university and one guest student from the University of Oldenburg. The general subject of the lab-course was electricity and galvanism around 1800.[1] Five different kinds of tasks were carried out during the week.

The first and most extensive was experimental work using mainly rebuilt instruments. The part on electricity ranged from experiments on attraction and repulsion with different materials, to the use of a large and complex electrical machine, which had been rebuilt at the University of Jena according to an electrical machine built in 1773 by Georg Christoph Schmidt.[2] Furthermore, two students tried to rebuild and understand Joseph Weber's electrical amplifier. With regard to galvanism, the students had to build their own Voltaic pile and use it to replicate Johann Wilhelm Ritter's experiment on the galvanic decomposition of water.[3] In general, the students were initially given quite confined and well-defined tasks, as commonly done in lab-courses in the natural sciences, but then we (the organisers) observed rather than directed the further course of experimentation.

The students had to keep minutes during the experiments, which constituted the second kind of task that was carried out. Writing up these minutes in neat form was done in the evening at home, since the students had to submit them the next morning. Obviously, a history-of-science lab-course is not a natural science lab-course, which caused – intended – irritation among the students. We will see later on how they reacted to these irritations.

The third kind of task consisted of text work on primary literature, mainly physics textbooks from around 1800. Using these texts in order to make sense of one's own experiments, rather than discussing the texts in their own right as conventionally done in text-based seminars, also caused irritation, again to be discussed later on.

The fourth kind of task consisted of rather conventional seminar papers presented by the students in order to provide the historical context of the experiments.

The fifth kind of task consisted of a process of reflection on one's own experiences, on the interplay between doing experiments on electricity and reading texts about electricity,[4] on keeping the minutes, on the status of replication, on the relation between past and present knowledge, and so on. This part, however, was mainly done after the lab-course.

Those students who required a certificate had to write a report on the lab-course. Their joint effort resulted in an 80-page report, which contains the daily minutes in the form of texts and even some of the notes taken during experimentation in the form of pictorial reproductions. These minutes, as well as their remembered experience, then served to reflect on the whole project. The process of self-reflection was itself documented, resulting in five meta-minutes, four highly personal ones on each individual's experiences and one joint one summarising the individual points. It is mainly this section of the report which merits the article at hand.

Section 2: An experiment on the use of replication

In order to introduce my particular questions, it is useful to distinguish three levels in any historiography of science (*see* Fig. 8.1). Past scientists (on level 2) are engaged with technical or natural things or entities on level 1, electricity in our case. The connection between levels 1 and 2 is achieved by experimenting. Historians of science (on level 3) are normally neither interested in nor immediately engaged with level 1, but only with people or texts on level 2 in as much as they are engaged with level 1. In contrast, the replication of past experiments does not stop short at level 2, but engages directly with level 1. In a replication, one interacts directly with the things and entities, i.e. with electricity rather than with electricity scientists in our case.

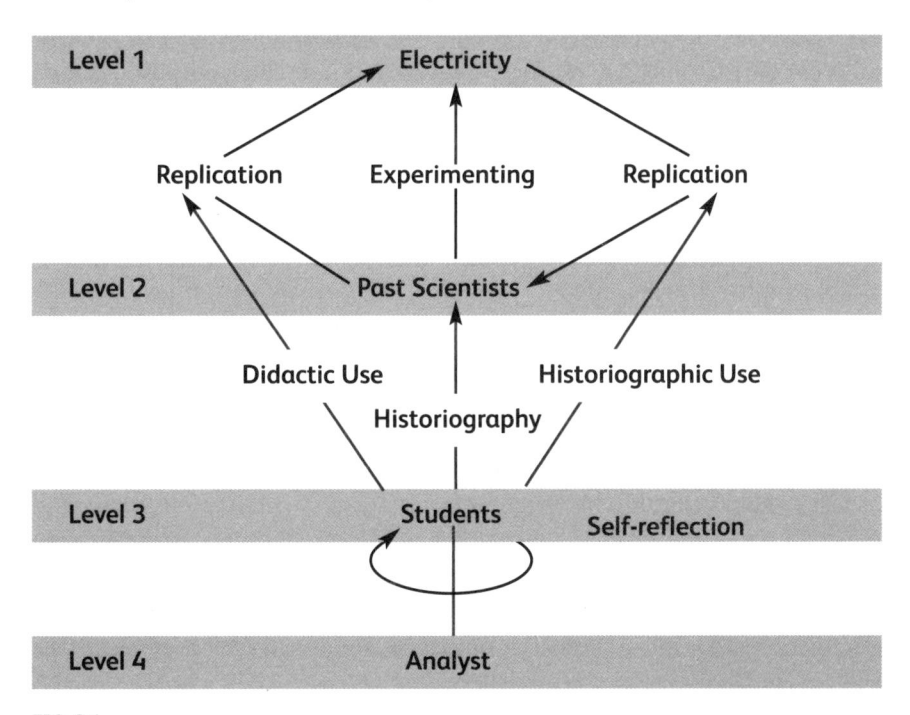

FIG. 8.1
Structure of levels of analysis in the electrical-historiographic-didactic experiment.

There are two different uses of historical replications, one didactic and the other historiographic. Although both have been applied in various places, I will focus on the Research Group on Physics Education and History and Philosophy of Science at the Carl von Ossietzky University of Oldenburg where both uses have been systematically developed and applied in many case studies over a long period of time. In Oldenburg the historiographic use developed from the didactic use, but, systematically, both can be neatly distinguished.

The use of replications for didactic purposes was felt a necessary supplement to the historical-genetic approach in teaching physics, which had been (unlike ahistorical methods) mainly text-based before.[5] The presumption of this method is that it is easier to learn physics from historical examples or from the long-term historical development than in an (apparently) ahistorical way. This approach is associated with the claim that an individual's process of learning a science more or less parallels the historical development of that science.[6] Neither the correctness of this claim nor the general presumption about the appropriateness of a historical approach for learning science is my concern here. What is important to note for my purpose is that in the didactic use of replication, history of science is used for learning science. To put it simply, level 2 is used for learning about level-1 items.

In the historiographical use, it is exactly the reverse. From the early 1990s onwards, members of the Oldenburg group explicitly renounced any ties to didactics. They neither saw any necessity for legitimising replication didactically, nor did they intend to use their replications for teaching purposes. In fact, some replications grew too complicated (both technically and theoretically) to be included in teaching at school or even at university. This is not the place to discuss replication as a research method for the historiography of science.[7] A lot has been written about questions such as by which means and to what degree a replication can be close to the original; whether this should be an aim at all; whether it is possible to regain tacit knowledge and, if so, what can reasonably be done with it; whether replication is a natural scientific method or a historiographical method; how to behave during a replication, etc. What has to be grasped for my present purpose is that – in whatever way – engaging with level 1 is a technique used to get to level 2.

Thus, both uses of replication are diametrically opposed to one another in terms of the relationship between the levels 1 and 2, between electricity and (past) science of electricity in the case at hand. To put it simply, in the didactic use one learns about electricity by means of the history of science, while in the historiographic use one learns about the history of science by means of electricity. In addition, in both uses one learns about and even experiences the experimental investigation of electricity.

Note that the same rebuilt apparatus can be used for both purposes.

Neither the material components of a replication, nor even the whole replication itself, determines its use.[8] This allowed us to leave open the question of whether the lab-course was didactic or historiographic. In fact, this open set-up was the crucial feature of the Jena lab-course.

One might argue that it was neither didactic nor historiographic. If one assumes that students of the history of science have to learn about past engagement with electricity rather than about electricity itself, the didactic option will fail to meet the requirements of our participants. And there will be no historiographic gains for them either, because the form of a lab-course keeps the students from coming close to those insights that were tacitly built into the program by the organisers.

On the other hand, one might take the lab-course to be both didactic and historiographic. If one entertains a strong historicist philosophy of science, which states that all knowledge about natural entities is historically situated, then there would be no such thing as electricity *per se* to be separated from the historical circumstances of its investigation. Not only do students *automatically* learn about electricity, they are actually compelled to do so as historians of science. Only this allows us to assess past investigations on electricity.[9] And if one wants to learn about electricity, this entails a study of the historical context of that knowledge. This means that acquiring modern knowledge of electricity would require learning about present-day scientific practice. Trying to learn about electricity by means of rebuilt instruments and old texts accordingly entails an inquiry into the history of science, and this approach more openly reveals the historicity of science.

In any case, the fact that the ultimate aim was left open from the outset allowed us to investigate experimentally the kind of use made by the students. The focal points of our interest were whether they used the replication didactically or historiographically, whether they experienced the replication in that way, and whether and how they reflected on these uses and experiences.

Section 3: Making sense of the minutes

In this section I will analyse what the students did. This section is based mainly on the minutes as reprinted in the report.[10] I will focus on two phases of the lab-course. In the first, in which I had expected (or hoped for) a historiographic use to be made of their experience of physics (transition from level 1 to level 2), the students did something fundamentally different whereas in the second instance, where I would have liked them to use historical knowledge in order to make sense of physical phenomena (transition from level 2 to level 1), they did not do anything like that at all.

The first instance was at the very beginning of the lab-course. We had extracted the first paragraphs of the respective chapters on electricity from several physics textbooks dating from around 1800. The introductory paragraphs from different authors are very similar, even in the wording, except for the particular materials and devices mentioned as examples. A typical example is that of Friedrich Albrecht Carl Gren (Gren, 1793: 657).

> If one rubs a dry glass tube or a sulphur stick, or amber, or a rod of sealing wax with a piece of flannel, one finds that small light pieces of paper, iron filings, small gold leafs, small cork balls, etc., are first attracted by these rubbed bodies, but then repelled by them.

In addition to the texts, the students were given most of these materials and asked to 'reproduce' the effects as described in the texts. Here are parts of the minutes of these experiments by two teams of students.[11]

Team A[12]

Glass tube (one end closed, other end spherical with opening = 25 mm) l = 415 mm, Ø = 40 mm
Rubbing with dry hand, duration $c.15\,s$

Sample	Effect
Wood sawdust	Negative
Pith ball	Negative
Cork	Negative
Gold leaf	Negative
Iron filings	Negative
Paper shreds	Negative

Glass rod l = 310 mm, Ø = 10 mm
Cat fur, duration of rubbing $c.10\,s$ (until a quiet crackling was audible)

Sample	Effect
Wood sawdust	On nearing the electrified rod up to 1–2 mm attraction and sticking, after 2–3s falling off
Gold leaf	From 50 mm proximity movement in the sample, then complete attraction, after 10s falling off
Pith ball	Effect of attraction upon direct contact, $c.2\,s$ sticking, then falling off
Cork	Effect of attraction upon direct contact, $c.5\,s$ sticking, then falling off
Iron filings	Very small effect, hardly noticeable

[...]
Glass rod l = 420mm, Ø = 20mm
Cat fur, duration of rubbing c.10s

Sample	Effect
Snuff	Reaction of attraction and almost immediate repulsion, assistant observed immediate repulsion in some places
Cork	Reaction of attraction and repulsion (in the following noted as back and forth)
Wood sawdust	Back and forth
Gold leaf	Back and forth
Pith ball	Back and forth
Iron filings	Back and forth

Glass rod l = 420mm, Ø = 20mm
Flannel cloth, duration of rubbing c.10s

Sample	Effect
Snuff	Negative
Wood sawdust	Negative
Gold leaf	From 20mm, attraction
Pith ball	Upon contact, weakly
Cork	Upon contact, weakly
Iron filings	Negative

[...]
Further observations. Cigarette ash is attracted by an electrified rod, a stream of water from a tap is likewise attracted.

Team B

First experiment: Glass rod with 20mm diameter rubbed with fur and brought close to the lycopod spores. No effect.
Repetition: Now at 3cm distance attraction of the spores and forming of tiny towers.
Repetition with crumbled cork: at 5cm distance 'dancing' of the crumbs.
Repetition with iron filings: brief, reaction at 3cm distance, lasting about 3s. Likewise 'dancing'.
Repetition with pith balls: from 6cm as just described, lasting longer. Strong repulsion.

FIG. 8.2 (left) Students try to 'repeat' the effects described in textbooks written c.1800.

FIG. 8.3 (above) Effect by nearing a rubbed glass rod to snuff.
(Courtesy of Collaborative Research Centre 482, University of Jena)

Repetition with gold leaf: reaction already at a great distance (8–10cm). Likewise brief 'dancing'.
Repetition with snuff: reaction of attraction and repulsion from 2cm. Building of stripes at right angles to the axis of the rod, 3cm long on both sides of the snuff heap, clearly perceptible.
Repetition: This time semi-circle shaped spreading above the heap. Repelled crumbs are re-attracted by the rod up to 6cm distance in the direction of the hand before touching the ground.
Repetition with flannel as rubbing device: no reaction with gold leaf as the most sensitive material.
Repetition with silk as rubbing device: only with gold leaf weak reaction at 6mm distance.

Attempts to produce sparks: darkened room; no spark upon touching the tap, the radiator, or the finger of another person. The combination of the thick glass rod and fur was chosen because this had given the strongest reactions.

To me, the most striking outcome of both groups' attempts is that it was not at all easy to reproduce the effects with all mentioned substances and in the clear-cut manner as described in the textbooks. This, I suspected, would raise doubts about the status of the texts. We had used the texts as a task for a lab-course. The students considered it to be an account of 'simple historical demonstration experiments' (E. Kusch, 65).[13] But what is the function of these short and oddly stereotyped introductory paragraphs for the logic of the textbook chapters on electricity?

Analysing the whole chapters on electricity reveals that the authors considered the laws of (in modern terms) conduction, electrostatic induction, attraction and repulsion to be the core knowledge of electricity.[14] Although inferred from many experiments with sophisticated apparatus, these four laws are all present in the apparently simple experiment on minuscule particles. (You may try to explain the effects based on these four laws.) Thus, the experiment is not simple, but simplified. It is the utmost possible condensation of the core knowledge of electricity into one experiment and into one paragraph. It would be interesting to know whether this experiment was actually performed around 1800 in the first lecture on electricity (as the students apparently presupposed), and if so, whether past lecturers generally succeeded, but this is hardly crucial for the understanding of the function of theses texts in the overall argumentation. Although placed at the beginning, this paragraph serves as a summary of the whole chapter. It is not a suitable beginning for *studying* electricity.

I see this interpretation corroborated by the students' difficulties in reproducing the effects in a reliable and demonstrable manner. The effects as described are neither impressive nor even understandable without further knowledge. For this reason, it was not necessary for textbook authors to describe the experiment in the form of an instruction that would allow everyone to perform it easily. The experiment was basically textual.

This is roughly the way that the experience of physics could have been put to use for further historical inquiry. To be sure, the students' irritating experience of a blatant contrast between the simplicity and clarity of the text and their own difficulties in producing the effects does not suffice for understanding the nature and function of the text, but it might have been at least an *incentive* for trying to do so. But the students did something different. True, at that point they had not been given the whole textbooks, but they did not ask for them either. In fact, they did not care about the historical situation at all. The striking difference between what they managed to produce and what was stated as if it was easy to produce, did not bother them – at least not at this stage. They repeated those effects that worked well, ignored those that did not work, and they even invented some new effects such as the divergence of a stream of water from a tap. In short, they escaped the purported lab-course situation of a pre-set task with an expected outcome. Instead of historical contextualisation, they chose to continue via free, i.e. ahistorical experimentation. I will come back to this in the following section.

The second instance concerns the experiments with the rebuilt electrical machine (*see* Fig. 8.4). Since there was only one machine, all students worked together. Here is part of the minutes.

> **Experiment 1:** Enrico (E) tries to pick up a charge from the condenser [*sic!*] of the electrical machine (EM) with his knuckle. Only after several increases in the number of revolutions – maximum 32 revolutions (Rv) – a gentle crackling

is audible and a gentle touch is perceptible. Nevertheless no correspondence between the number of revolutions and the charge is apparent.

This experiment is executed by the other participants as well. The effects, in particular the ones by touch, get more and more weak. Again the number of Rv of the EM is increased. Mostly, only a gentle rustling or crackling is audible.

Robert (R) suggests changing the direction of rotation in his experiment. This has a perceptible effect in his experiment. Nevertheless it is not reproducible. [...]

Experiment 2: One person stands on the insulating stool and holds the head of the conductor in his hand. Then the charging by the EM takes place. Subsequently, the charged person touches another participant (finger to hand or finger to finger).

Christian (C) is charged. On touching E, a slight itching is felt. E gets charged. On touching C, a stronger itching is felt. Heiko (H) gets charged. On touching Michael (M), a stronger itching occurs. [...]

Experiment 5: C stands on the stool, one hand at the condenser [*sic!*], the other holds the modern Leyden jar (MLF). Test whether the jar is charged even if it is not at the conductor. Several tests. R turns at the EM, but this has no or only a very weak effect. Reflection on whether it makes a difference, whether R keeps his hand at the crank when he touches the MLJ. Test reveals that it has no consequences.

Experiment 6: Replication of a Leyden jar (RLJ). Its neck is connected to the conductor by a copper wire. The jar gets charged. Jan (J) touches the outside of the jar and the sphere on the bottle neck with the loose copper conductor (CC). The first time, there is a loud bright spark. The second time, a weaker spark. Several repetitions with different numbers of revolutions. Spark always occurs, strength depends on the number of revolutions.

Belly of the RLJ is earthed at the radiator by means of a copper wire. 10 Rv, clear spark. 20 Rv, clear spark. 30 Rv, very clear spark, also very loud. Transfer of earthing from the radiator to the protective contact of the socket. 20 Rv, very clear. [...]

Experiment 8: Repetition of experiment 1. M at 5 Rv. Clear shock. Breidbach (B) at 5 Rv. Clear shock. H at 7 Rv. Clear shock.

Apparently, the difficulties at the beginning were caused by boundary conditions (humidity of the air) and by the machine (warming-up period, the collector points no longer scratch over the glass plate of the EM). Maybe changed handling of the EM, as well.

In order to analyse this passage of the minutes, it is useful to distinguish the notions of *phenomenon*, *observation* and *experience*. By *phenomenon*, I understand anything happening at or with the electrical machine that can be observed. It does not matter whether a phenomenon occurs by itself or is artificially produced. By *observation*, I understand the conscious perception of a phenomenon. An *experience* is the conscious perception of something happening to oneself. For the present contexts, experiences are observations of phenomena with respect to personal preconceptions and expectations.

To me, the most interesting issue to be drawn from the minutes is how the students reacted to unexpected experiences. Unexpected experiences mostly consisted in recognising that a phenomenon is weaker than expected or fails to appear altogether. When the students observed only small effects (gentle shocks and light crackling) in experiments 1 and 5, three kinds of reactions took place.

The first consisted in venturing hypotheses about the causes of failure, thereby circumventing rather than explaining it. The experimenter keeping his hand at the crank was one possible reason for failure, which was tested and ruled out.

The second way to react was by changing something. Letting another person do the experiment was always the first choice. In other cases, these two ways of reacting merged: is changing the direction of rotation in experiment 1 based on a hypothesis that the direction is important, or is it just a case of blindly changing anything? Experimenting is fast, sometimes faster than thinking. In this case, at least, the apparently stupid change resulted in a stronger effect, if only temporarily.

FIG. 8.4
Students experimenting with the rebuilt electrical machine.
(Courtesy of Collaborative Research Centre 482, University of Jena)

The third way consisted in trying to standardise procedures. The most important tool for this consisted in noting the number of revolutions. Although no clear-cut relationship between this number and the strength of the effect could be established, the number of revolutions was the constant reference. Quantification served as a substitute for mastery.

If everything worked well (as in experiments 2 and 6), there was no attempt for an explanation either. This is remarkable: literally nothing was done with good experiments! Here, the students obviously behaved as if they were in a science lab-course, where the immediate goal is to get rid of annoying tasks. Producing the expected result means by definition that this task has been achieved and that one can proceed to the next one.

Only when the same experiment gave significantly different results from before (as in experiment 8 in comparison to experiment 1), the students felt inclined to venture possible explanations, such as 'boundary conditions', 'warming-up period' or a changed 'handling' of the machine. These, however, are circumscriptions of possible fields of influences rather than possible causes themselves. These vague conjectures seem to have been sufficient; the phenomena itself had somehow been produced after all. It is worth noting that the students had been given and had used the textbooks the morning before the experiments. Nevertheless, they did not consider clarifying both success and failure by referring to the literature of around 1800. Two fundamentally different reasons for this are given in the meta-minutes.

Section 4: Making sense of the meta-minutes

In my analysis of the students' meta-minutes, I will focus on three subjects: the course of experimentation, the minute-keeping, and the students' assessment as to whether they did real replications and for what purpose.

The course of experimentation

The impression gained from the minutes of the first experiments – that all students had difficulties in producing the effects mentioned in the text and then tried something new – is confirmed by the students themselves. They state frankly that they focused increasingly on strong and spectacular effects, leaving aside those effects that were difficult or impossible to produce (C. Reiss, 58; M. Müller, 62). C. Reiss even derived a general three-stage pattern for all experiments on electricity:

> First came trials to proceed systematically, followed by a phase of result-oriented action and finally ending in exploratively playful 'experimenting around'. (C. Reiss, 58)

For scientific research, one might expect explorative research first and systematic research later, but here the initial systematic way of experimenting is not a sign of maturity, but may partly stem from unfamiliarity with the devices and substances and partly from a confusion of what is expected of them in a usual science lab-course. Anyway, the students evaded this structure, left aside the historical text and did something new, characterised as 'playful'. New, however, does not mean arbitrary. A standardisation emerged; the crackling of the rubbing device, for example, was used as a criterion for sufficient electrification (M. Markert, 59; M. Müller, 62). And playful does not mean inattentive:

> During the experimental actions, self-observation was almost impossible. Producing the phenomena and building up series of experiments absorbed all attention. A successful or failed procedure immediately triggered problem-solving strategies, in which the question of why one did exactly this or that was irrelevant at the moment of action. Today's knowledge and skills mostly played a secondary role. During the production of the phenomena, each person in the team became an experimentalist or physicist of past times himself. (M. Müller, 63)

Obviously, a coherent interplay between people, things and effects, characterised as 'gradual enculturation' (M. Müller, 63), had emerged. I will come back to this later.

Strikingly, the students regard the experiments with the electrical machine as fundamentally different from those 'rather explorative' (C. Reiss, 57), 'elementary' (M. Markert, 59), or 'simple' (E. Kusch, 65) experiments described in the textbook entries. They emphasise that

> ... by the possibility of using the number of revolutions for quantifying one's own practice, it was possible to allow for a better error-analysis of negative results. (C. Reiss, 58)

It has been noted above that error-analysis has not been done at all. To me, the quantification of anything, even of something of such secondary importance as the number of revolutions, served to establish a coherent – or at least repeatable – experimental practice. This may have been (felt to be) necessary precisely because the students did *not* possess the theoretical background for the explanation of the, at times awkward, behaviour of the machine and additional equipment. As a reason why they did not try to explain these effects, they state that it was too difficult for them to enter into contemporary theories of electricity (C. Reiss, 58).

Keeping the minutes

According to the students, the manner of keeping the minutes developed largely alongside the recorded action itself. Starting with the meticulous and – immediately recognised as such – impossible demand of recording everything, the choice of what to write down was soon governed by the same criteria as the experimenting itself: positive and spectacular effects (C. Reiss, 57–58). Both kinds of work mutually reinforced each other. Only the spectacular was recorded, and only the recorded was worth remembering.[15]

The interplay between experimenting and keeping the minutes created remarkable effects, both social and epistemic ones. The social effect was a temporary division of labour. Although M. Müller and R. Hampe had planned to interchange their roles, they remarked that

> … soon a hierarchy emerged in the two-person team, if only briefly, by which the group dynamics were influenced and the actions were directed in a certain direction. (M. Müller, 62)

Surprisingly, they do not state whom they regard superior in the hierarchy, the experimentalist or the minute-keeper. This is an important question, since it touches on the reason for credibility for scientific effects. One instance is remarkable enough to justify quoting at length:

> In our 2nd series of experiments on the 2nd day, in the 3rd and 4th experiment, a reproducible and complex phenomenon occurred which we documented. By the end of the day we had more series of experiments at hand, which were secured and stood in clear-cut contradiction to the 2nd series of experiments. From then on, the first phenomenon could no longer be reproduced. With only the final minute of the whole day, the two experiments would have been dealt with as 'vague defects'. But thanks to the permanent minuting, we already had a written document at hand, which pointed to this phenomenon. Since it had occurred regularly during the morning, we could have performed further experiments on it in order to limit possible errors and maybe to disclose hitherto unobserved properties of electricity. But here the limits of recording apply since we afterwards suspected the phenomenon to depend on the speed of nearing or on the position of the rubbed rod relative to the pith ball, but these had of course not been part of the minutes so far, because we had not regarded them as significant for the experiment. (M. Markert, 61)

What is the status of this effect? If the students had been participants in a usual science lab-course, the answer would be something odd and irrelevant, since it is not mentioned in textbooks or previous protocols, which both fix the desired outcome in advance.[16] If they had been participants in an ordinary text-based seminar either on electricity or on the history of research on electricity, this effect would not even have occurred. Thus whether there was something 'real' here (be it historically, scientifically or electrically) partly depends on what sort of business the students saw themselves to be engaged in.[17] In this particular case, they did not dare to pin down their status and the status of the effect. In general, however, they had a clear awareness of what they had done.

The type of replication
We, the organisers, obviously succeeded in not predetermining what use should be made of the replications:

> Since the aim during the lab-course was not totally clear, i.e. whether it was about the experiencing of experimental practice around 1800 as such or about obtaining the described results by past means, [...] a mixing of these fields of knowledge could not be avoided. (C. Reiss, 57)

How then do the students assess retrospectively what they actually did and what they were supposed to do? All participants agree in stating (or presupposing implicitly) that they should have done – or would like to have done – replications for historiographic purposes, thereby employing, however, a rather restricted concept of replication. And all agree that what they actually performed were not real replications at all (M. Müller, 64). Thus the whole assessment is tainted by the deficiency of their own practice with regard to a (supposed) ideal of replication, which they assumed they were expected to fulfil (M. Markert, 60). In particular, they complain about having to use partly modern equipment (M. Markert, 60), about too little time to get into past ways of thinking (C. Reiss, 58), about single experiments extracted from larger experimental programs (all, 72), about too little contextualisation (M. Markert, 60), about being told about the method of replication only at the end of the lab-course (C. Reiss, 57) and about their own divergence from the ideal (M. Markert, 60; M. Müller, 62 and 64, E. Kusch, 65; all, 71 and 72). Only in the reflection does the historiographic demand become clear.

> The series of experiments on Monday afternoon had the common aim of producing two different kinds of electricity by means of the available material and are therefore no replication. (M. Markert, 60)

Whatever the students meant by 'historiographic replication', they experienced what they had done as something different. Retrospectively, the happy, productive and coherent phase of experimenting (like a 'physicist of past times', M. Müller, 63) is subdivided into what was close to the historical original and what was not. This has consequences for assessing what they did and why.

For example, the students give different explanations for not having used the textbooks of around 1800 to explain strange effects. Rather than to complain about their inability to understand them in a reasonable time (in fact, pages 36–53 of the report prove that the students understood them very well), they now blame the fact that the strange effects encountered during the lab-course were not sufficiently close to the time of 1800 – as if electricity itself has since changed!

> Using contemporary textbooks on the second day of the lab-course to explain our observations is therefore highly problematic, since which part of our experiment can be judged as sufficiently 'correctly' replicated to allow the textbook statements to be applied to them? (M. Markert, 60)

The students were obliged to discuss the rubbing experiments of the first day based on textbooks, but retrospectively, at least, they see this as misguided. For methodological purity, they refrained from using modern knowledge, as well (M. Müller, 62 and 64). (As if with that knowledge everything would be easy to explain!)

Nevertheless, the quibbles with various anachronisms is not due to careless preparation and/or carrying out of the lab-course, but a fundamental feature of the method of replication as such. This, however, was not grasped by the students, as the following quote reveals:

> Should we not succeed with the retrospective clarification of these difficulties, we can hardly venture statements about replication as a method and its use with regard to our lab-course. (M. Markert, 60)

What the students have failed to grasp is that the retrospective clarification of difficulties, in particular those resulting from anachronisms, is not a *precondition* for a replication. It is *practising* replication all along. In fact, this kind of retrospective judgement is the most important part of a replication. Anachronism is not a problem to be avoided (either by achieving full resemblance of the experiment to the original or by totally ignoring the criterion of closeness to the original); anachronism is the dynamic principle that makes replications useful for the historiography of science in the first place.

If the students do not characterise the lab-course and their own practice as historiographic, what do they regard it as then? Strikingly, they never consider the possibility that the whole lab-course had the didactic purpose of teaching them about electricity using a historical approach. In a way, however, this is consequential. If the experiments do not really belong to the past, because they do not fully resemble past experiments, there can be no learning about anything by means of past experiments.

A third possibility could be that the lab-course allowed them not only to acquire established knowledge (about history or about electricity), but to find out something new about electricity. In this case the lab-course would be neither historiographic, nor didactic, but scientific. The students would not only *learn* something *about* scientific research, they would *practise* it themselves. The only instance where knowledge about electricity itself (rather than its past or present inquiry) was in the focus of interest is the above mentioned moment when they pondered whether they had discovered a new effect. But even there, as was already said, they shied away from bluntly labelling their own practice as 'natural science research'. They characterised it as '*almost* having the character of a scientific enterprise' (C. Reiss, 57; emphasis added) or '*if you like to put it like this*, in the following we experimented scientifically' (E. Kusch, 65; emphasis added). Only when they happened to fully depart from the 'available sources' (all, 70) did they give up circumspection: 'Here experimentation was carried out in the manner of a natural scientist' (all, 70) (if without success, since no real scientific finding emerged).

The fourth possibility is a natural, or usual, science lab-course. Although they sometimes behaved as if it was such (recall the lack of interest in expected results), no one seriously took our enterprise as being intended as such.

They experienced the lab-course as something else. All students agree that the lab-course allowed them to learn about experimental practice and to experience it themselves. Thus, according to the experience of the students, the use of replications is basically neither historiography, nor learning about electricity using a historical approach, nor research on electricity, nor a usual science lab-course, but something different, difficult to pin down in a fitting term. The difficulty to specify this kind of enterprise can be inferred from the fact that they mostly characterise it negatively. M. Markert, for example, while appreciating his insights into experimental practice, contrasts it with accurately located experimental practice of the past:

> I characterised the lab-course as successful, because for me the salient aspect was getting acquainted with research practice. But actually, the lab-course was about research practice around 1800 and I experienced practice only as far as my experience could have been applicable to research around 1700, 1900, or 2000 and not to a particular status of research. (M. Markert, 59)

On the contrary, C. Reiss compares it with present-day science when he characterises his own practice as 'quasi natural science' (58). All participants, however, agree that the most crucial thing is one's own experience. And this, finally, allows them to overcome their own rigid position of total adherence to either the past or the present. Referring to Sibum (1998), their conclusion is:

> Finally it can be said that in all performed experiments, we succeeded only in producing phenomena. We gained historical insights only during the

concluding discussion by taking into account the methodological approaches. [...] While Sibum assumes that past 'gestural knowledge' can be reconstructed by one's own practice, we came to the conclusion that this allows only one's own 'gestural knowledge' to be acquired. Therefore, one's own experiment becomes a source, which has to be interpreted, and which allows statements concerning historical research practice only by means of contextualisation. (all, 72)

The conclusion is still written in the spirit of deficiency – note the fourfold use of 'only'. Nevertheless, the lesson drawn is of prime importance. Replication does not mean that by using rebuilt instruments – even if they are extremely close to the original – one dives into the past. What replication can do is no more and no less than allowing for one's own experiences which can then be put to various kinds of uses – historiographic research being one of them. In my opinion, this important insight alone is worth the whole lab-course. If we had given them clear-cut advice on how to use a replication historiographically, it might have given some minor results. Not doing this, however, allowed the students to undergo this fundamental methodological experience by themselves.

Section 5: Self-reflection on level 4

From the students, we expected intensive and serious self-reflection. They should not only write about experiments and texts about electricity and about electricity itself, but also about their own engagement with them. It is apt that I, the analyst, apply a similar degree of self-reflection to my engagement on level 4 which consisted of analysing at once the participants' simultaneous engagement with electricity and texts and their reflection upon this.

What have I done on level 4? During the lab-course and with regard to writing the report, I mostly remained removed from levels 1 to 3. I preferred to watch the game rather than play it. It may appear that this is an idle, non-interventionist enterprise. In fact, it is not. First of all, it should be noted that it was I who placed myself on the meta-level. I am not authorised by the students on level 3, let alone by the persons and things on levels 2 and 1. This act of self-positioning gives power. At the cost of not being immediately engaged with electricity, texts on electricity, or the self-reflection among the students, I am in a position of overview of all of these levels at once.[18]

The position above the whole gives the pleasure of insights which cannot be gained from either level below. And, after all, it saves the level-3 insights from sinking into oblivion. At the same time, I feel uneasy about this position. Placing myself on the meta-level is far from idle. On the contrary, it is even aggressive, because it simultaneously declares others as inferior and makes oneself invulnerable by stating that one is above all. Whatever participants on either level below would say, the analyst on the meta-level can take this as part of the game rather than taking it seriously at eye level.[19] In my case, the students have not even been informed that any meta-analysis of their report is being carried out.

Would I, after all, be pleased to be analysed by a meta-analyst on level 5, trying to make sense of my stumbling attempts to make sense of the electrical-historiographic-didactic experiment? Maybe not. Thus the iterative creation of meta-levels seems unreasonable, both epistemologically and politically. Instead, I suggest mutual commentating. In particular, I offered the students the chance to comment on my comments. This does not place them at level 5, but neither does mutual commentating level out the steps between them and me. In fact, in my part of this report (sections 1–5), I am one step above them, and in their comments (in section 6), they are one step above me. (For those who have difficulty imagining how both parties can be simultaneously upstairs of the other, I recommend recalling M. C. Escher's painting *Ascending and Descending*.)

Here is what they wrote.

Section 6: Making sense of ourselves
(by Enrico Kusch, Michael Markert, Matthias Müller and Christian Reiss)

Contrary to Jan Frercks' worries concerning our reaction to the meta-analysis of our 'stumbling attempts', we were actually quite thrilled about the paper he wrote. As students of the history of science, we usually experience the situation the other way round, as we analyse the 'stumbling attempts' of other people. This paper gave us the unique opportunity to experience the feeling of being analysed.

During the discussion of this paper, we realised that Frercks' and our interpretation of certain events during the lab-course differed fundamentally. In particular this difference concerns the interpretation of the events surrounding the 'simple demonstration experiments'. In our view, and also as far as we remember, the reason for the way we carried out the experiments is not to be found in our misunderstanding of the introductory paragraphs on electricity. We were aware of the fact that the experiments described in these chapters were neither simple to reproduce nor meant as a kind of starting point for electrical experimentation.

The open concept of the lab-course is the main criterion we hold responsible for the way in which our experimentation developed from a historiographical account into an ahistorical one. Due to the absence of a task or goal and based on our 'methodological purity', we soon reached a point where a historiographical reproduction of the described experiments didn't make sense any more. So we consciously stopped the lab-course part and started fooling around, for we were aware of the fact that the theoretical discussion would follow the next day, as we knew the timetable of the lab-course.

Given this instance and the lack of methodological discussion during the lab-course (the methodological discussion in our report mainly took place in retrospect), our interpretation had a more epistemological orientation. As Frercks has correctly stated, we 'mostly [...] characterise[d] it negatively'. But this has to be understood against the background of the lab-course being, on paper, a historiographical one. It was just our attempt to save this aspect. Referring to Sibum's performative historiography as a method of Experimentelle Wissenschaftsgeschichte, a different procedure of preparation and processing of the experiments would have been conceivable.[20]

We want to close this section and this article with an affirmation of such a lab-course, which can also expose questions regarding epistemology and philosophy of science, for example on the role of the experiment as a source of scientific knowledge, how the experiment relates to publication and the status of experiments in textbooks. According to the conclusion in our report, carrying out experimental actions oneself can help to answer these questions.

Both in the actual experimentation and, to an equal degree, in the substantial reflection of our experiences, we learned much more about the nature of experience in general, and of experience in scientific contexts in particular, than we would have done in a solely text-based seminar.

Notes

1. Before the start of the lab-course, the participants were expected to have read at least Schmidt (1778), Ritter (1800) and the introduction of Weber (in press) on electricity and galvanism around 1800, and Rheinberger and Hagner (1997), Riess (1998) and Sichau (2002), pp. 1–72, on methodological issues.
2. For the process of rebuilding the machine, see Weber/Frercks (2006). The details about rebuilding the glass components are to be found in Weber et al. (preprint). The intention with rebuilding the machine is somewhat different from the Oldenburg approach. Rather than to come close to past experimental practice by supplementing textual sources with one's own experience during the replication, the main point of interest in Jena is the relation between the machine itself and its textual and pictorial representation. The core question is to what degree the latter allowed for a replication by contemporaries. The description of Schmidt's machine is in Schmidt (1778).
3. One student experimented on the electrical properties of a heated turmalin, but this experiment was not accounted for in the report.
4. For reasons of simplicity, I will speak of electricity rather than of electricity and galvanism, although around 1800 it was still contested whether both were basically the same.
5. I take a representative sample of texts on this use of replication to consist of Schulz and Riess (1988), Riess (1995), Heering (2000), Höttecke (2000), Sichau (2000) and Riess (2000).
6. All practitioners emphasised that they intended to gain not only scientific knowledge, but also knowledge about scientific practice (the 'nature of science'), ranging from philosophical problems such as the role of

experiments in science to the social, technological and political influences and relevance of science. These topics, however, can easily be taught alongside contemporary science. The only reason to choose examples from the history of science for this aim is to demonstrate that science is historical rather than timeless.

7 Fundamental texts are Sibum (1995), Heering (1998) and Riess (1998). A concise account of the conventional version of the method is contained in Heering (2005). Critical assessments from outside Oldenburg are Pestre (1994) and Gaudillière (1994). Further developments of the method are contained in Sibum (2000), Frercks (2001) and Sichau (2002). A fairly representative sample of case studies is Heering (1992), Heering (1994), Sibum (1995), Voskuhl (1997), Staubermann (2000), Frercks (2000), Frercks et al. (2001), Beneken et al. (2001), Heering (2002), Staubermann (2004) and Staubermann (2006). Many more case studies resulted in master, diploma, or PhD theses, which have been published only in German or not at all.

8 This applies at least to the restricted notion of replication as argued for in Frercks (2001), which comprises the reconstruction of the instruments (including necessary historical research for this) and one's own experiment. Others, for example Heering (2005), include the historical contextualization into the concept of replication, which, in my opinion, makes it doubtful whether something else can still be done with one's own experiment other than putting it to historiographic use.

9 I will not enter the discussion about whether one can, should or must not use later established knowledge, in particular present-day knowledge, for the historiography of science.

10 This report, written by Enrico Kusch, Michael Markert, Matthias Müller and Christian Reiss is entitled 'Abschlussarbeit zum Praktikum der experimentellen Wissenschaftsgeschichte vom 30.08. – 03.09.2004'.

11 With the help of J. Harrison, I have tried to retain the at times sloppy wording of the report in the English translation.

12 Although work was done partly in teams of two persons or alone, I will not compare the different groups.

13 All the following quotes are from the report. Those rendered with a name stem from the individual meta-minutes.

14 See Frercks (2004).

15 Is the retrospective assignment of the development of experimenting an artefact of recording?

16 The irrelevance of the results of experiments performed in natural science lab-courses has a disciplinary and an ideological function. The disciplinary function consists in making the students clear that they are not yet real scientists. The ideological function consists in implicitly stating that this kind of experimenting is merely didactic, thus saving a field of established knowledge from being questioned endlessly; see Frercks (2009).

17 This can be inferred from the fact that the appertaining sections of the records, the only ones indicating at something scientifically interesting, have not been included in the report.

18 In this respect, the position resembles Latour's 'centres of calculation', see Latour (1987).

19 The simultaneous feelings of envy of the scientists being immediately engaged with things and knowing much more about them than the analyst and of arrogant distancing from the 'actors' or 'participants' by spotting down on them from the meta-position runs through that major portion of sociology of science that refuses to get drawn into the game. Collins (1984) argues for a temporal immersion into science on the part of the analyst. The most sophisticated reflection on the problem of (self-)reflection in the sociology of science is Ashmore (1989).

20 See Sibum (2000).

Bibliography

Ashmore, M. (1989): *The Reflexive Thesis: Wrighting Sociology of Scientific Knowledge* (Chicago & London: University of Chicago Press).

Beneken, H., J. Frercks, P. Heering, A. Makus, A. and F. Riess (2001): 'Forces, Distances and Laws: 'An Experimental Competition about Electricity among an Englishman, a Frenchman, and a German', in *Laboratorium* (H. U. Obrist and B. Vanderlinden, eds) (Köln: Dumont), pp. 237–40.

Collins, H. (1984): 'Researching Spoonbending: Concepts and Practice of Participatory Fieldwork', in 'Social Researching: Politics, Problems', in *Practice* (C. Bell and H. Roberts, eds) (London: Routledge & Kegan Paul), pp. 54–69.

Frercks, J. (2000): 'Creativity and Technology in Experimentation: Fizeau's Terrestrial

Determination of the Speed of Light', in *Centaurus*, 42, pp. 249–87.

Frercks, J. (2001): *Die Forschungspraxis Hippolyte Fizeaus: Eine Charakterisierung ausgehend von der Replikation seines Ätherwind-experiments von 1852* (Berlin: Wissenschaft und Technik Verlag).

Frercks, J. (2004): 'Disziplinbildung und Vorlesungsalltag: Funktionen von Lehrbüchern der Physik um 1800 mit einem Fokus auf die Universität Jena', in *Berichte zur Wissenschaftsgeschichte*, 27, pp. 27–52.

Frercks, J. (2009): 'Epistemisches Theater: Die Dialektik von Forschung und Lehre bei Vorlesungsvorführungen in der Chemie um 1800', in *Versuchsanordnungen 1800* (J. Vogl and S. Schimma, eds) (Zürich and Berlin: diaphanes 2009), pp. 17–38.

Frercks, J., P. Heering, J. Hennig, A. Makus and F. Müller (2001): 'Places of Experimental Action: Performance, Measurement and Construction', in *Laboratorium* (H. U. Obrist and B. Vanderlinden, eds) (Köln: Dumont), pp. 240–44.

Gaudillière, J. P. (1994): 'Lavoisier, Priestley, le phlogistique et l'oxygène: De l'étude de controverse à la réplication pédagogique', in *Aster*, 18, pp. 183–215.

Gren, F. A. C. (1793): *Grundriß der Naturlehre, zum Gebrauch akademischer Vorlesungen entworfen*, 2nd edition (Halle: Hemmerde and Schwetschke).

Heering, P. (1992): 'On Coulomb's Inverse Square Law' in *American Journal of Physics*, 60, pp. 988–94.

Heering, P. (1994): 'The Replication of the Torsion Balance Experiment: The Inverse Square Law and its Refutation by Early 19th-century German Physicists', in *Restaging Coulomb: Usages, controverses et réplications autour de la balance de torsion* (C. Blondel and M. Dörries, eds) (Firenze: Leo S. Olschki), pp. 47–66.

Heering, P. (1998): *Das Grundgesetz der Elektrostatik: Experimentelle Replikation und wissenschaftshistorische Analyse* (Wiesbaden: DUV).

Heering, P. (2000): 'Getting Shocks: Teaching Electrostatics with Historical Experiments at Secondary School Level', in *Science & Education*, 9, pp. 363–73.

Heering, P. (2002): 'Analysing Experiments with Two Non-Canonical Devices: Jean Paul Marat's Helioscope and Perméomètre', in *Bulletin of the Scientific Instrument Society*, 74, pp. 8–15.

Heering, P. (2005): 'Weighing Heat: The Replication of the Experiments with the Ice-calorimeter of Lavoisier and Laplace', in *Lavoisier in Perspective* (M. Beretta, ed.) (München: Deutsches Museum), pp. 27–41.

Höttecke, D. (2000): 'How and What Can We Learn from Replicating Historical Experiments? A Case Study', in *Science & Education*, 9, pp. 343–62.

Latour, B. (1987): *Science in Action: How to Follow Scientists and Engineers Through Society* (Cambridge, MA: Harvard University Press).

Pestre, D. (1994): 'La pratique de reconstitution des expériences historiques, une toute première réflexion', in *Restaging Coulomb: Usages, controverses et réplications autour de la balance de torsion* (C. Blondel and M, Dörries, eds) (Firenze: Leo S. Olschki), pp. 17–30

Rheinberger, H.-J. and M. Hagner (1997): 'Plea for a Scientific History of the Experiment', in *Theory in Biosciences*, 116, pp. 11–31.

Riess, F. (1995): 'Teaching Science and the History of Science by Redoing Historical Experiments,' in *Proceedings of the Third International History, Philosophy, and Science Teaching Conference*, vol. 2 (F. Finley et al, eds) (Minneapolis: University of Minnesota), pp. 958–66.

Riess, F. (1998): 'Erkenntnis durch Wiederholung: Eine Methode zur Geschichtsschreibung des Experiments', in *Experimental Essays: Versuche zum Experiment* (M. Heidelberger and F. Steinle, eds) (Baden-Baden: Nomos).

Riess, F. (2000): 'History of Physics in Science Teacher Training in Oldenburg', in *Science & Education*, 9, pp. 399–402.

Ritter, J. W. (1800): 'Volta's Galvanische Batterie, nebst Versuchen mit derselben', in *Magazin für den neuesten Zustand der Naturkunde mit Rücksicht auf die dazugehörigen Hülfswissenschaften*, 2, pp. 356–400.

Schmidt, G. C. (1778): *Beschreibung einer Elektrisir-Maschine und deren Gebrauch*, 2nd edition (Berlin and Stralsund: Gottlieb August Lange).

Schulz, R. and F. Riess (1988): 'Zur Rechtfertigung des historisch-genetischen Ansatzes im naturwissenschaftlichen Unterricht', in *physica didactica*, 15, pp. 32–59.

Sibum, H. O. (1995): 'Reworking the Mechanical Value of Heat: Instruments of Precision and Gestures of Accuracy in Early Victorian England', in *Studies in History and Philosophy of Science*, 26, pp. 73–106.

Sibum, H. O. (1998): 'Die Sprache der Instrumente: Eine Studie zur Praxis und Repräsentation des Experimentierens', in *Experimental Essays: Versuche zum Experiment* (M. Heidelberger and F. Steinle, eds) (Baden-Baden: Nomos), pp. 141–56.

Sibum, H. O. (2000): 'Experimentelle Wissenschaftsgeschichte', in *Instrument – Experiment: Historische Studien* (C. Meinel, ed.) (Berlin & Diepholz: Verlag für Geschichte der Naturwissenschaften und der Technik), pp. 61–73.

Sichau, C. (2000): Practising Helps: Thermodynamics, History and Experiment, in *Science & Education*, 9, pp. 389–98.

Sichau, C. (2002): *Die Viskositätsexperimente von J.C. Maxwell und O. E. Meyer: Eine wissenschaftshistorische Studie über die Entstehung, Messung und Verwendung einer physikalischen Größe* (Berlin: Logos Verlag).

Staubermann, K. (2000): 'The Trouble with the Instrument: Zöllner's Photometer', in *Journal for the History of Astronomy*, 31, pp. 323–38.

Staubermann, K. (2004): 'The Oldest Collection of Astronomical Plates: Recreating Photographic Practice of the 1880s', in *Journal for the History of Astronomy*, 35, pp. 447–56.

Staubermann, K. (2006): 'Exercising Patience: On the Reconstruction of F. W. Bessel's Early Start Chart Observations', in *Journal for the History of Astronomy*, 37, pp. 19–36.

Voskuhl, A. (1997): 'Recreating Herschel's Actinometry: An Essay in the Historiography of Experimental Practice', in *British Journal for the History of Science*, 30, pp. 337–55.

Weber, H. (in press): *Die Elektrisiermaschine im 18, Jahrhundert* (Berlin: VWB Verlag).

Weber, H., O. Breidbach, K. Heide and H. Grimm (preprint): 'Glas und Elektrizität: Zur Glastechnologie in der Konstruktion der Weimarer Elektrisiermaschine von 1773'.

Weber, H. J. and Frercks (2005): 'Replication of Replicability: Schmidt's Electrical Machine', in *Archives des Sciences: Publication de la Société de physique et d'histoire naturelle de Genève*, 58, pp. 113–22.

PETER HEERING

Materialised skills

INSTRUMENTAL DEVELOPMENT AND PRACTICAL EXPERIENCES

Introduction

THIS PAPER ATTEMPTS to discuss some aspects of the solar microscope. The aim of this paper is two-fold. First, I will describe some of the experiences I had with a solar microscope. The description should not only enable the reader to develop an understanding of the instrument's performance, but also give an idea of the meaning the device had to some of the historical actors. Second, I will analyse the instrument labelled as the solar microscope – as I will show there is not just the solar microscope; several developments of this instrument can be identified. Bringing these observations together with the conclusions from the first part of my paper, some interpretation for the development of the instrument can be described, thus enabling a more thorough discussion of the technical development of the device. Whilst these two aspects are discussed explicitly, there is also a third, more implicit level in this paper. This study is based on making my own experiences in re-enacting the demonstrations of the historical actors. In my conclusion, I will show that some of the findings made in this analysis are related to the methodology employed.

The solar microscope

Solar microscopes can be characterised as projection microscopes; they use sunlight to create an image of a microscopic specimen on a screen or the wall opposite to the instrument. Technically speaking, a mirror reflects (parallel) sunlight onto a condensing lens. The microscopic specimen is placed directly in front of the focal point of the lens. The position of the specimen is determined by the diameter of the light cone: the specimen is in a slider and should be completely illuminated, yet the light cone should also have the diameter of the opening in the slider so that a maximum of light is used. The specimen is transparent; consequently, the result of this first part of the optical part is an illuminated transparent object. On the other side of the specimen, the projection

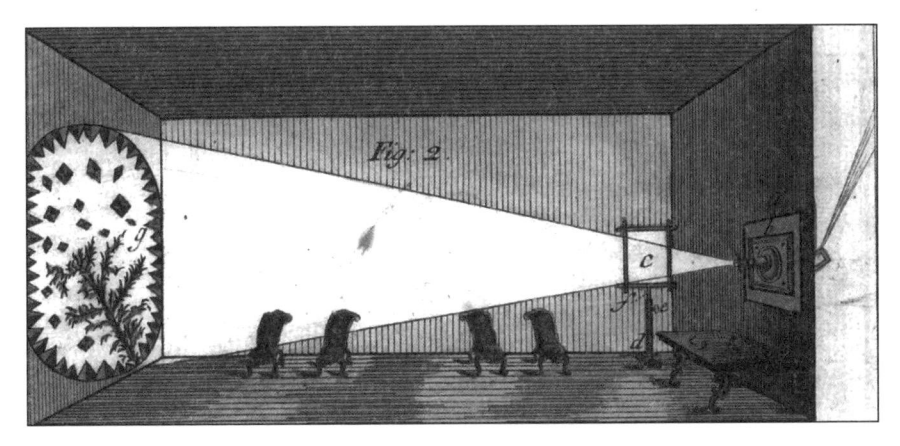

FIG. 9.1
Projection situation with a solar microscope (Ledermüller, 1762).
(© Deutsches Museum)

optics is placed: in the simplest case just a single lens, but in most instruments a lens system. Speaking purely technically, the instrument is not a microscope but a sort of slide projector for microscopic objects: it creates a real image on a screen instead of a virtual image in the eye of the observer as common microscopes do. This technical peculiarity results in certain particular options for the use of the instrument, as I will discuss in this paper, these options are one reason for the success of the instrument.

The instrument is used in a darkened chamber and according to the historical actors, the quality of the darkening is of high importance. To give but one example, the microscopist Henry Baker pointed out that 'when this Microscope is employed, the room must be rendered as dark as possible: for on the Darkness of the Room, and the Brightness of the Sun-shine, depend the Sharpness and Perfection of your Image' (Baker, 1769: 23f.). The solar microscope is placed in the shutter of the window in such a manner that the mirror is outside the room (Fig. 9.1).

Martin F. Ledermüller's plate is very telling as it does not only show the instrument in the shutter of the window and the projection (in this case of crystals) on a wall, it also shows an additional screen [c] that is meant to draw the image. (Due to the light intensity, the drawing could be made on the reverse of the screen where the drawing hand would not create a shadow.) There is also a table on which spare objects, other sliders as well as other projecting lenses, could be left, and four chairs. These chairs are deliberately included in the image as they refer to a major aspect of the use of the solar microscope – it served as a device to how an audience microscopic specimen that could be looked at collectively.

Solar microscopes started to appear in the price lists of instrument makers as well as in publications in the 1740s, and they quickly became very popular. This popularity can be explained in part by the characterisations used to market the instrument, e.g. Edward Nairne claimed that 'Of all Kinds of Microscopes the Solar may be justly reckoned the most entertaining' (Nairne, n.d.: 4). Yet it is not just the instrument makers who praised this device (and whose statements could be read as advertisements), but also practitioners such as Joseph Priestley.

> The sun's rays being directed by the looking glass through the tube upon the object, the image or picture of the object is thrown distinctly and beautifully upon a screen of white paper, and may be magnified beyond the imagination of those who have not seen it. (Priestley, 1772: 742)

Taking these statements literally, one might question oneself why the solar microscope is nowadays not that well-known. However, looking at nineteenth-century publications reveals that the instrument received an entirely different reception. As the British microscopist Goring stated in 1827:

> The image of a common solar microscope may be considered as a mere shadow, fit only to amuse women and children …. The utmost it can do is to give us the shadow of a flea, or a louse as big as a goose or a jackass. The swinish vulgar will always be gratified by such spectacles, because they have no idea that a microscope of any kind is to do more than exhibit objects very much dilated in point of bulk. (Goring quoted in Bradbury, 1967: 159)

Goring's statement is in direct contrast to the descriptions of eighteenth-century instrument makers and practitioners. Moreover, the solar microscope was used in the mid-nineteenth century for producing photographs of microscopic objects. Yet, despite the praise from the eighteenth-century actors, most historians have followed statements from nineteenth-century microscopists. The discrepancy between these characterisations can be seen as one of the starting points of my work on solar microscopes.

The practice with the solar microscopes

In order to develop a better understanding of the practice with the solar microscope, I intended to use the replication method (for a discussion of this method *see* Heering, 1998; Sichau, 2002; Breidbach et al., 2010). In doing so, I had the opportunity to work at the Deutsches Museum in Munich where I could examine the instruments from their collection and work with two original devices. Moreover, another instrument was reconstructed at the Carl-von-Ossietzky Universität Oldenburg. One of the instruments I used in Munich

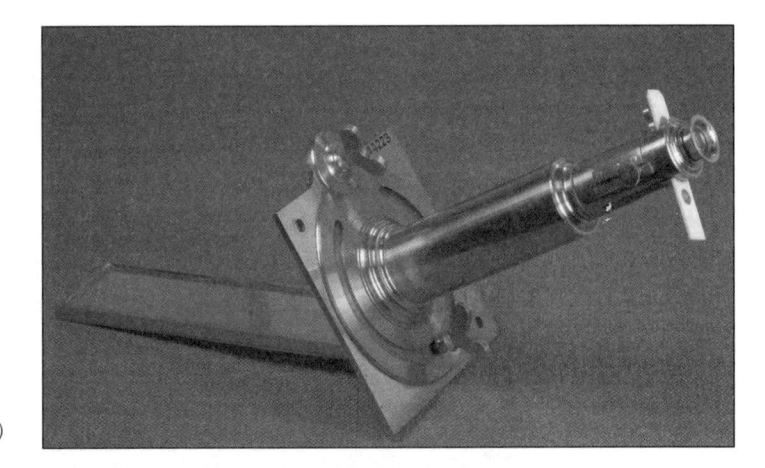

FIG. 9.2
Dollond's solar microscope kept at the Deutsches Museum in Munich.
(© Deutsches Museum)

FIG. 9.3
Junker's solar microscope kept at the Deutsches Museum in Munich.
(© Deutsches Museum)

was made at the Dollond workshop around 1780 (Fig. 9.2), the other by the field chaplain Junker (Fig. 9.3).

It quickly became evident that the quality of the projections correspond more to the description of the eighteenth-century practitioners than to Goring's criticism. The solar microscope was suited for making impressive demonstrations, to – as Priestley put it – throw images 'distinctly and beautifully upon a screen of white paper' which 'may be magnified beyond the imagination of those who have not seen it'. Even pictures of the projected images are a meager and unsatisfactory representation (Fig. 9.4).

Such an observation naturally raises questions: even though the images were extremely impressive, can they serve as the only explanation for the success of the solar microscope in the eighteenth century? And what can explain the rejection of the instrument by nineteenth-century microscopists?

One answer can be developed by taking into consideration the status of microscopy in the eighteenth century. As Marc Ratcliff (2009) pointed out, microscopical practice underwent significant chances during the eighteenth century. At the same time, he argues convincingly that in the middle of that

FIG. 9.4
Projection with a Dollond solar microscope at the Deutsches Museum in Munich.
(Peter Heering)

century the necessity for ensuring the observers against erroneous interpretations of the images became more pressing. In this respect – and I shall come back to this argument shortly – the solar microscope had advantages as it enabled a collective observation.

This ability has already been identified by Alice Walters. However, she makes her argument in a significantly different perspective:

> … social dangers handicapped other instruments. By their very nature, most common microscopes invited solitude, and so were easy to associate with impoliteness. (Walters, 1997: 140)

This creates a problem, as specific social standards became established during the Enlightenment and these standards also affected practices in the natural sciences.

> The most challenging obstacle to overcome in the effort to associate science with politeness was dispelling contemporary caricatures of those interested in natural philosophy as solitary, pedantic, asexual, melancholic bores, more comfortable with the microscopic creatures found in pond water than with the polite ladies and gentlemen of society. To counter this image, polite science explicitly associated the study of natural philosophy with conversation, the fundamental activity of polite society. (Walters, 1997: 126)

In this respect, as she observes, the solar microscope offers an opportunity to bring microscopical practices in accordance with social standards:

> Indeed, solar microscopes strongly encouraged common observation …. The popularity of these instruments in the eighteenth century undoubtedly derived

in large part from their ability to illuminate the religiously and aesthetically inspirational microscopic world in polite social settings.
(Walters, 1997: 141)

This argument appears to be very convincing; however, it misses one aspect. To illustrate this aspect, yet looking again to the advertisements of one of the instrument makers can be useful.

> Besides this particular Property it hath, that the Numbers of People may view an Object at the same Time, and may point to different Parts thereof, and by discoursing on what they see, may understand each other better, and more probably find out the Truth, than when they are obliged to look one after the other. (Adams, 1747: 11)

Having Walters' argumentation in mind, it is evident that George Adams is addressing the potential of the instrument to enable a discourse of the audience – a polite social activity. Yet there is another aspect that Adams addresses, and this seems to be highly important. He points out explicitly that the instrument might enable the audience to find out the truth – and this aspect corresponds nicely to the argument made by Ratcliff. In this respect it is remarkable that several authors labeled the solar microscope as a 'camera obscura microscope' (*see* Nairne, n.d.). According to Jonathan Crary's analysis, the camera obscura was 'synonymous with the production of truth and with an observer positioned to see truthfully' (Crary, 1990: 32). Consequently, the label on the instrument indicated in this context that the solar microscope was a device to see the microcosm truthfully.

Beyond these two interpretations of the particular meaning the solar microscope (or the projected images) had within the reference frame of the eighteenth-century Enlightenment, there is another dimension that can be ascribed to the instrument. As Jutta Schickore pointed out, according to the understanding of researchers in the eighteenth century

> [t]he observer's mental activities were understood to be one major factor conditioning the outcome of microscopical investigations. … If the observers exerted strict control over their mental operations, they could prevent erroneous results. (Schickore, 2001; 128)

This appears to correspond to the criteria made by Ratcliff, yet a different perspective can be developed: in order to be able to exert control over the mental operations, learning processes are necessary. And this is another potential the solar microscope opens, actually for two reasons: on the one hand, the images can be looked at simultaneously, thus an instruction on how to see can be realised; on the other hand, the image is a real image on a wall, consequently novices to microscopy do not have to learn how to look at microscopic images.

These peculiarities of the observational situation became particularly evident during the re-enactment of the demonstrations when an audience was present in the darkened room. The audience started to discuss the images and tried to reach a consensus about what they were seeing.

Yet another aspect became striking even when I was alone in the darkened chamber. The perception of the microscopic specimen is significantly different to the solar microscope in comparison to the common microscope. This is more evident when two eighteenth-century accounts of microscopical observations are compared. The Augsburgian instrument maker Brander praised his experiences with a common microscope as follows:

> What can be more pleasant and more appealing but when we – even without stepping out of our parlour – seemingly catch sight of a new world, or at least of completely new inhabitants, that were up to now entirely unknown to us?' (Brander, 1769: introduction)

Catching sight of a new world (or the inhabitants) is a metaphor that is used frequently in eighteenth-century texts on microscopy. However, what is also implied in looking at these beings is that the user looks through the instrument into the microscopic world. In some sense, the microscope mediates between microcosm and macrocosm, but also separates both worlds. This is different when using a solar microscope: here the image appears in a darkened room on a screen, and in a way it can be argued that macroscopic and microscopic beings are in the same room. This impression of the observational situation is implicitly created by an account published anonymously in the *Neues Hamburgisches Magazin*:

> To see this well-known small animal, I mean the flea, alive in the above mentioned size [of a lion or an elephant], arouses particularly for the fair sex no little pleasure. They laugh at his quaint shape, in which it appears. They point a finger at it, and, when they exclaim sarcasms against this miserable, they taste the sweetness of revenge for the mischief it did to them. Now they finally see how terrible their hereditary enemy is, whom they have known that long.' (Anon., 1781: 460)

The remarkable aspect from this account (even though it is probably an advertisement) is that the audience appears to address the flea – just as if the animal is in the room, just like in a zoo.

To sum up, it can be argued that the solar microscope is able to create very bright images that appear attractive to the audiences. In this respect, eighteenth-century claims appear to be realistic, whilst the nineteenth century-

Materialised skills: the technical development of solar microscopes

It was not only the attitude of the audience and the different manner of perception of microscopic specimen that became evident during the experimentation with the instruments. Another aspect was the different procedural knowledge that was required to use the devices. In order to discuss these differences, it is necessary to be aware of the operation one has to carry out in order to show images to an audience. Basically there are three things to be done: the adjustment of the mirror to compensate the seeming motion of the sun (which has to be done approximately every five minutes), focusing the image, and changing the specimen by moving the slider or by replacing one slider with another one.

As it turned out, the Dollond instrument was fairly straightforward and somewhat easy to use – it did not take much time to learn how to operate the different parts of the solar microscope. In this respect, the practice with the solar microscope was significantly different to that claimed by some historians of science, according to whom the solar microscope was:

> [a] difficult instrument to use, because it required strong sunlight at an appropriate angle, constant adjustment of the mirror outside and a completely darkened room. (Hankins, 2003: 39)

Things were slightly different with the Junker device. Here, some time was needed to develop the skills required to operate the instrument adequately. To make the differences slightly more understandable, it is necessary to take a closer look at the operations required.

The mirror adjustment is realised by two independent operations: one has to turn the mirror to follow the seemingly east–west movement of the sun, and to change the angle of the mirror against the horizon in order to compensate the change in the height of the sun. With the Dollond instrument, two screws can be seen above and below the tube of the instrument. The upper one is connected to a toothed wheel that turns a second, larger toothed wheel on which the mirror is fixed. The lower screw turns a worm gear that raises or lowers the mirror. As mentioned, it is necessary to readjust the mirror about every five minutes, and with the Dollond instrument it was fairly easy to determine the moment when the readjustment of the mirror was necessary: in the upper part of the projected image a yellowish gleam appeared – this resulted probably from a reflection of the mirror's brass frame and could be taken as an indication that it is about time to readjust the mirror. This was very easy; the mechanism allowed simple and controlled movement of the mirror, thus it became possible that the readjustment of the mirror was not even noticed by the audience.

Likewise, it was very easy to change the specimen: the slider moved in a check rail, and moreover, the slider had above each specimen a small dent that corresponded to a pin in the instrument. As a result, when moving the slider to show the image of the next specimen, the slider locks in place so that the specimen is in the ideal position.

And finally, the focusing unit was also very user-friendly; it consisted of a cograil that changes the distance between the lens and the specimen (*see* Fig. 9.5). Again, this mechanism allowed a very precise and easy adjustment of the position of the lens, thus enabling the user to focus the image.

Things are slightly different with Junker's device; the technical quality is not as high as the technical quality of Dollond's device. When projecting with the instrument it quickly becomes evident that the quality of the optical parts does not meet the quality of Dollond's instrument (even though it was manufactured later). Likewise, using the instrument is not as easy: the mirror inclination is changed with a cord that is fixed to the mirror. Its length can be modified by turning a wooden pin, yet the friction between the pin and the wooden ground plate makes it difficult to carry out a minute adjustment. Moreover, the pin can easily tilt, requiring more force to make it move again, resulting in a less precise adjustment of the mirror. It is also difficult to turn the mirror, done by turning the wooden plate on which the tube is fixed in the wooden base plate of the instrument. Again, there is wood moving on wood, the friction between the two parts is difficult to overcome, and a controlled movement (and as a result a controlled readjustment of the mirror) is difficult to be realised. Furthermore, as the image is not as bright and clear as the one with Dollond's instrument, and as there is no indication when and how far to move the mirror, it becomes necessary to move the mirror further than the

FIG. 9.5
Focusing mechanism of Dollond's solar microscope.
(Peter Heering)

FIG. 9.6 Dollond's mechanism of mirror adjustment.

FIG. 9.7 Junker's mechanism of mirror adjustment.
(Peter Heering)

ideal position, then move in the other direction, return again. With this sort of iteration it is possible to find the best adjustment for the mirror, but not without the audience noticing that the mirror position is changed. One reason for this, apart from the difficulties of producing a minute and controlled movement, is the manner in which the tube is connected to the instrument. In Junker's solar microscope, the tube is directly fixed to the plate to which the mirror is attached. Consequently, when the mirror is turned, the slider and the ocular are turned as well, as a result, the image on the screen is moving. However with Dollond's instrument, the tube is fixed to a brass plate that covers the toothed wheel on which the mirror is fixed (*see* Fig. 9.6 and Fig. 9.7).

Likewise, it is more difficult to change from one specimen to another: there is no mechanical means to adjust this movement. The slider can be freely moved not only in the horizontal direction, but also to some extent in a vertical one. As a result, one has to learn how to move the slider appropriately, finding the best position, again in an iteration of moving the slider forward and backward. And finally, the focusing is also less comfortable with Junker's instrument: the lens is simply screwed into the holder, and by unscrewing it slightly, its distance to the specimen can be modified (*see* Fig. 9.8). A side effect of this technical solution is that the lens itself is rotating in its fitting. As the lens is (at least in the instrument I used) not perfectly central to the geometrical axis of the tube, the image is projected at a slightly eccentric angle. As a result, the image moves when the lens is turned, which could result in a displacement of some 30cm at a distance of 4m (which results in an image of

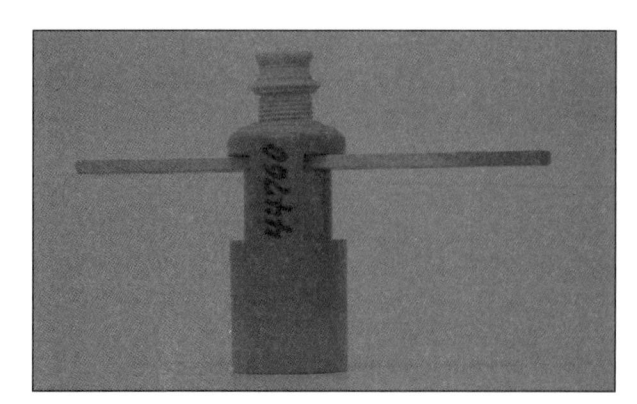

FIG. 9.8
Focusing mechanism of Junker's instrument.
(Peter Heering)

about 1.6m in diameter). Consequently, in this case, even the screen has to be re-adjusted.

To summarise, it can be argued that Dollond's instrument can be labelled as much more user-friendly in comparison. There are, when compared to Junker's instrument, few skills required to operate the device, skills that can easily and quickly be developed. With Junker's device it is somewhat different: even though it can be used after some initial experiences, more skills need to be developed, and it needs more experience to use his solar microscope for the projection.

In some sense it can be argued that this is not surprising. Dollond's workshop can be taken as being one of the leading manufacturers (*see* Sorenson, 2001) for optical instruments. Consequently, one should expect a superiority of Dollond's device. Moreover, as Junker himself points out explicitly, his instrument is meant to be used in an educational context, and he even offers a discount if a sovereign decides to equip several schools with Junker's devices (Junker, 1791). Consequently, he tries to make his device as cheap as possible in order to enable as many schools as possible to acquire the instrument.

Yet there is an explanation why Dollond's instrument has better optical qualities, and even why his instrument is made out of wood and cardboard (which were common materials for these instruments: *see* Wiedeburg 1758) and not brass. With respect to the technical realisation, Junker's instrument appears to be somewhat old-fashioned, while Dollond's instrument is characteristic of instruments that were made particularly by London instrument makers from the 1770s onwards. Thus the question one might raise at this point is not why Junker's instrument is less sophisticated, but why Dollond and other makers developed their devices in a manner that made them easier to be used.

Of course, one might argue that instrument makers always try to modernise devices. However, the question of why British instrument makers pushed the development of the solar microscope in this particular direction remains. A possible answer can be developed from the first part of this paper. Here, the argument was made that one reason for the solar microscope becoming popular during the second half of the eighteenth century was its correspondence to social standards. According to these standards, the instrument was used by amateurs and enabled discourse about the projected images. Moreover, a huge advantage when compared to common microscopes was the suitability of the instruments to see microscopic objects without the necessity to develop the ability to look into a microscope. In this respect, the development of the instrument can be seen in a similar line: the instrument can be used without having to spend time and effort in order to develop the required skills. Taking into account that the salon was one social setting where the instrument was used, such a development can be taken as an adequate adaption of the instrument towards the requirements of the market. Persons of a higher social status, as well as amateurs interested in microscopy, can easily use the sophisticated instrument marketed by Dollond. On the other hand, Junker's instrument is meant (among others) for teachers to be used in a classroom – this means that the user is working with the solar microscope in a somewhat professional setting. In such a context, users may take the time to acquire the skills necessary to operate the instrument, and here financial aspects might be more important.

Yet the Dollond instrument is not by all standards superior in its user-friendliness. This became evident when sliders other than those that belonged to the instrument were used. Contrary to prior experience, these sliders (those that came with the Junker device, as well as others I used for making my own preparations) had slightly different measures. As a consequence, it was significantly more difficult to move them into the correct position as well as to change from one slider to another. In a way, the above described characteristics of Dollond's instrument also result in less flexibility of what can be projected with the instrument. It was the other way round with Junker's instrument where it was no problem to use other sliders. The skills developed in working with the original sliders were adequate to use the instrument with other materials as well, the skills enabled the user to be more flexible with respect to which sliders were to be used.

To avoid misunderstandings, two points should be made explicit. I am far from arguing that this interpretation is the sole cause of the technical development of the solar microscope – on the contrary. Other reasons probably played a role, yet to my understanding, one aspect probably has been that instrument makers were seeking to market their devices, and this should mean that they tried to adjust the instrument to the needs of potential customers. And the distinction between a comparably more user-friendly instrument, such as Dollond's, and a somewhat more demanding instrument, such as Junker's, is not to be understood in terms of a professional versus an amateur's instrument.

One could, of course, argue that Dollond's instrument is aimed more at the amateur as it is less flexible in its use; yet this would be a misinterpretation. The flexibility can be gained through the effort to develop certain skills, after some practice, and I have been able to use Dollond's instrument also with other sliders.

As Ratcliff (2009) argued very convincingly, the notion of eighteenth-century microscopy as being a non-research activity and merely a leisure activity for the salons, is not an adequate description of the situation. Even though instruments were used in social settings and used as icons to demonstrate the owner's enlightenment, it would be an oversimplification to limit the uses of these instruments to such purposes. Although Ratcliff made this argument for microscopy in general, it also holds for the solar microscope.

Conclusion

It can be said that the solar microscope was popular because it enabled a perception of the microcosm in a unique manner and in a particular social setting. The instrument is powerful enough to produce convincing images that appear to meet the claims of the eighteenth-century practitioners. In this respect, the practical part of my case study was vital: only through the re-creation of these projections, an understanding of the peculiarities of this access to the microcosm became possible. Moreover, it was the experiences in the darkened room (mine, those of my guests, and mine with my guests) that helped me to understand the potential of these projections. Despite the criticisms of the nineteenth-century microscopists, the solar microscope turned out to be an instrument that has a powerful educational potential as well as an ability to produce an access to the microcosm. In this respect, and this is one important aspect of this methodological approach, my own experiences enabled me to come to different interpretations of the source material.

It has to be made explicit that the images produced and the experiences that were made in the darkened room are not to be understood as sources – on the contrary. It is evident that these experiences cannot be projected onto the historical actors. Modern individuals have completely different prior experiences; consequently the amazement some people experienced in the darkened room is not necessarily transferable to the former actors. Yet these experiences help to make the accounts of these actors understandable. Even experiencing the problem of how to communicate this experience (which turned out to be impossible to document properly) helped me to understand why the former practitioners claimed that one had actually to see the images.

Another aspect that became evident is the difference between the two solar microscopes and the requirements for using them; what I described as 'user-friendliness' is a criterion that can be developed through practise with the instrument. Yet, in order to develop such a criterion, it was necessary to carry out the experiments in a self-reflexive manner. By this term I mean that one has to observe how one's own interaction with the instrument is realised, what kind of skills become necessary in order to operate the device, and how these skills are developed and modified. This appears to be an area which is (at least almost) inaccessible with the traditional methods of the historiography of science; yet these methods are vital in order to be able to embed the experiences in the historical analysis. In other words, it is necessary to operate the instrument in a self-reflexive manner, otherwise no observations would be made except for being able to claim that after a while it was possible to manipulate the device adequately. However, it is insufficient just to operate the instrument in this self-reflexive manner as the resulting observations might tell us more about the (modern) user than the historical situation. Here it is necessary, in order to use this acquired knowledge for the historical analysis, to correlate the findings with the sources – this could be done by generating new questions out of these experiences as well as by looking again at the material from a different point of view. These different perspectives, these newly-developed questions, are basically a central outcome of the interaction with the apparatus. Thus in this respect, the particular manner of working with the apparatus becomes relevant for the historical analysis.

Acknowledgements

I am indebted to the Deutsches Museum in Munich which enabled me in its Scholar in Residence programme to work with the original instruments. It was not just the opportunity to work with its solar microscopes, but also the friendly and open atmosphere in the institute that enabled me to clarify my thoughts and to develop an understanding. In particular I would like to thank Ulf Hashagen and Christian Sichau (who was also responsible as a curator for the instruments I used). Moreover, I would like to thank the workshop at the Carl-von-Ossietzky Universität Oldenburg which carried out the reconstruction of the solar microscope by Benjamin Martin. This reconstruction was based on an instrument kept at the Universiteitsmuseum Utrecht that was made available for us by Klaus Staubermann. Tiemen Cocquyt, Hans Holtorf and Holger Koch took responsibility for the technical supervision of the reconstruction, and Jan Deiman supplied us with the optical data of the lenses. Moreover, I had the opportunity to examine solar microscopes at the Optisches Museum Jena, the Museum Boerhaave Leiden, the Universiteitsmuseum Utrecht, and the Mathematisch-physikalische Salon Dresden. I am indebted to the curators who enabled me to work with their collections: the examination of these instruments played a crucial role in identifying technical developments in the actual design of solar microscopes.

Bibliography

Adams, G. (1747): *Micrographia Illustrata, or, The Knowledge of the Microscope Explain'd: Together with an Account of a New Invented Universal, Single or Double, Microscope, Either of which is capable of being applied to an Improv'd Solar Apparatus* (London: printed for the author).

Anon (1781): 'Versuch, über die Vortheile des Sonnemikroskops' in *Neues Hamburgisches Magazin*, 119, pp. 457–79.

Baker, H. (1769): *The Microscope Made Easy*, 5th edition (London: Dodsley).

Bradbury, S. (1967): *The Evolution of the Microscope* (New York and Oxford: Pergamon Press).

Brander, G. F. (1769): *Beschreibung zweyer zusammengesetzten Mikroscope* (Augsburg: Eberhard Kletts sel. Wittwe).

Breidbach, O., P. Heering, M. Müller, M. and H. Weber (eds) (2010): *Experimentelle Wissenschaftsgeschichte* (München: Wilhelm Fink Verlag).

Cavicchi, E. (2009): 'A Witness Account of Solar Microscope Projections: Collective Acts Integrating across Personal and Historical Memory', in *British Journal for the History of Science*, 41, pp. 369–83.

Crary, J.(1990): *Techniques of the Observer: On Vision and Modernity in the Nineteenth Century* (Cambridge, MA: MIT Press).

Fleck, L. (1986 [1947]): 'To look, to see, to know', in *Cognition and Fact: Materials on Ludwik Fleck* (R. S. Cohen and T. Schnelle, eds) Boston Studies in the Philosophy of Science series, vol. 87, (Dordrecht, Reidel), pp. 129–51.

Hankins, T. L. (2003): 'How the Magic Lantern Lost its Magic', in *Optics & Photonics News*, 14(1), pp. 34–40.

Heering, P. (2002): 'Analysing Experiments with Two Non-canonical Devices: Jean Paul Marat's Helioscope and Perméomètre', in *Bulletin of the Scientific Instrument Society*, 74, pp. 8–15.

Heering, P. (2005): 'To see or not to see: Jean Paul Marats öffentliche Experimente und ihre Analyse mit der Replikationsmethode', in *NTM International Journal of History & Ethics of Natural Sciences, Technology & Medicine*, 13(1), pp. 17–32.

Heering, P. (2008): 'The Enlightened Microscope: Re-Enactment and Analysis of Projections with Eighteenth-Century Solar Microscopes', in *British Journal for the History of Science*, 41, pp. 345–68.

Heering, P. (2009): 'The Enlightened Microscope: Working with Eighteenth-Century Solar Microscopes', in *Illuminating Instruments* (P. J. T. Morris and K. Staubermann, eds) (Washington DC: Smithsonian Institution Scholarly Press), pp. 19–38.

Junker, F. A. (1791): *Ueber das Sonnenmicroscop* (Magdeburg: for the author).

Ledermüller, M. F. (1762): *Nachlese seiner Mikroskopischen Gemüths – und Augen-Ergötzung* (Nürnberg: Adam Wolffgang Winterschmidt).

Nairne, E. (n.d.): *The Description of a Single Microscope; And of an Apparatus Applicable to the same, in order to make it a Solar Microscope, which will equally serve for a Camera Obscura, and also for viewing Prints in Perspective, Greatly Improved for those Uses* (London: for the author).

Priestley, J. (1772): *The History and Present State of Discoveries Relating to Vision, Light, and Colours* (London: J. Johnson).

Ratcliff, M.: (2009): *The quest for the invisible : microscopy in the Enlightenment* (Farnham and Burlington, VT: Ashgate Publishing Group).

Schickore, J. (2001): 'Ever-Present Impediments: Exploring Instruments and Methods of Microscopy', in *Perspectives on Science*, 9(2), pp. 126–46.

Sichau, C. (2002): *Die Viskositätsexperimente von J.C. Maxwell und O. E. Meyer. Eine wissenschaftshistorische Studie über die Entstehung, Messung und Verwendung einer physikalischen Größe* (Berlin: Logos Verlag).

Sorrenson, R. (2001) 'Dollond & Son's Pursuit of Achromaticity, 1758–1789', in *History of Science*, 39, pp. 31–55.

Stafford, B. M. and F Terpak (2001): 'Revealing technologies/magical domains', in *Devices of Wonder: From the World in a Box to Images on a Screen* (Los Angeles: Getty Research Institute), pp. 1–142.

Walters, A. N. (1997): 'Conversation pieces: Science and Politeness in Eighteenth-Century England', in *History of Science*, 35, 123–54.

Wiedeburg, J. E. B. (1758): *Beschreibung eines verbesserten Sonnen-Microscops* (Nürnberg: Georg Bauer).

KLAUS STAUBERMANN

Aim at the stars, reach the Nicol

THE ZÖLLNER PHOTOMETER

IN 1861 A young physicist called Karl Friedrich Zöllner entered a photometer in a competition for a prize offered by the Vienna Academy of Science. The prize was intended for the design of an astro-photometer and numerous photometric measurements with the instrument. The prize was never awarded, neither to Zöllner nor to any of the other candidates. Still, by the end of the nineteenth century Zöllner's photometer had become one of the most influential devices in astrophysics.

A review of Zöllner's prize application attracted the attention of astronomers at the Berlin Academy Observatory who had been in charge of coordinating a major star-chart programme for the Academy of Sciences in Berlin. Star-chart surveys during the nineteenth century formed a significant part of the work of astronomical observatories. New modes of labour division had been introduced at observatories across Europe and measurement procedures were standardised. However, whereas positional measurements were largely standardised, brightness measurements were still considered subjective. This is why Academies such as in Vienna or Berlin wished for an instrumental standard for brightness measurements. Shortly after he was declined the Vienna prize, Zöllner was invited by the director of the Berlin Academy Observatory and co-ordinator of the Berlin Academy's star-chart project, Johann F. Encke, to present his instrument. Zöllner demonstrated his photometer as well as dismantled and reassembled it, which made such an impression that it was reported to the director of Leipzig Observatory, Carl Bruhns.[1] Having been a former student of Encke and observatory technician in Berlin, Bruhns had been appointed professor and director of the newly founded Leipzig Observatory in Saxony shortly before he met Zöllner. Bruhns was still in the process of equipping his new observatory when Zöllner's photometer was pointed out to him. Following the demonstration of his photometer in Berlin, Bruhns invited Zöllner to join him in his work at Leipzig and in 1867 Zöllner was appointed the first professor of astrophysics at Leipzig University.

During the course of the next years Zöllner's instrument became a standard in astronomy, used by observatories all around the world. The reconstruction project presented in this paper aims to understand how Zöllner, although he himself was not a trained astronomer, designed a photometer that became an international success in this discipline. I will argue that Zöllner's personal and professional background, interests and skills are manifested in his instrument and that, by rebuilding the instrument and practising with it, I can show how Zöllner's interests and skills led to the development of the photometer. This paper describes my replication of the photometer and the reworking of the observation practices.

Karl Friedrich Zöllner was born in Berlin on 8 November 1834. Not unusual for the 1850s, he studied at both Berlin University and the Königliches Gewerbeinstitut (Royal Technical Training Institute). Berlin University could only provide a limited number of laboratory places for its students, so that some of them, like Zöllner, trained at the better equipped technical schools. Zöllner focused his studies on physics, but his main interest was in light and colours.[2] His interests were closely related to his personal background: his father had established a dye business in Schöneweide near Berlin. Dyers had always been interested in questions of colour, contrast and illumination and it was well known to them that light, especially artificial light, had an effect on the appearance of colours and contrast.[3] In the early nineteenth century, with the rapid development of the dyeing industry especially in France, the appearance of dye colours had become a topic of scientific interest. For example, the French chemist Michel E. Chevreul, having been appointed the Director of Dyeing at the national gobelins and tapestry workshop in 1824, investigated both chemical and psychological questions connected with dyes, colours and colour impression. Chevreul's work stimulated an influential stream of innovative theory and practice of colouring both in France and elsewhere.[4] With the industrialisation of the dyeing industry, similar developments took place in Germany and the production of dyes and control of the production process became increasingly a topic of scientific investigation.[5] By the mid-nineteenth century most of the dye-producing and tissue-printing companies, such as that of Zöllner's father, had their own small well-equipped laboratories and often employed chemists who were in charge of the mixture and standardisation of dyes.

Measuring and standardising colours was one of the problems industrial dyers had to engage with. Most colourists had a set of standardised tissue available for comparison with the produced samples. Instruction books for colourists often included tissue samples of the published dye recipes. Although colour chemists had tried to standardise dyes by means of tests and instruments these could only be used for controlling parts of the production process.

Colour analysis was already an academic subject at Berlin University when Zöllner tackled the colourists' problem of inventing suitable measurement devices. His most influential teacher here was Wilhelm Dove in Berlin. Dove had just invented a device for the purpose of demonstrating different colours by means of polarisation. Polarisation and its theoretical foundation by scientists such as Étienne-Louis Malus, Thomas Young and Augustin-Jean Fresnel in the early nineteenth century was already an established scientific topic at Berlin University, with applications in crystal and mineral analysis. Polarimeters were widely used in laboratories at universities and industrial schools. These included photometers invented by physicists such as François Arago and Jacques Babinet.[6] Not surprisingly, Zöllner's photometer in 1857 incorporated many of the design features of Babinet's comparison and polarisation photometer of 1853: a mirror for the comparison of two light sources and an analyser consisting of two polarising Nicol prisms.[7] Especially Nicol prisms had become standard devices in laboratory analysis, both at universities and in industry, by the middle of the nineteenth century. Instruments employing Nicol prisms included the saccharimeter, a device for measuring sugar content developed by the chemist Eilhard Mitscherlich, one of Zöllner's professors and a family friend in Berlin.[8] Laboratories like Dove's teaching laboratory would have been furnished with several types of Nicol analysers and it is likely that Dove introduced Zöllner to Babinet's photometer.

Dove had published a popular text book on colours in which he introduced colour as a means for material analysis.[9] Moreover, he had described the phenomena that occur as a result of combining of different types of optics, colours and light sources. Zöllner, while studying with Dove, had designed an apparatus based on Dove's idea that allowed the comparison of intensities of complementary colours by projecting images onto a screen.[10] Zöllner's original idea was to employ two so-called magic lantern projectors which then would act as standardised light sources, and would send light through a coloured filter glass onto a glass screen. One lantern shone from behind, through the filter glass, into the observer's eye; the other shone from the side and had its light reflected onto the front of the plate. The intensity of the second image, polarised by reflection on the glass plate, could be changed by observing it through a polarising prism. Zöllner hoped that with standardised light sources, such as galvanically glowing platinum wires, the light absorption and thus the colour of dyes in test tubes could then be compared and measured.[11] Such an instrument would have met the requirements for standardised colour measurements in the dyers' laboratory. However, Zöllner had relied on Dove's assumption that two complementary colours, for example red and green, would produce grey. But the phenomenon of complementary colours was in fact more complicated and Zöllner's 'colorimeter', as he had called it, was never produced. Nevertheless, two questions had occurred to Zöllner when designing his photometer: how to produce a reliable light source, and how to eliminate the subjective eyesight of different observers. He worked on both problems during the next two years and chose them as a topic for his doctoral dissertation in 1858 and for his studies at the laboratory of the physicist Gustav Wiedemann at Basel University. Wiedemann, not much older than Zöllner, was a friend of the Zöllner family in Berlin and had just equipped an experimental laboratory in Basel. Wiedemann suggested that Zöllner investigate the relation between light emission of glowing wires and current intensity. It was probably Wiedemann who pointed Zöllner at the prize for astronomical photometry offered by the Academy of Science in Vienna, which encouraged Zöllner to design his astro-photometer.[12]

Textual and pictorial records allow only a limited understanding of historic practices. This is why the historiographical approach of replication and practice is introduced here: to help us understand Zöllner's photometer, its success, and its decisive role both for the development of astro-photometry and for Zöllner's professional career. Publications, both articles and books, by Zöllner are numerous. Apart from these publications, a few letters by Zöllner are known to exist. The original photometer, which still exists in parts, is kept at the Deutsches Museum in Munich, and a few later copies of the photometer can still be found in museums and observatories around the world. This allows us not just to study the original instrument, but also to practice with the replicas. Performing with a historic instrument is not uncommon in disciplines such as astronomy. Astronomers still employ old instruments and carry out measurements with them, mostly to judge the quality of old data obtained with these instruments. This approach is different from other scientific disciplines, which are often less historically continuous in their attention to historic instruments and measurements. One reason for astronomers' interest in historic instruments and practice is the singularity of astronomical events, for example the appearance of comets or novae or the changes in brightness of variable stars. The replication of the Zöllner photometer described here is based both on Zöllner's publications and existing letters, and the remaining photometers kept in Munich and elsewhere. I will describe the process of analysing the original instrument, its replication and finally the performance with it.[13]

I have already pointed out that Zöllner's photometer was designed for comparing an artificial light source with star light by means of polarisation, based on a design by Babinet and Zöllner's work with Dove. The measurement procedure involved matching the brightness and colour of the artificial star with a real star observed through a small telescope. A combination of three polarising prisms and a rock-crystal plate allowed the brightness and colour of the artificial image to be altered and the alteration to be measured. Through a glass plate in the telescope, the observer could see both the image of the real star

ZÖLLNER PHOTOMETER

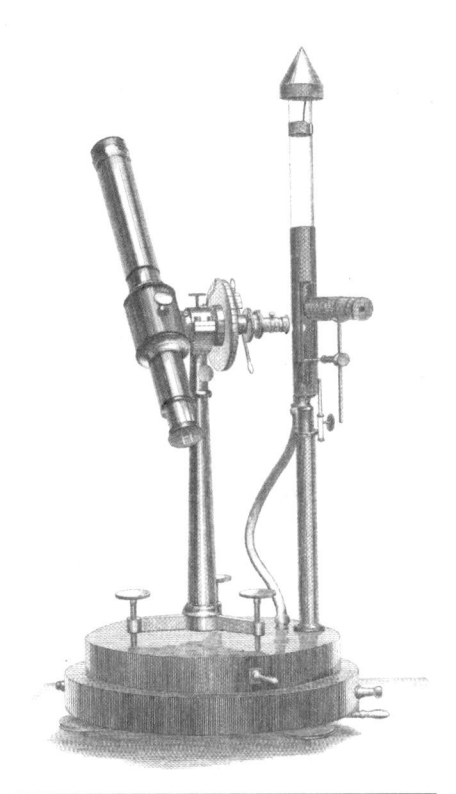

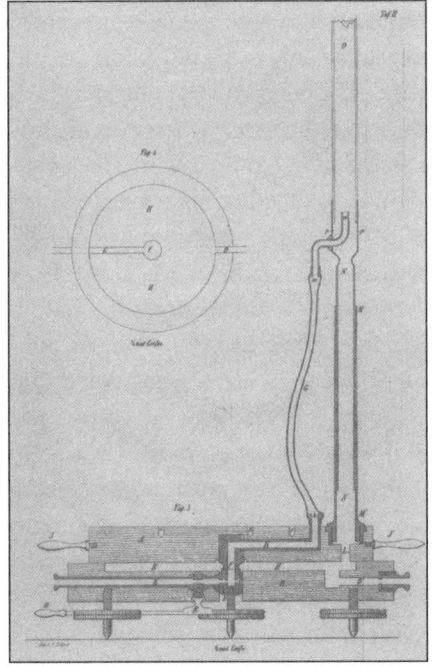

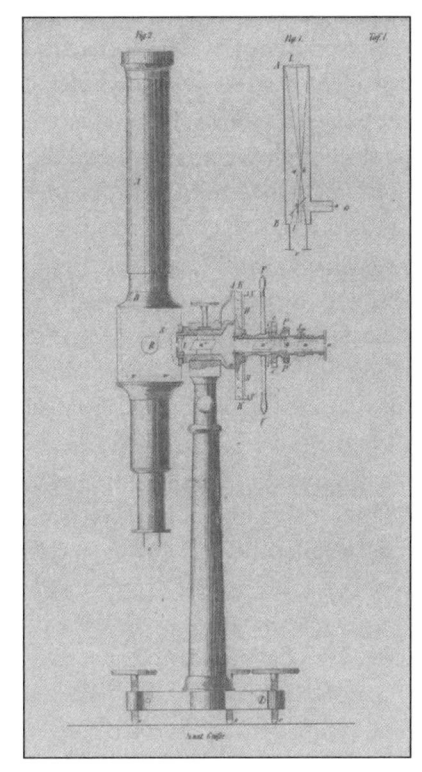

FIG. 10.1 (above left)

Zöllner published a prize essay about his photometer in 1861, shortly after it had been rejected by the Academy of Science in Vienna. It contained this image of the photometer.

FIG. 10.2 (above right)

This drawing from the prize essay illustrates the working of the photometer. Light enters the photometer from the right, travels through a prism, a crystal plate and two further prisms before it reaches the glass plate, where it is reflected into to the eye-piece and the observer's eye.

FIG. 10.3 (right)

This separate drawing demonstrates how gas enters the base of the burner and flows through the centre axle before it enters the burner. Air flow is secured by a second opening (on the right) and the air space in the base is intended to keep the air flow constant.

(Zöllner, 1861)

213

and the reflection of the artificial star. The analysis of the original instrument was carried out at the Deutsches Museum and the optics workshop of the European Southern Observatory in Garching. In the Garching optics workshop the object lens, the eyepiece lens, the prisms and the crystal plate of the photometer were analysed and their optical properties determined. The first item to be measured was the object lens which collects the star light and projects it into the eye of the observer. According to Zöllner he had used three object lenses of different diameter, of which only one still survives. These lenses, and with them the tubes, could be exchanged depending on the brightness of the objects to be observed. Object lens No. I had a focal length of 10.5 par. Zoll (1 par. Zoll = 2.71 cm) and a diameter of 1.5 par. Zoll; object lens No. II had a focal length of 20 par. Zoll and a diameter of 2.3 par. Zoll; and object lens No. III had a focal length of 38 par. Zoll and a diameter of 2.6 par. Zoll. A first measurement of focal length and diameter of the existing lens showed that it did not match any of the three lenses mentioned by Zöllner. However, with a focal length of 20 par. Zoll but a diameter of 2 par. Zoll it was close to object lens No. II, the lens which Zöllner used for most of his later observations, and one can assume that these lenses were similar if not identical. In the published drawings of the Zöllner photometer, the object lens is drawn as a single lens. A reflection test revealed that the object lens has four optical surfaces, indicating that it must be a doublet. Furthermore, it could be shown that the object lens was not cemented in but that the two separate parts were adjusted by the frame. For such an adjustment the frame is fixed around the lens with a small chisel and a hammer. Measurement of aberration, chromatism and transmission of the lens indicated that the lens must have had an optical quality comparable to today's amateur telescopes, although the transmission of the lens is ten per cent below today's values due to different polishing techniques. Compared to other mid-nineteenth-century lenses the object lens found in the photometer represented a standard of medium-sized quality lenses widely manufactured at this time.[14]

An analysis of the eyepiece lens showed similar results in quality to those from the main lens. This lens was used to magnify the image projected onto it by the object lens. Here again the focal length is different from Zöllner's published values. This is not surprising, since Zöllner had already suggested that different observers should use different eyepieces corresponding to their different eyesight. However, since the focal length of the eyepiece was 65 mm shorter than Zöllner's published value, another reason must have played a role. Reducing the focal length increases magnification, a factor which might have been more important for a historic observer. A replica of the eyepiece lens was purchased from a local optics workshop which could offer a lens comparable to the original. For the object lens the case was different, since I could not find a non-cemented doublet on the market. However, an optics workshop was found which was prepared to replicate the lens according to the measured mechanical and optical properties, including the optical curvatures and thickness of the lens. It must be mentioned that the replication of the lenses is not as difficult as one might assume, considering their age. This is due to two facts: first the technique for manufacturing these lenses has not changed considerably, and second the replicated lenses are not 'critical'. This means that their optical properties did not include extreme dimensions or proportions, e.g. the ratio of diameter and focal length. Finally, for a photometrical measurement all objects are observed in the centre of the field, usually the optically best corrected part of the lens.

Finding a suitable rock crystal plate was not a problem either. Zöllner had introduced a circular polarising rock crystal plate, first to eliminate the different colour sensitivity of different observers and later to determine the colour of the stars. It was described as 'cut perpendicular towards the optical axis, left-turning rock crystal plate'.[15] After an inquiry at the Mineralogisches Institut of Münster University in Germany about the possible replicablity of the plate, it was found that it would not be a problem to produce a plate corresponding to Zöllner's values for diameter and thickness and to mount it into the photometer replica. The thickness was slightly increased for practical reasons, but the colours shown by the replication were similar to the ones shown by the original photometer at the Deutsches Museum. Polarising crystal plates have almost disappeared from modern science today, whereas they were common devices in mid-nineteenth century optics. Work on polarisation had been stimulated by an increasing interest in mineralogy related to mining and scientific travel. Work with polarising media was an established part of physics teaching in the mid-nineteenth century, and instruments for demonstration of polarisation could be found both in lecture theatres and private cabinets. Demonstration of these phenomena was certainly part of the physics lectures attended by the young Zöllner at Berlin University.[16]

The rock crystal plate was then placed between two Nicol prisms which allowed a colour to be selected once the polarised light had been circularised by the crystal plate. Turning the first prism and the plate against the second prism allowed the observer to select the colour. Turning the two prisms and the crystal plate against a third prism changed the brightness of the image. The prisms used by Zöllner were called 'Nicol' prisms, named after the Scottish natural philosopher William Nicol.[17] Nicol had found that calcspar had not only the quality of being a transparent double refracting medium, but that it could be used to produce a polarised image. Nicol's idea was to cut the calcspar in two halves at a particular angle and cement the two halves together again with a substance called 'canada balsam'. The cutting and polishing of the optical surfaces at a particular angle would refract the polarised image in such a way that the image would leave the prism exactly on the opposite side. To avoid further reflections within the prism, the lateral surfaces were left rough or painted black.[18] Reproducing these prisms was not an easy task as Nicol had already pointed out at the time. Nicol had been a skilled experimenter known for his experience in cutting and slicing microscopic samples. Cutting calcspar required techniques and tools which were available to nineteenth-century practioners, but are hardly in use today. Calcspar was a standard material in the nineteenth century and used in large quantities, but does not play any significant role in today's science.[19] For the photometer replica, no optician could be found who was able to re-manufacture the Nicol prisms in a quality similar to the ones reached 150 years ago within the time and financial scale of the replication project. When Zöllner ordered the Nicol prisms for his instrument, they were standard devices in laboratories and available 'off the shelf'. Many of them were in use during the nineteenth century and some still exist, e.g. in remaining later copies of Zöllner photometers. That means it is possible to use original Nicol prisms, just as they were available in the middle of the nineteenth century, in a present-day replica. Two observatories were prepared to loan their original Nicol prisms from later Zöllner photometers to be used with the replica: Archenhold Observatory in Germany and Uppsala Observatory in Sweden.

Although obtaining and using Nicol prisms was not difficult for Zöllner, the problems in restoring such prisms are worth noting since they show how local material culture can vanish over time. When the prisms from Archenhold Observatory arrived, it was found that only one of the three prisms could be used with the photometer replica, since the other two showed spots on their cemented surfaces. The closer the original prisms were in relation to the artificial light source, the more the cement had been affected. In both prisms the spots could be found where the light beam of the artificial light source had passed through the surfaces. It is likely that the prisms from the Berlin photometer were used with an electric light source. The intensity of the light beam might have affected the condition of the cement. I decided to use the third prism, which had not been affected, the one furthest from the light source.

The prisms provided by Uppsala Observatory also showed signs of damage: all their surfaces had suffered from scratching. Since the cleaning of optical surfaces is a difficult and costly affair, I decided to restore only the two prisms which could not be used from the Berlin set. A first inquiry with a Berlin-based company that specialised in the restoration of optics gave an estimated price of £200 (in 1995) for the cleaning of one optical surface. This amount was beyond my project budget. Thankfully, the Belge workshop which had already carried out the replication of the object lens offered to help me with the restoration for a price of £200 for restoring all four optical surfaces. However, this workshop did not have much experience with historic prisms and the material characteristics of calcspar. This made the restoration challenging in that it demonstrated how a twentieth-century optician would approach a historic optical device. We were aware that calcspar is a fragile mineral: I had already tried to cut a piece of calcspar from Brazil, which had proved unsuitable for optical use due to its opacity and colour. To absorb the vibrations of the polishing machine we mounted the prisms in gypsum. Despite taking this precaution, one small edge of a prism broke off. The full damage done to the prisms became visible only after they were taken out of their mounts: the already brittle cement had not stood up to the vibrations of the polishing machine, and the two prism halves had fallen apart. The optician, who was not aware of the refracting index of the canada balsam and its importance for the polarisers, cemented the prisms with a modern cement. Now both light beams were reflected out of the prism, which rendered the prisms useless for photometry. No knowledge existed as to how the modern cement could be dissolved. Finally, we decided to put the prisms in a highly concentrated cleaning solution, hoping that the cement would dissolve before the calcspar. This proved successful. After the cleaning of the internal surfaces, the prism halves were cemented together with canada balsam. The resulting optical and mechanical qualities were now comparable to the nineteenth-century originals. The prisms were finally set in cork frames to avoid further damage and mounted into brass frames as had been done 150 years ago.

Nicol prisms were established devices in nineteenth-century astronomy. This lasted until the 1930s when the last Zöllner photometers were built; after this time Nicol prisms were hardly produced and their material culture declined. The demand for calcspar had led to the exhaustion of the known geological sources. It has been pointed out that for an acceptance of such devices by experimenters they need to be 'unproblematic objects'.[20] Nicol prisms in nineteenth-century astronomy were not considered as 'problematic'. They were continually used but were hardly ever questioned. In fact, in the history of the Zöllner photometer the reliability and accuracy of these prisms was challenged only twice, but neither criticism received any significant response from astronomers. The first occasion on which the accuracy of the prisms was challenged was in 1865 when Zöllner's examiner for his inaugural thesis, the Leipzig physicist Wilhelm Hankel, objected to the underlying accuracy that Zöllner was assuming for his measurements.[21] This question was repeated by the director of Leipzig Observatory, Bruhns, some ten years later. Bruhns discussed the photometer at a meeting of the German Astronomische Gesellschaft in Leiden where both Zöllner himself and observers who used the Zöllner Photometer were present. However, all argued that 'the shortcomings attributed to Zöllner's Photometer were groundless and could not be found in the copies being used'.[22] This points to a fact which has been investigated by the historian of science Kathryn Olesko in her work on nineteenth-century scientific practice: though scientists in the German states were concerned with the accuracy and reliability of their instruments, the application of these criteria very much depended on individual scientists and disciplines.[23] In astronomy, polarisers remained mainly unchallenged, whereas in physics they never became fully accepted.

Further optical components used in the Zöllner Photometer included two lenses which were inserted between the third prism and the screen and between the first prisms and the artificial light source. The first lens was a convex lens introduced for projecting the artifical image onto the screen. Zöllner himself gave an inaccurate value for the focal length ('30–40mm'), but the value could be determined in the original instrument from the distance between lens and screen. The convex lens between the artificial light source and the prism was meant to focus the artificial image and make it appear more 'star-like' by reducing its size. However, it is not clear when Zöllner introduced this lens, where exactly he placed it, and when he decided to use it for his observations. Moreover, the introduction of this lens with its extra focal length required fundamental changes in the design of the instrument which would have changed several of its other features. Most importantly, this lens was not found in the original instrument at the Deutsches Museum, and it could not be seen where it was inserted. Therefore it was decided not to replicate it, but to see first if the artificial images appeared star-like or not.

The last optical component to be replicated was the glass screen reflecting and projecting the artificial image into the eye of the observer, while at the same time allowing the natural image to be seen by looking through the plate. This also allowed the observer to see both images against the same background, so that different background brightnesses would not affect the measurements. The glass plate was

FIG. 10.4

The Zöllner photometer as it was found at the Deutsches Museum in Munich. Note that the original gas burner has been replaced with a kerosene burner and counterweight, and a large card board tube has been added.
(Klaus Staubermann)

FIG 10.5 The head of the workshop that built the replica, the Deutsches Museum's curator, and the Museum's conservator examine the original photometer and match it with historic drawings. This exercise was decisive in securing the quality of the replica.
(Klaus Staubermann)

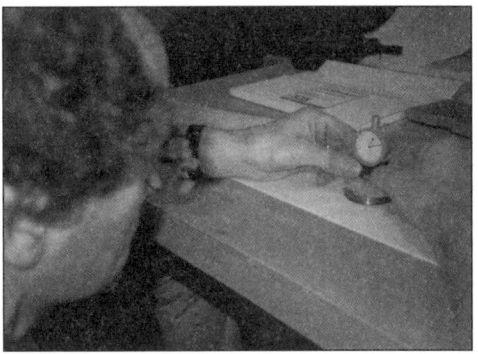

FIG. 10.6 The lenses of the photometer were measured at the European Southern Observatory in Garching (here the curvature). The tests were useful as they revealed that the quality of the historic lenses did not differ too much from those today.
(Klaus Staubermann)

5mm thick and of a quality comparable to present plates. Therefore a standard glass plate from an optics supplier was chosen for the replica.

Apart from its optical components, the photometer itself was mainly manufactured from brass, one of the most commonly used materials in nineteenth-century instrument making. This includes the foot which can be levelled by means of three screws, the lower part of the tube which can be moved around a horizontal axis, an adjustable counterweight to compensate the weight of the tube and the two wheels for turning the prisms and crystal plate. Both wheels have a graduation for reading the turning angle; the small one divided into 100 degrees and the big one divided into 360 degrees, combined with a 'Nonius' (vernier) allowing for reading of one tenth of a degree.

An investigation of Zöllner's published technical drawings of the photometer showed that they could not always be relied upon: matching the drawing of the photometer itself and that of the gas burner showed that the position of the first prism and the diaphragm would have been located somewhere inside the gas flame.[24] This was one of the reasons why it became essential to measure the original photometer at the Deutsches Museum. To avoid possible communication errors which might affect the replication, it was decided that the engineer in charge of the replication of the photometer should measure the original photometer himself.[25] This has been shown to be a great advantage since when reproducing the replica the engineer could refer to his own notes and measurements.[26] Independently taken measurements of the instrument confirmed that the accuracy of the reproduced replica was comparable to that of the original instrument. The only non-brass parts were the three different tubes which were made from layers of paper. This helped reduce the weight of the tubes and kept the photometer balanced when moving the tubes. For the replica, only one tube has been replicated from black cardboard paper.

Although most parts of the original instrument still exist, some needed to be substituted. It was not always clear from the existing drawings where or how these parts were inserted. These parts included a counterweight, which needed to be added in order to compensate for the weight of the tube. Apparently Zöllner only became aware of the need for such a weight when he replaced the initial tube of the photometer by a larger one. In the remaining instrument at the Deutsches Museum, the counterweight is missing and only from the instability of the instrument itself and from drawings of later instruments could its existence be detected. Furthermore, one out of the three screws for fixing the scale wheel opposite the tube is missing. This indicates where the counterweight could have been added. I built the counterweight from brass which now compensates the weight of the replicated tube and does not disturb the observer when taking his measurements.

That the different tubes were intended to adjust the instrument to the varying brightness of the stars is apparent from Zöllner's description of the instrument. Due to the design of the instrument, the possibility of adding large tubes was limited.[27] One method of overcoming this problem is manifested in the changed design of the photometer: Zöllner added an opening on one side of the tube allowing for the insertion of an eyepiece. Looking through the eyepiece, the observer could see the artificial star through the glass screen and the real star projected onto the screen from the front. By applying this method, less light from the real star would be lost because it was only reflected and fainter stars could be observed.[28] I had noticed the opening when investigating the original photometer at the Deutsches Museum, but its function only became clear later while practising with the replicated instrument. Only when I applied the photometer on stars fainter than the ones originally measured by Zöllner did I become aware of the need for such a device and the idea behind the replicated opening. This helped me to trace the idea to another inventor, the Moscow observer Witold Ceraski. Zöllner had applied the change later, either based on his own experience or after he had read about Ceraski's suggestion.

One of the essential parts of the photometer was its wooden base which acted as a gas regulator at the same time. Mounted on it was the burner, a gas burner with a separate gas and air supply, the latter in order to increase the brightness and stability of the flame. The upper part of the base with the photometer on it could be turned around a vertical central axis, moving the burner at the same time and keeping the gas supply constant. The gas supply could be regulated by means of a tap, or cock, on the bottom of the foot. The burner could be adjusted in height to ensure sufficient light would enter the photometer. Furthermore, a small telescope was added to the burner to calibrate the height of the flame. A glass cylinder with a brass cap protected the flame from wind. Finally, the base was hollow in order to function as an air tank which would supply the flame with a constant amount of air. A pipeline could be connected to the tank and its other end kept in an area protected from wind.

The wooden base was one of the most challenging parts of the replication: it does not exist anymore, nor does Zöllner give sufficient information for its replication in his drawings or writings. The most difficult problem to tackle was the gas cock on the bottom side of the base. Zöllner's drawing gives a side-view of the cock, which does not explain the functioning of the device. The workshop which replicated the stand suggested using a modern valve instead of the cock. Since the functioning of the cock and its use with the instrument was not clear yet, I tried to find out about its possible functioning by rebuilding it. From the drawing it could be seen that the gas entered the burner through a large hole on the side of the cock. However, a hole of a size which would allow a free flow of the gas would have made the structure of the cock very fragile, especially considering that the cock also served as centre axis of the base. A smaller hole would have made it difficult to control the gas supply properly. I decided to design a cock with a larger number of small openings, which would allow a free flow of gas independent from the relative position of the cock's handle and still give the axle a solid structure.

Furthermore, Zöllner's drawings only showed the brass components of the gas cock. To prevent the gas from leaking into the air tank or the environment I added a rubber washer. These details also help us to pinpoint the date Zöllner produced his drawings: they were surely drawn after the construction of the instrument, since they included information about the construction which must have resulted from the building process (like the gas cock). On the other hand they must have been drawn before the instrument was put into practice, since they do not show features described by Zöllner after he began his observing. When I replicated the base, it also became clear that the instrument had been intended for outdoor use. I neglected to varnish the wood inside the air-tank because I considered the inside of the tank sufficiently protected from the environment. This proved to be a mistake, since immediately after taking the photometer out into the cold and humid night air the tank began to show signs of wear and the foot became difficult to rotate. This problem was solved by adding a thin layer of felt between the upper and the lower part of the foot. Although Zöllner could not prepare himself for all the problems related to the new environment, he still tried to take precautions.

Otto Sibum and other historians re-working historic experiments have frequently pointed out that it is impossible to rebuild an instrument similar to

FIG. 10.7
The replica, apart from the optics and the gas supply, was built by the workshop of Oldenburg University. This workshop over the years has built a reputation for constructing authentic scientific instrument replicas. The instrument to the very right is a replica of Fizeau's apparatus for the measurement of the speed of light (1850).
(Klaus Staubermann)

FIG. 10.8
The completed replica ready for observing at Cambridge Observatory.
(Jan Frercks)

the original. However, before one compromises on the similarity of a replication, one needs to investigate the functioning of the part in question. A part which is essential for the functioning of the whole instrument should be replicated as closely as possible to the original. Compromises regarding design and materials can be made where they do not affect the historic practice with the replica. One example for a possible compromise is the collimator telescope used by Zöllner to calibrate the gas flame. The calibration telescope was a precision device and a laboratory standard in its time. However, trials with various collimator telescopes showed that one could compromise by edging the millimetre scale of the calibration telescope into the glass cylinder surrounding the gas flame. For the observer it does not make a difference if the scale is visible inside the telescope or on the cylinder. The changed design of the replicated telescope and scale did not affect practice with it, but saved me a considerable amount of time and money.

Zöllner had experimented with different types of light sources and had tried to establish galvanic glowing wires as a standardised light source during his doctoral work. However, his experiments at the time had shown that gas still provided the most reliable and constant illumination. This gas, called 'town gas', was produced from burning coal. Zöllner's laboratory in Basel, where he carried out his photometric experiments on galvanic glowing wires and also tested his new photometer, was connected to the town's gas supply. This was not the case in Schöneweide near Berlin, where in 1859 he began observing. However, the building, his late father's dyeing factory, had its own gas supply. According to Zöllner it used bituminous coal imported from England, and stored enough gas in the tank to supply his gas burner for one night of observation.

No records exist about Zöllner's gas production equipment, but there is general information about gas production and the composition of nineteenth-century town gas. Interestingly, neglecting the power supply is common practice by experimenters. When, for instance, the French physicist Hippolyte Fizeau described his experiments on how he determined the velocity of light in 1849, he took it for granted that his readers would know what limelight was and how it could be made bright enough to be seen over a distance of 16 km. From contemporary sources on nineteenth-century town gas, we know that its illuminative power was around ten times higher than that of a candle. Furthermore, analysis of the gas had been carried out during the nineteenth century on which a reproduction could be based. Since gas production equipment was not available to me and contemporary town gas in Britain produces flames different both in colour and brightness because of its use for heating, it was decided to mix the gas from separate components found in nineteenth-century town gas.[29]

Due to the high price of industrial gas I decided to order the three main components of nineteenth-century town gas first: hydrogen, methane and nitrogen. Although this mixture had a comparable calorific value, it was found that it was not illuminative enough. Actually, the flame appeared much fainter than a candle's flame. Therefore, a fourth component, ethylene, was proportionally mixed with the hydrogen, methane and nitrogen – and it produced the right colour and brightness: the colour of the flame Zöllner had described 'as red as the planet Mars near the horizon'.[30] The replicated gas does not contain as many impurities as the original gas might have contained. This does not affect the brightness of the flame much, but might have affected its stability. However, as we shall see later, air turbulence has the strongest influence on the stability of the flame when the instrument is taken outside.

The reproduction of the material culture of the photometer did not cause any fundamental problems and largely matched the existing textual evidence. The difficulties in producing devices like the Nicol prisms were caused by the contrast between their wide use in the nineteenth century and their limited use today. The material culture from which they originated is not lost, but has become more difficult to access than it was 150 years ago. However, this example shows how closely natural resources and scientific practice are related. One can argue that the Zöllner photometer was the result of mid-nineteenth-century engineering practices. It was the outcome of established materials and techniques combined to be applied on a defined problem. The process of replication, which took me one and a half years, provided me with one fundamental insight: by reworking the photometer it turned from an astronomical instrument into an industrial device. But let me first describe the performance with the replicated photometer.

The essential task after reproducing the replica is the performance with it. Performance here means carrying the replicated photometer out of the workshop under the night sky. Setting the instrument up outside requires first of all a table or a surface small enough to allow moving around the instrument while observing stars in different directions, and high enough to allow a comfortable position for the observer. A suitable observing site also needs a gas supply which must be protected from the weather. Zöllner had chosen his late father's factory outside Berlin as an observatory site. The gas supply was that of the factory, which at night he had exclusively for his own use. His observatory was placed on the roof of the factory, a one- or two-storey building. The roof of the room where the photometer was kept could be partly removed. The site itself was overlooking the river Spree and the Spreewald pine tree forests. The river would have caused fog which could have restricted Zöllner in observations near the horizon. The replicated photometer was set up on a site as comparable as possible to Zöllner's, the two-storey building of the old Cambridge Observatory. This site is located a few kilometres from the town centre with the river Cam nearby. Furthermore, the building is protected from the light of the town and traffic by surrounding trees. Since a removable roof comparable to the one used by Zöllner was not available, I decided to keep both the photometer and the gas equipment in the dome on top of the observatory, and to carry the photometer out onto a balcony around the dome for observations. This procedure takes only a few minutes and the photometer stays constantly connected to

FIG. 10.9

This close-up of the replica gas burner shows the arrangement of glass cylinder, calibration telescope, and diaphragm. The smoke caused by the flame has largely blackened the upper part of the glass cylinder.

(Marina Frasca Spada)

the gas supply. Weather in Cambridge is not only very changeable, it is also generally quite windy. With the instrument kept outside the building for measurements, the flame is more susceptible to wind than in Zöllner's original set-up. Another influence to consider was the humidity of the surrounding air; the environment in Cambridge was probably similar to the conditions close to the river Spree in the 1860s: both places are near the banks of rivers and affected by fog and humidity in the mornings and evenings.

About the weather conditions which Zöllner faced between December 1859 and September 1860, we have a detailed record both from Zöllner himself and the meteorological office in Berlin.[31] It is not surprising that these meteorological observations were carried out by the same Carl Bruhns, who soon after became the director of the new Leipzig observatory and Zöllner's employer. Throughout the nineteenth century, and before meteorological and astronomical recording was considered inseparable, Zöllner had argued that his observations were limited due to bad weather at that time. This is partly supported by the fact that both autumn and winter of 1859 and the year 1860 seem to have been extraordinarily rainy and cold. On the other hand, Zöllner did not observe on many nights which had been reported as clear, and also did not make use of the whole night even when weather conditions permitted. Zöllner himself reported that he had to finish observing early during some nights due to seasonal changes and early morning fog, but he also tended to

FIG. 10.10

The author in the process of preparing a night of observing. To the left the gas pipe leading to the gas cylinders within the building. The lens is still covered for protection from dust and the gas cylinder gas been removed for lighting the flame.

(Marina Frasca Spada)

FIG. 10.11

The author demonstrates how to calibrate the height of the flame. Obviously, this would have been done after dark. Once it was dark the observer had to rely on his ability to move and act in the dark.

(Jan Frercks)

complete observing around one o'clock in the morning even when the sky was apparently clear. No other explanation could be found for this fact than Zöllner's personal attitude. Therefore, it is not surprising that the Vienna Academy criticised his limited number of measurements within the time given to the applicants for the prize work. But Zöllner's attitude is not difficult to comprehend. When I did the re-observing I found it meant spending many hours at the observatory, waiting for clear skies. My social life came to a halt when I tried to combine observing, studying – and sleep. Often the sky cleared up around sunset, before it clouded over again about one hour later. For me this meant cycling to the observatory every night to set up the equipment, only to have to pack it away again after an hour or so. Cycling to the observatory took me twenty to thirty minutes; for Zöllner, who lived in Berlin, travel must have taken even longer. Furthermore, Zöllner was a physicist who was used to a laboratory environment and who probably was neither used to nor prepared to spend much time observing at night. At the time Zöllner did his observing, he was also writing articles and carried out physiological experiments, which included the design of new instruments. So it is scarcely surprising that he abandoned his observations in the middle of the night.

Confronted with a new environment, Zöllner had to solve several problems: moonlight illuminated the sky, something which he had not had to contend with before, in his laboratory; the cold froze his equipment, mostly affecting the gas supply; humidity led to the condensation of water, both on the lens and on the glass cylinder of the burner; and, finally, wind affected the stability of the flame. Apparently Zöllner learned to control all these factors. When the replica was placed on the balcony, similar problems occurred. The humidity of the air in February, at the beginning of the observations, was so high that water immediately condensed on the hot glass cylinder surrounding the flame. The water ran down on the inside of the cylinder into the flame, causing a flickering to the point of extinction. The lens did not show any traces of condensed water, which was probably due to the fact that the dome where the photometer was kept during daytime was not heated. The condensation problem was only solved when temperatures rose higher and the air became less humid in spring. Another problem was the wind which affected the flame. Although wind does not directly affect the flame, gusts cause fluctuations of the air supply and therefore of the intensity of the flame. Controlling the unprotected flame when lighting the gas can be a rather unpredictable enterprise: since the pressure of the gas in the pipe can only be guessed from the high-pitched noise emitted at the tip of the burner, much care needs to be taken when lighting the gas. The flame can become as high as 30cm, and a gust can easily blow it against the instrument or the observer. On one occasion the flame was blown against the photometer's cardboard tube by an unexpected gust, and part of the tube was charred. These risks can only be overcome by careful precautions, like always standing at the far side of the instrument when igniting the gas flame. Zöllner's idea of a remote air-supply for the burner via a pipe was tried but did not solve any problems. This was partly due to the limited diameter of the pipe, and partly to the fact that air also fluctuates around the flame, coming in through the upper opening of the cylinder, which needs to remain open. Putting wind shields around the whole equipment to protect the photometer in an effort to recreate the conditions in Zöllner's observatory was more successful. It became clear that Zöllner had learned to both expose and protect his photometer from the new environment.

Moving in the dark is another challenge faced by the observer. Although the equipment can be set up during dusk, further adjustments need to be made in actual darkness. A candle might help, but it dazzles the eye for some time. Only repeated experience of moving around the location can help the observer to get used to working in the dark. However, a small amount of light is produced by the gas flame itself, and by reflecting it with a white sheet of paper the observer can read the scale of the polariser and the colorimeter. A typical observation starts as follows: set the photometer and its base up on the outside table, remove the cover from the object lens, open the gas cylinders to a fixed pressure controlled by a manometer, remove the protective glass cylinder, light the gas, put back the glass cylinder on the burner and adjust the height of the flame for the next half an hour or so. Interestingly, with calm conditions and if protected, the flame showed a significantly better stability than expected from Zöllner's records and from the objections of nineteenth-century observers before they had employed the photometer. Though I had previously decided to adjust the flame after each measurement, I soon found that this was not always necessary. Not only the photometer but also the gas supply became 'stable' during the measurements.

Stabilising the gas supply today is probably easier than it was for Zöllner in the late 1850s. Zöllner had to control a complete gas production apparatus, whereas I could order my gas in industrial cylinders. However, installing the cylinders at the observatory was not an easy task either. This was partly for safety reasons: the weight of the large cylinder was 62.5 kg. At this time, a person employed by the observatory was only allowed to lift a maximum weight of 20 kg. Hence, four people would be needed to carry the cylinder upstairs, too many for the small staircase leading to the dome where the photometer was kept. The problem was only overcome when three people of unknown identity carried up the cylinder in the absence of the health and safety officer. On the other hand, no objections were expressed about my operating with an open flame on top of a historic semi-wooden building housing the observatory's impressive library. It can be argued that health and safety awareness and concerns change over time as well.[32]

Taking measurements is relatively trouble-free once the instrument has been set up. Before the first observation, the tube containing the object lens needs to be focused so the natural star appears sharp. (For the artificial stars this could already be done in the workshop.) The glass screen might have changed position when moving the instrument, in which case it can be re-adjusted by means of a screw connected to its frame. Then the photometer is pointed at the star to be measured. The horizontal movement is carried out by turning the base of the photometer around its central axis. The vertical movement is carried out by turning the tube with the integrated screen and eyepiece around the optical axis of the artificial image. For objects observed near the zenith a rectangular reflection prism can be fixed behind the eyepiece to allow the observer a more comfortable position. The instrument is ready for measurement when a natural star can be seen in the centre of the field between two artificial stars of apparently equal brightness.

First, the observer chooses a colour for the artificial star which is equal to the colour of the natural star by turning the wheel containing the crystal plate and the first prism, the colorimeter. The colours of the artificial stars are brighter than those of most real stars. This is disturbing in the beginning, and practice is needed to get accustomed to these colours. The observer selects the colour closest to the colour of the real star. This involves turning the colorimeter back and forward several times to select the most fitting colour. The same is done for brightness measurements: the Nicol prisms in their wheel are turned backwards and forwards until a similar brightness is found.[33] Once this position has been set, the corresponding angle can be read from the scale. As I have already pointed out, a white sheet of paper can be used to reflect light on the scale for better reading.

Choosing the right diaphragm for producing the point-like artificial star image seemed insignificant at first, but it turned out to be crucial for the quality of the measurement. The choice to be made was to find a diaphragm which would give a stellar appearance of the artificial star. Since I did not know which diameter would provide the most star-like appearance, I produced three diaphragms with apertures of 0.1, 0.3 and 0.5 mm. The images produced by the diaphragm with the smallest aperture proved to be too faint, and the ones produced by the 0.5 mm too large in appearance. I figured that a convincing appearance of the artificial star was more important for obtaining accurate measurements than its actual brightness, since the latter could be partly corrected by the choice of tubes and the use of the Nicols. Zöllner used three different telescope tubes and varied their aperture. He did not outline this practice, but indications of it can be found in his measurement records. Since I had only one tube available, I varied the aperture by the use of cardboard diaphragms I placed in front of the lens. Making these choices required skills which I seemed to develop with practice.

The choice of lenses and diaphragms also determines to what extent the observer can control his perception. Here, it is important that a sufficient brightness interval for manipulating both the real and the artificial star is provided. This means that the observer needs freedom to increase and decrease the brightness of both stars in order to find the best match. A too limited interval can lead to a poor measurement in the sense that a brighter star is measured as too faint or a faint star as too bright. This error occurs when the comparison star's brightness cannot be increased or decreased equally. It seemed from my observations that when the artificial star cannot be made bright enough, the real star will be estimated as too faint. When the artificial star cannot be made as faint as required, the real star will be estimated as too bright. Apparently the observer subconsciously tries to compensate for the limited control interval, which then leads to an under- or over-estimation of the real star's brightness. This possible error later became know in astronomical photometry as 'perception error' (Auffassungsfehler, 1899).

We know that Zöllner carried out his measurements by turning and reading the scale eight times, four times on each side of the scale, for each star. Later he reduced this to four measurements for each star. The same was done with the replica: since no systematic deviation was found when taking readings at opposite sides of the circle, the number of measurements was reduced to four. The turning angle of the brightness scale was read, and then the scale turned to an arbitrary position to repeat the same procedure again three more times. Time was allowed to let the eye adjust to the stars' brightness after it had been dazzled by the light of the flame when reading the scale. If the flame had been controlled by means of the small telescope and the millimetre graduation on the glass cylinder, even more time would be allowed to let the eye adjust to the darkness again. Since changes in the height of the flame might have affected its brightness, after each adjustment of the flame's height a new sequence of measurements was started.

The idea behind a sequence of measurements was to measure the brightness of one star and to use this as a reference for all other stars of the same sequence. The number of stars in a sequence depends on the stability of the flame and the movement of the photometer. Any movement of the photometer could affect the quality of the flame and therefore measurements were limited to the same area of the sky. Zöllner was not an astronomer and he often omitted to choose new reference stars for different measurement sequences. Furthermore, when testing his instrument, Zöllner did not move beyond drawing an interpolation curve for the values he had measured and did not see the need to carry out an error analysis or derive any conclusions from his interpolations. Zöllner saw his measurements more as a practical test of the instrument than as part of a possible stellar brightness catalogue, as wished for by the Vienna Academy of

Science. His measurements included many incompatible reference stars which made a later use of his data by other astronomers almost impossible.[34] He carried out observations on moonlit nights which affected the consistency of his measurements. Also, he did not take into account the extinction of the earth's atmosphere sufficiently (he tried to compensate by choosing stars not too close to the horizon). Furthermore, Zöllner published the intensities which he derived from the rotation angle of the prism but did not give the actual values of his measurements, which makes it impossible to fully understand his data. Finally, as I have pointed out earlier, no observation notebooks are in existence anymore. It must be sufficient to remark that sometimes my measurements seemed more accurate than Zöllner's, sometimes less, but always within the limits of the deviations assumed by Zöllner for his own measurements.[35]

The fact that the reworked measurements did not seem to deviate much from the originals can partly be explained by the reconstruction procedure. When I replicated the instrument, I questioned every part of the it. Questioning the functioning of every part of the instrument made it easier to put it into practice. This would have been different for a user who was not accustomed to the functioning and modes of practice of the photometer. Rebuilding Zöllner's photometer and redoing his observations was crucial for the understanding of his observational practice. Reworking Zöllner's photometry also helps us to explain how and why Zöllner approached the sky the way he did. Before I replicated the photometer I perceived it, like many historians before me, as an astronomical instrument and focused on features such as lenses and the mount. It was the practice with the instrument which changed my perspective as a historian. Considering what caused 'trouble' while building the replica of the photometer and practising with it, my focus changed from the 'telescopic' or astronomical features to the features connected with producing the artificial image and how to control the gas flame. The object lens, considered at first to be the most important part of the photometer, turned out to be of only limited historiographic significance. On the other hand, by practising with the photometer it turned from an astronomical instrument into a laboratory device. This can be considered the most important insight gained during the replication process. Zöllner, during most of the twentieth century, had been perceived as an astronomer because of his contributions to astro-photometry and his first professorship in astrophysics. Rebuilding Zöllner's photometer and redoing his observations changed this perception. Both Zöllner's observational practice and his instrument originate from industrial laboratory practices. It was his family background in the dyeing business and his interest in industrial light and colour measurements that drove him to designing new types of instruments. The Nicol polarimeter was an established laboratory device in his time and widely used at universities and in industry. Zöllner's achievement was to engineer a device that could be moved out of the laboratory and under the night sky and generate standardised photometric measurements of the stars.

The story of Zöllner's photometer must be considered as part of a broader development which took place in Germany and elsewhere during the second part of the nineteenth century.[36] Laboratory researchers designed highly specialised instruments for both industrial and academic uses. Although these instruments were designed for highly specific purposes, their design was flexible enough to respond to a wide range of applications and practitioners.[37] Multiple design features enabled the application in diverse environments. What these new types of instruments had in common was their intended use for standardisation and control, both in science and in industry. As the polarimeter had helped chemists and dyers to standardise and control production and manufacturing processes, it also helped to standardise astronomical measurements and control observers' perception.

Notes

1 Koerber (1899), p. 22.
2 Herrmann (1974), p. 3.
3 Gage (1995), p. 14.
4 Kemp (1990), p. 306.
5 Bernoulli (ed.) (1850), p. 225.
6 Müller (1897), p. 243.
7 Babinet (1853), p. 774–75.
8 Schütt (1992).
9 Dove (1853).
10 Dove (1852), p. 69.
11 Zöllner (1857), p. 381.
12 Koerber (1899), p. 3. However, the award seems to have been public knowledge at that time and was announced in general newspapers; see for e.g. *Illustrierte Zeitung*, vol. 24, no. 608, 24 February 1855, p. 138.
13 I am indebted to Oldenburg University for their generous support of this replication, which made this research project possible.
14 Rucks (1996), p. 954.
15 Zöllner (1861), p. 14.
16 Dove (1853).
17 Morrison-Low (1992), pp. 123–31. National Museums Scotland keep a wide range of nineteenth-century Nicol prisms.
18 Nicol (1829), pp. 83–84; and W. Nicol (1840), p. 238.
19 Simon Schaffer has already pointed to the difficulties involved in the replication of prisms which in his study made it impossible for Italian physicists to replicate Newton's prism experiments. Schaffer (1989), p. 67.
20 Schaffer (1989), p. 99.
21 See Hankel: *Habilitationsgutachten Dr. Zöllner, Johann Carl*, Universitätsarchiv Leipzig, PA 1093.
22 *Vierteljahresschrift der Astronomischen Gesellschaft*, vol. 10 (1875), p. 234.
23 Olesko (1991), p. 221.
24 The matching was done by photocopying all drawings on foils in the same scale. Then the different drawings could be combined easily.
25 The need to have the workshops' engineers 'on site' for a successful construction of an instrument was already known in the nineteenth century. Simon Schaffer has pointed this out in his study of Maxwell's work on the replication of scientific experiments. Schaffer (1992), p. 34.
26 Communication with the historians replicating experiments at Oldenburg have revealed that problems related to the replication process can occur due to misunderstandings between the historian and the engineer supervising the

27. However, later users redesigned and connected their photometers to telescopes large enough to carry the photometer.
28. Information on Fizeau's experiments I owe to Jan Frercks, Lüneburg.
29. Bunsen and Roscoe (1857), p. 43.
30. Zöllner (1861).
31. Bruhns: 'Meteorologisches Tagebuch der Sternwarte Babelsberg, Sternwarte Babelsberg', Oktober 1859 to November 1860, Berlin-Brandenburgische Akademie der Wissenschaften, Akademiearchiv, Signatur 679.
32. After an inquiry with the health and safety representative I was supplied with a small ladder which would enable me to climb from the dome down to the roof of the main building of the observatory. From there I could either jump into the rose beds five metres below or wait for the arrival of the fire brigade.
33. For a report of this procedure see: Lord Rosse (1876), p. 92.
34. One does find a few stars which Zöllner observed more than once and which enable us to connect at least a few of his measurement sequences. We have tried to reduce Zöllner's measurements for current photometric research based on his and others observations: Sterken and Staubermann (eds) (2000).
35. Every measurement was preceded by the choice of a diaphragm balancing the brightness of the real and artificial stars. The pressure of the (replicated) town gas was set to 0.5 bar and the pressure of the ethylene to 0.75 bar. I usually waited half an hour after the gas had been lightened to make sure the flame is stable. Zöllner's revised measurements based on my practising with the replicated photometer can be found in Sterken and Staubermann (eds) (2000).
36. D. B. Herrmann and D. Hoffmann have convincingly argued that the development of astro-photometry must be seen in the broader context of nineteenth-century industrial gas-lighting technology. Herrmann and Hoffmann (1976), pp. 94–104
37. Shinn (2001), pp. 29–48.

Bibliography

Babinet, J. (1853): Note descriptive du photomètre industriel, in *Comptes Rendu*, vol. XXXVII.

Bernoulli, Chr. (ed.) (1850): *Neue Encyklopädie der Wissenschaften und Künste*, vol. 1, 4.

Bruhns, C.: 'Meteorologisches Tagebuch der Sternwarte Babelsberg, Sternwarte Babelsberg', Oktober 1859 to November 1860, Berlin-Brandenburgische Akademie der Wissenschaften, Akademiearchiv, Signatur 679.

Bunsen, R. and H. Roscoe (1857): 'Photochemische Untersuchungen II', in *Poggendorff'sche Annalen der Physik und Chemie*, vol. C.

Dove, H. W. (1852): *Über den Einfluß der Helligkeit einer weißen Beleuchtung auf die relative Intensität verschiedener Farben, Bericht über die zur Bekanntmachung geeigneten Verhandlungen der Königl*, Preuß (Berlin: Akademie der Wissenschaften zu Berlin).

Dove, H. W. (1853): *Darstellung der Farbenlehre und Optische Studien* (Berlin: G. W. F. Müller).

Gage, J. (1995): *Colour and Culture* (London: Thames and Hudson).

Hankel, W.: *Habilitationsgutachten Dr. Zöllner, Johann Carl*, Universitätsarchiv Leipzig, PA 1093.

Herrmann, D. B. (1974): 'Ein eigenhändiger Lebenslauf von Karl Friedrich Zöllner aus dem Jahre 1864' (Berlin).

Herrmann, D. B. (1982): *Karl Friedrich Zöllner* (Leipzig: B.G. Teubner).

Herrmann D. B. and D. Hoffmann (1976): 'Astrofotometrie und Lichttechnik in der 2. Hälfte des 19. Jahrhunderts', in *NTM-Schriftenreihe Geschichte der Naturwissenschaften, Technik und Medizin*, 13, 1.

Illustrierte Zeitung, vol. 24, no. 608, 24 February 1855.

Joerges, B. and Shinn T. (eds) (2001): 'Instrumentation between Science, State and Industry', in *Sociology of the Sciences Yearbook* (Dordrecht: Kluwer Academic Publishers).

Kemp, M. (1990): *The Science of Art* (New Haven and London: Yale University Press).

Koerber, F. (1899): *Karl Friedrich Zöllner* (Berlin).

Morrison-Low, A. D. (1992): 'William Nicol, FRSE c.1771–1851, Lecturer, Scientist and Collector', in *Book of the Old Edinburgh Club*, new series, vol. 2.

Nicol, W. (1829): 'On a method of so far increasing the Divergency of the two Rays in Calcareous-spar, that only one Image may be seen at a time', in *Edinburgh New Philosophical Journal*, vol. VI.

Müller, G. (1897): *Photometrie der Gestirne* (Leipzig: Engelmann).

Nicol, W. (1840): 'Ueber eine Verbesserung in der Construction des Kalkspathprisma mit einfachem Bilde', in *Poggendorff'sche Annalen der Physik und Chemie*, vol. IL, p. 238.

Olesko, K. M. (1991): *Physics as a Calling. Discipline and Practice in the Königsberg Seminar for Physics* (New York: Cornell University Press).

Rosse, Lord Charles P. (1876): 'Zöllner's Astro-Photometer', in *Conferences Held in Connection with the Special Loan Collection of Scientific Apparatus* (London: published for the Lords of the Committee of Council on Education by Chapman and Hall [1877]).

Rucks, P. (1996): 'Prüfung astronomischer Optik mit Laserinterferometrie', in: *Sterne und Weltraum*, vol. 12.

Schaffer, S. (1989): 'Glass Works: Newton's Prisms and the Uses of Experiment', in *The Uses of Experiment: Studies in the Natural Science* (D. Gooding, T. Pinch and S. Schaffer, eds).

Schaffer, S. (1992): 'Late Victorian metrology and its instrumentation. A Manufactory of Ohms', in *Invisible Connections* (R. Bud and S. Cozzens, eds) (London: SPIE Optical Engineering Press).

Schütt, H. W. (1992): *Eilhard Mitscherlich* (München: Oldenbourg Gruppe).

Shinn, T. (2001): 'The Research-Technology Matrix: German Origins, 1860–1890', in Joerges and Shin (eds) (2001).

Staubermann, K. (2000): 'The Trouble with the Instrument', in *Journal for the History of Astronomy*, 31, pp. 323–38.

Sterken, C. and K. Staubermann (eds) (2000): *Karl Friedrich Zöllner and the Historic Dimensions of Astro-Photometry* (Brussels: Vrije Universiteit Brussel (VUB) Press).

Vierteljahresschrift der Astronomischen Gesellschaft, vol. 10 (1875).

Zöllner, F (1857): 'Photometrische Untersuchungen', in *Annalen der Physik und Chemie*.

Zöllner, J. C. F. (1861): *Grundzüge einer allgemeinen Photometrie des Himmels* (Berlin: Mitscher & Röstell).

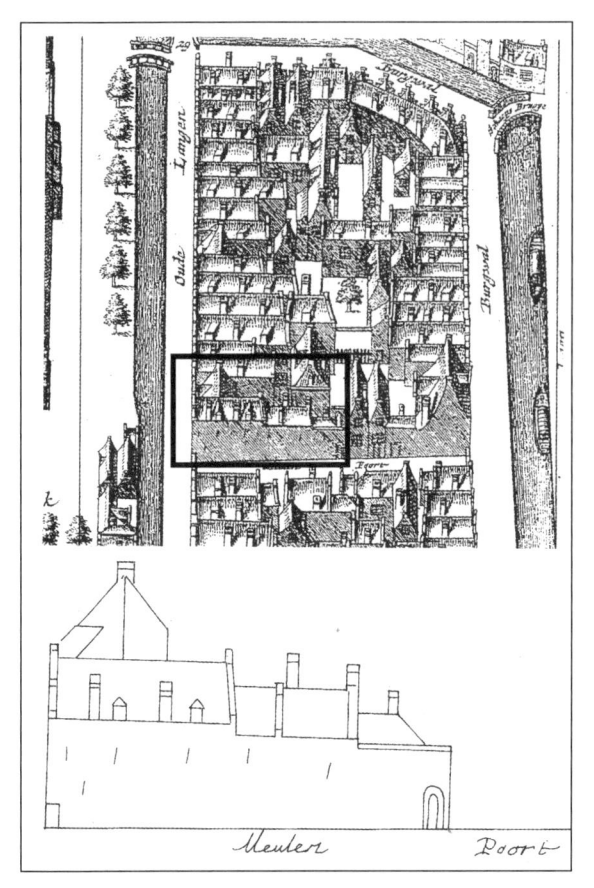

FIG. 11.1 Detail of Dirck van Bleyswijck's pictorial plan of Delft, 1675–78, showing the sidewall of Maria Thins' house on the Oude Langendijk.

(Houtzager, H. L., et al. (1997): *De Kaart Figuratief van Delft* [Rijswijk: Elmar B V])

FIG. 11.2 Drawing by Abraham Rade-maker of 1678 showing houses on the Oude Langendijk. A fraction of Maria Thins' house appears at the extreme right.

(Gemeenter Archief Delft)

PHILIP STEADMAN

Reconstructing the painting studio of Johannes Vermeer

MOST OF THE contributions to this volume have to do with the reconstruction of machines, instruments or scientific experiments. My subject is rather different. I will describe the reconstruction of a *room*, as it existed in the seventeenth century in the city of Delft in Holland. (Although one might say that this room was at the same time an instrument.) The room was used as a studio by the painter Johannes Vermeer, from some time in the mid-1650s up to the time of the artist's death in 1675. It was in a house belonging to Vermeer's mother-in-law. The house was demolished in the early nineteenth century; but there are several sources of information on which to base a reconstruction, of which the most important are Vermeer's paintings themselves. Reconstructing this studio is not just a matter of minor art-historical or biographical interest: an analysis of the paintings, in combination with a knowledge of the architecture of the room, can throw important new light on the great painter's working methods.

In 1653, when he was aged 21, Vermeer married Catharina Bolnes and the couple went to live with her mother Maria Thins in her large house on the Oude Langendijk, in the centre of Delft.[1] There are to my knowledge just two contemporary images of the house. It can be seen from the air in a pictorial plan of Delft, made by Dirck van Bleyswijck in the 1670s (Fig. 11.1). The front of the house looked across the street and canal towards the town's Market Square and the New Church. In this view we see the long sidewall of the house that bordered a narrow alley called the Molenpoort. And part of the front façade is visible at the extreme right of a drawing by Abraham Rademaker of 1678 (Fig. 11.2).

Vermeer died suddenly in 1675 at the age of 43. War with France had ruined the Dutch economy and the market for paintings; Vermeer had got deeply into debt; and his wife Catharina was obliged to declare herself bankrupt. Her financial affairs were put in the hands of a trustee nominated by the city's Aldermen. The bailiffs made an inventory of the contents of the house, in preparation for their sale.[2] From this inventory we know the location of Vermeer's

RECONSTRUCTIONS

FIG. 11.3 Reconstruction of Maria Thins' house.
(Warffemius, A. (2002): 'Jan Vermeers huis: Een poging tot reconstructie', in *Delfia Batavorum Elfde Jaarboek 2001* (Delft: Historische Vereniging Delfia Batavorum), Fig. 8, p. 70)

studio, since it describes a room on the first floor containing easels, palettes and canvases. The room was on the street front and faced north, the ideal aspect of course for painting by natural light.

In 2001 Ab Warffemius, an architectural historian from Delft, published a reconstruction of Maria Thins' house.[3] Besides the images of Figures 1 and 2 and the bailiffs' inventory, Warffemius used a plan of the footprint of the house from a nineteenth-century cadastral map. Because of its date and its purpose in levying taxes, we can expect this plan to be dimensionally accurate. The building was big in comparison with other seventeenth-century houses in the city and had an unusually wide frontage. Warffemius was able to glean some information about the building's internal layout from the positions of the chimneys, dormers and gables in Van Bleyswijck's view. He also recognised that the short black lines on the sidewall are the cross-pieces of iron tie bars, which would coincide with the lines of internal walls. (All this of course assumes that the map is reasonably reliable on such details.) What is more, Warffemius appreciated something important about the format of the inventory. The rooms in it are *numbered*, and Warffemius realised that the bailiffs had assigned these numbers in order as they worked through the house, starting from the front entrance.

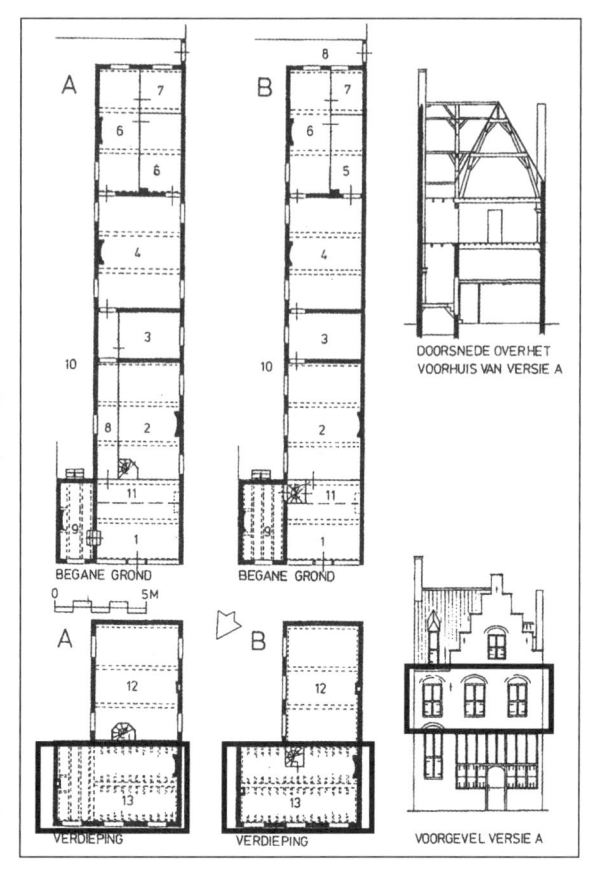

Fig. 11.3 shows Warffemius's measured drawings, including the street facade, a cross-section of the back extension, and two alternative plans, which differ only in relatively minor details. The room numbering is as in the inventory. Warffemius has relied not just on the documentary sources, but also on

his general knowledge of the typical construction of Delft houses of the period. Vermeer's studio is room 13 on the first floor at the front of the house. Notice that it has three large windows, one of which is visible in Rademaker's drawing. Warffemius assumes three windows since this is standard for other seventeenth-century houses of this width. The room has main beams in the ceiling running across the width of the plan, and rafters supported on these beams at right angles to the window wall. The height of the room according to Warffemius is 2.1m and the length 6.6m. It will be appreciated that this length is given by the frontage of the house, which Warffemius derived from the cadastral plan. All these details are important to what follows.

Now let us turn to Vermeer's paintings, many of which as it transpires show the interior of this very room. Only some three dozen pictures by Vermeer are known today: it seems he worked slowly, and his life was short. At the start of his career, Vermeer produced large canvases with Biblical and mythological themes – *Jesus in the House of Martha and Mary* (now in Edinburgh), *Diana and her Companions* – as well the genre subject *The Procuress*. These compositions are filled with human figures, there is not much by way of architectural background, and the compositions have little perspectival structure.

Then around 1657 – just about the time he moved to Maria Thins' house – everything changes. The canvases become generally much smaller, and Vermeer's

FIG. 11.4

Johannes Vermeer, *The Music Lesson*, c.1662–65, oil on canvas, 73 x 64cm.

(The Royal Collection © 2011, Her Majesty Queen Elizabeth II)

FIG. 11.5
Johannes Vermeer, *Allegory of the Faith*, c.1671–74, oil on canvas, 114.3 x 88.9cm.
(© The Metropolitan Museum of Art/Art Resource/Scala, Florence)

attention shifts to three new types of subject matter. There are small portrait heads of young women, of which he painted six. There is the town of Delft itself, of which he made two pictures – the magnificent panoramic *View of Delft* and the more modest *Little Street*. And most numerous, accounting for around two-thirds of Vermeer's entire *oeuvre*, there are the domestic interiors with sometimes two or three figures, often just lone women absorbed in their own occupations, reading and writing letters, trying on jewellery, playing music. The figures move further away from the viewer, and the architecture of the room becomes as much Vermeer's preoccupation as his human subjects and their psychological relationships. In these pictures we see the mastery of perspective and above all the miraculous treatment of light and shadow for which Vermeer is so much admired.

When one studies these interiors as a group, one has to be struck by the impression that perhaps as many as ten pictures seem to show the very same room. *The Music Lesson* gives the widest and most comprehensive view (Fig. 11.4). In every case we look directly at a blank far wall, without doors or windows. We never see any wall to the right. In this and two other pictures we see a part of the ceiling, with similar rafters running across the picture. These rafters are always perpendicular to the window wall (as in Warffemius's reconstructed studio). The height of the room appears to be about the same in all three cases. There is a wall with windows to the left; and when we cannot see that wall, as in *Allegory of the Faith* (Fig. 11.5), the light still comes from that direction.

In the majority of cases the floor is covered in black and white marble tiles, set diagonally – although these are laid in a number of different patterns. In this group of ten pictures, the windows are the most distinctive feature of the architecture. Again we see them most clearly in *The Music Lesson*. Each window opening has four lights. The upper fixed lights have simple rectangular pat-

FIG. 11.6

Johannes Vermeer, *The Glass of Wine*, c.1658–60, oil on canvas, 65 x 77 cm.

(© bpk/Gemäldegalerie, SMB/ Jörg P. Anders)

terns of leading. The lower lights are side-hung casements, with leading in a complicated pattern of squares, circles and half circles. One casement is shown half open in *The Glass of Wine* (Fig. 11.6). There are external shutters on these casements, which are closed over the far window in this picture and in some other paintings. *The Glass of Wine* is an early picture in the group, and in this painting and two others there are smaller ceramic floor tiles, in brown and black, again set on the diagonal.

With all these pictures, because they show tiled floors, it is possible to reconstruct the three-dimensional geometry of the architecture and furniture, by working the standard methods of perspective construction backwards, so to speak. By analysing the pattern of floor tiles, one can find the theoretical viewpoint of each picture. One can then work out the three-dimensional geometry of the room and furnishings from their two-dimensional images in the painting. Fig. 11.7 shows my plan and side view of the space of *The Music Lesson*.[4] Such techniques of 'reverse perspective' are related of course to the methods of photogrammetry. They are fairly standard and have been used by art historians to reconstruct the three-dimensional scenes in many naturalistic paintings.

In general in such a process of reconstruction it is not possible to determine the exact *scale*, only the *relative* dimensions of different features. Speaking strictly geometrically, the pictures might just as well show dolls in dolls' houses as real people in real rooms. But with Vermeer the situation is different, because he depicts many *real* objects, the originals or copies of which survive today in museum collections. Assuming he shows them at their real sizes – allowing for perspective foreshortening – these objects can serve as measuring scales. I will mention just a few examples.

In *Allegory of the Faith* (Fig. 11.5) the terrestrial globe is recognisable as one of an edition produced around 1660 by both Jodocus Hondius and Willem Blaeu. The crucifixion on the back wall is a copy (with minor modifications) of a real painting by Jacob Jordaens, a picture that possibly belonged to Vermeer's family. There is a large canvas of *Christ on the Cross* mentioned in the inventory of Maria Thins' house: this could be it.[5] The chair in the foreground of *The Glass of Wine* has a very distinctive design decorated with brass lions' heads on the uprights. My colleague Marc van Leusen found an example in the attic of the Prinsenhof Museum in Delft and made measurements.

All of the many maps and globes in Vermeer's paintings have been identified and copies tracked down.[6] Vermeer's versions are extraordinarily faithful. Many 'painted paintings' hanging on the back walls, like the Jordaens crucifixion, have been recognised as real works by Vermeer's contemporaries. Examples of other pieces of furniture besides the chairs have been found. The virginals in *The Music Lesson*, for example, are the work of the celebrated Antwerp firm of Ruckers.

Above the virginals in *The Music Lesson* hangs a mirror. In it we can see reflected the woman's head, the black and white tiles, and the table covered with its carpet. Fig. 11.8 shows a detail. Beyond these is a shadowy group of items of equipment, which on closer examination turn out to include an easel with a canvas on it, a mysterious box of some kind, and a stool. At the very top left-hand corner of the mirror is a small patch of yellow, which must be a part of the back wall of the room, behind us the viewers, behind Vermeer as he painted. Disappointingly there is no sign of the artist himself, not even his boot or the sleeve of his jacket. As Benedict Nicolson says, 'the painter appears to have been swamped by a wave of diffidence and at the crucial moment to have fled the scene'.[7]

From what it reflects, it is possible to work out the precise angle at which the mirror hangs, and from this to reconstruct that

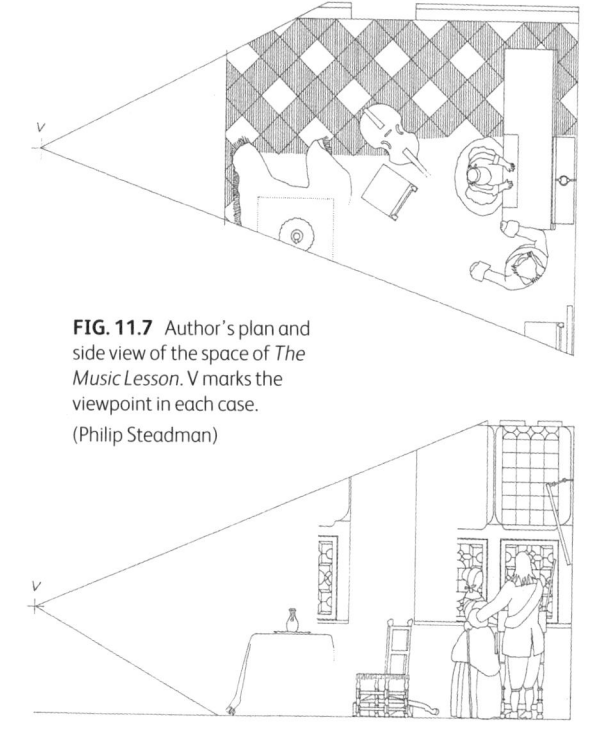

FIG. 11.7 Author's plan and side view of the space of *The Music Lesson*. V marks the viewpoint in each case.
(Philip Steadman)

JOHANNES VERMEER'S PAINTING STUDIO

FIG. 11.8
Detail of the reflection in the mirror in *The Music Lesson*.
(The Royal Collection © 2011, Her Majesty Queen Elizabeth II)

part of the room visible only in the mirror. Fig. 11.9 reproduces my bird's eye view of the complete room. This shows the easel, the stool and the position of the back wall. The tile pattern matches up nicely across the join between the area seen directly in the picture and the area seen only in the mirror. The theoretical viewpoint (marked V) is at Vermeer's eye level had he been sitting on the stool. The viewpoints of all these interiors are at the eye-height of a seated observer.

I have made reconstructions of the spaces in all ten paintings by this method of reverse perspective, using the real objects to determine the scales.[8] It turns out that – with some discrepancies – the dimensions of the room and its windows *are* similar in all cases. The patterns of floor tiles change: however, it transpires that there is an underlying *grid* to which all the patterns conform, including the smaller ceramic tiles whose edges turn out to be exactly half the dimension of the marble tiles. The real objects – the furniture, the maps – can all be made to come out close to their known sizes, again with a few exceptions. I had estimated the dimensions, from the paintings, of the chair with the lion's heads, before my colleague found the actual example in the Prinsenhof. He sent me measurements

FIG.11.9 Author's bird's eye view of the space of *The Music Lesson*, including the part of the room visible only in the mirror. The theoretical viewpoint is at V.
(Philip Steadman)

of the height, the width, and the height of the seat, and I was within a centimetre in each case. It is very difficult – I suggest – to escape the conclusion that the ten pictures in this group show one and the same room – with some variations in the floor tiles – and that Vermeer has set up actual furniture and props in the room to create *tableaux* for his different compositions.

Armed with the dimensions from the perspective analysis, I have had a scale model made of the room. I have also had model furniture made, and dolls dressed to play the parts of Vermeer's sitters. At the back of the model there is a large photographic plate camera. The camera lens can be moved to the correct viewpoint for each painting and a photograph taken. Figure 11 in art section shows the result for *The Music Lesson*. It will be appreciated that the fidelity with which the photograph matches Vermeer's original provides an independent test, in effect, of the accuracy of the perspective reconstructions. The model, what is more, makes it possible to study some of the patterns of light and shade in the scene – something a purely geometrical perspective analysis does not address.

Notice a number of features of the lighting in the painting that are nicely captured in the photograph. See the shadow on the far wall below the window, the shadow of the window frame, and the shadows of the virginals' legs. Notice also the small bright patch on the wall just below the far window, which is illuminated by light reflected back off the varnished end of the virginals' case. The photographer Trevor Yorke and I did not have to work too hard to achieve these results – in fact they more or less fell out. We made the assumption that Vermeer would have been painting by north light, and simulated this with lamps behind two large diffusers.

The distinguished Vermeer scholar Arthur Wheelock has said that the light in *The Music Lesson* is obviously direct sunlight.[9] If so, there must have been some very strange astronomical phenomenon occurring above Holland in the 1660s. See how the mirror casts two shadows on the wall, one dark, the other less dark. Were there two suns over Delft? No, the shadows are cast by north light coming from the two windows. The darker area is shadowed from both windows, the lighter area shadowed just from the far window. The double shadows made by the mirror are caught in the photograph, but they are much wider than in the painting. This is a detail where Vermeer *has*, it seems, departed from actual appearances and introduced an inconsistency. We know the angle at which the mirror must hang in order to reflect what Vermeer shows. In the model the mirror is set at this angle, leaning sharply out from the wall. Judging by his painted shadows, however, Vermeer has preferred to depict a mirror that hangs much closer to the vertical.

Is this then the very studio in Maria Thins' house on the Oude Langendijk, in which Vermeer worked in the 1660s? I should emphasise that when I

JOHANNES VERMEER'S PAINTING STUDIO

published my results in my book *Vermeer's Camera* in 2001, I had not seen the reconstruction by Warffemius, which was published in a Delft historical yearbook the following year. He and I worked quite without each other's knowledge, and were only able to compare results once we had both published. We have already seen how the directions of the rafters in the paintings match Warffemius's reconstruction. The length of the room, as calculated *The Music Lesson* is 6.6m, precisely the length determined by Warffemius. The height of the room as estimated from the paintings on the other hand is 2.85m, considerably taller than Warffemius's dimension of 2.1m. I am reluctant to criticise Warffemius on this point, but it should be said that he depends for vertical dimensions largely on Rademaker's drawing, which might not be entirely reliable. We know that other drawings by Rademaker of buildings in Delft that remain unchanged today can be quite inaccurate in detail. (We have no way of judging the width of the room from the evidence of the paintings, since we never see a wall to the right, opposite the window wall.) How about Warffemius's three windows? In the paintings we only ever see two windows directly, although there could in principle be space for a third, out of sight to the left.

As it happens, *The Music Lesson* is not the only picture in this group to include a mirror. *Allegory of the Faith* has a *mirrored ball* hanging from the ceiling giving a panoramic if shadowy view of the back of the room, including the rear wall behind Vermeer's viewpoint. Fig. 11.10 shows a detail. Along with the apple and the crushed snake on the floor, this ball serves the allegorical programme of this rather unsatisfactory composition. Since it reflects the entire

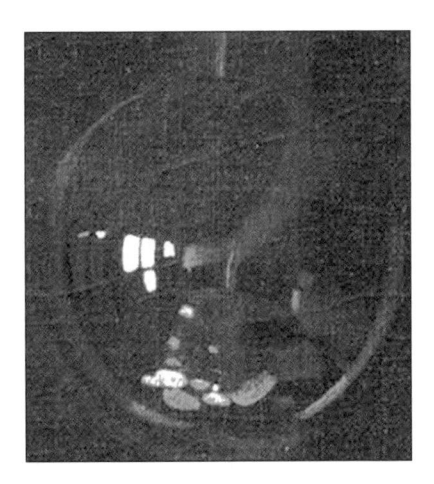

FIG. 11.10
Detail of mirrored sphere in *Allegory of the Faith* (see **Fig 5**).
(© The Metropolitan Museum of Art/Art Resource/Scala, Florence)

FIG. 11.11
Photograph of a miniature mirrored ball in the model of Figure 11.14, dressed as for *Allegory of the Faith* (cf. Fig. 11.10).
(Trevor Yorke)

243

world around it, the ball symbolises the capacity of the human mind to comprehend the vastness of God's creation.[10] Our present interest is however more prosaic: it is in what, precisely, we can see in this reflection.

It is not so easy to make out the detail in reproductions or even in the painting itself. To help in interpreting the reflection, I have set up the scene in the scale model, complete with a miniature mirrored sphere. My colleague Trevor Yorke took a photograph (Fig. 11.11). By comparing photograph with painting, it becomes clear what both show. The white patches at the bottom of the ball are the woman's face, her white dress and her open Bible. To the left we see a bright inverted L shape. This is one of the windows, with an external shutter closed on one of the casements. Further to the left still, we can just make out two small patches of light, near the ceiling, and in the painting itself one can see that below these are bluish rectangular areas. This must be a second window, almost completely covered by blue curtains: indeed it must be the window nearest the seated woman, and furthest away from us the viewers. Now look to the right of the bright L-shaped window with its closed shutter. Here is a *third* patch of light, in the upper half of the wall. This looks like a *third* window, at

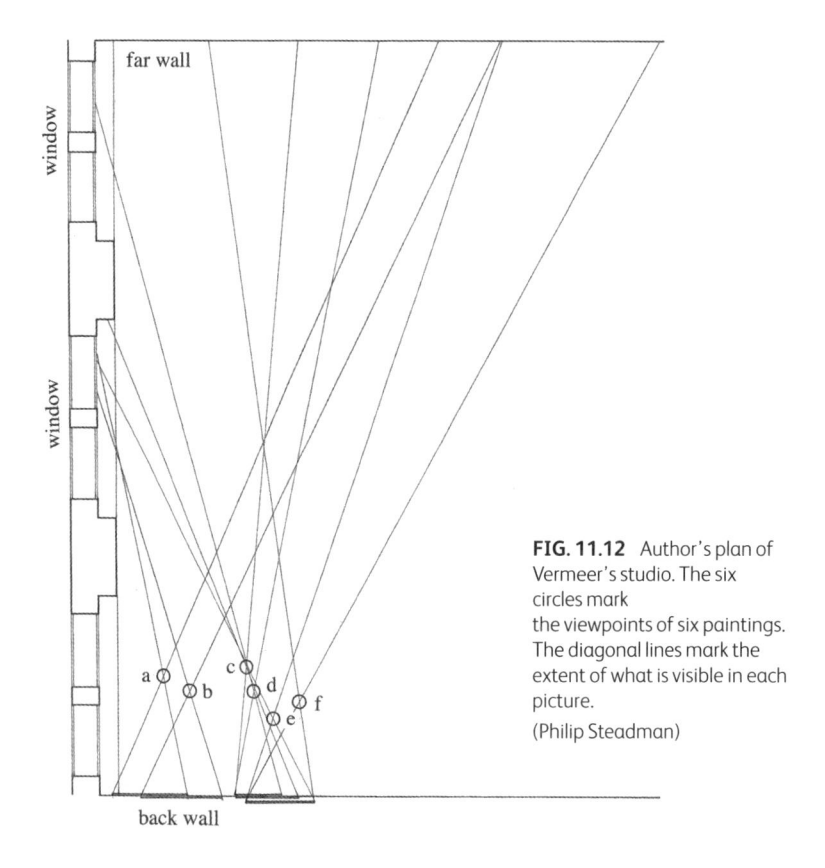

FIG. 11.12 Author's plan of Vermeer's studio. The six circles mark the viewpoints of six paintings. The diagonal lines mark the extent of what is visible in each picture.
(Philip Steadman)

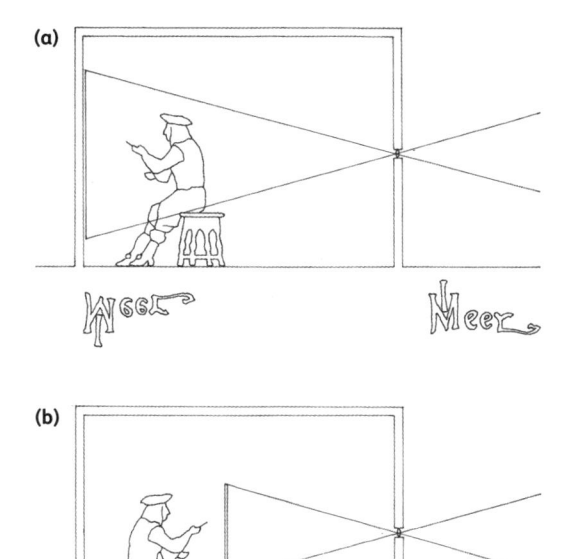

FIG. 11.13 Two designs of cubicle-type camera obscura, as described in seventeenth century literature. In (**a**) the image is projected on the far wall of the booth. In (**b**) the image is interrupted by a translucent screen, which the artist studies from the opposite side. In both cases the image is inverted. In (**a**) the image is also laterally reversed, which it is not in (**b**).
(Philip Steadman)

the back of the room, that we never see in any other painting. The shutters are closed on both casements. To the right again is yet another bright patch that must be part of the back wall of the room.

So on this admittedly rather slender evidence, it does seem that there are indeed three windows in the painted room. There is the anomaly between Warffemius's and my estimates of the height of the room, as discussed earlier. Nevertheless, I would propose that – given the correspondence between my analysis and Warffemius's reconstruction of the windows, the rafters, and the precise length of the room – there can be little doubt that what we see in these ten paintings are views of Vermeer's actual studio on the Oude Langendijk.

I promised at the outset that this knowledge of the architecture of Vermeer's room would throw new light on his working methods. Fig. 11.12 contains the key finding from all my perspective analysis. It shows a plan of the room, with the far blank wall, the window wall at the left, and the back wall in the position found from the mirror reflection in *The Music Lesson*. The small circles mark the positions of the theoretical viewpoints of six paintings.[11] The diagonal lines mark the extent of the room that is visible in each picture. A similar diagram can be drawn in side view. Now let the diagonal lines be carried back through the viewpoints to meet the back wall. In each case they describe a rectangle on that wall. *In all these six cases, the rectangle is the exact same size as Vermeer's canvas.* What might be the cause of this strange geometrical result? It can hardly be due to chance.

For more than a hundred years, historians have suggested that Vermeer might have made use of the camera obscura, the predecessor of the photographic camera.[12] The camera obscura can take many forms, small or large, but one type of design described by several authors in the seventeenth century consists of a darkened booth or cubicle with a lens in one wall (Fig. 11.13). The observer or artist sits inside the booth. Light entering through the lens forms an image, upside down, either on the opposite wall, or on a screen. We have descriptions of cameras of this general type being used from the Renaissance. Leonardo da Vinci was the first person to describe possible applications in art. He says: 'You will receive these images on a white paper placed within this dark room rather near to the hole; and you will see all the objects in their proper forms and colours, but much smaller; and they will be upside down by reason of that very intersection [of the rays of light in the aperture].'[13]

Leonardo recommends making a screen from thin translucent paper, so that the image can be viewed from the opposite side. In this arrangement, the observer does not get in the way of the light. The first arrangement where the image is simply projected on the far wall is, however, still perfectly practical. An artist can study the image, and can trace it onto paper. He can even put his canvas under the image and paint. In Leonardo's time camera obscuras did not have lenses, just small holes, and the images were as a consequence extremely faint. Some time around the 1580s it seems, cameras were made for the first time with glass lenses instead of pinholes, which gave much brighter images and made them practical for use by draughtsmen and painters. By Vermeer's time, in the mid-seventeenth century, there are several printed accounts of how to exploit the camera in art, and there are written descriptions of a small number of Dutch painters actually using the device.[14] There are, however, no documents to indicate that Vermeer himself was a camera user, and any evidence for him using the device must come from the paintings themselves.

It was the American graphic artist Joseph Pennell who first made this suggestion in an article in the *British Journal of Photography* of 1891.[15] Pennell drew attention to what he called the 'photographic perspective' of Vermeer's *Soldier and Laughing Girl*. The two figures sit just a few feet apart, across the corner of the table. Nonetheless, the image of the officer's head is about twice the dimension of the girl's head. The effect is perfectly correct in geometrical, perspective terms. It happens because the viewpoint of the picture is very close to the soldier. Today we are used to seeing such effects in snapshots. But for a seventeenth century painting it is quite unusual. Vermeer's contemporaries, painting similar scenes, would have made the two sitters more nearly equal in size. Pennell thought that Vermeer might have arrived at this photographic perspective through using the camera lucida 'if it was invented in his time, for it gives exactly the same photographic scale to objects'.[16] The camera lucida is a quite different instrument from the camera obscura – although it serves a similar purpose – and in fact was *not* known in Vermeer's time. But Pennell's point would apply equally if Vermeer had been using the camera obscura.

In the 1940s the historian A. Hyatt Mayor took the argument some steps further, by drawing attention to Vermeer's treatment of *highlights*, the reflection of sources of light off shiny surfaces like glazed pottery, varnished wood, or polished metal.[17] In *Girl with a Red Hat* the girl rests her arm on the back of a chair with decorative lions' heads. The chair is the type seen in *The Glass of Wine* (Fig 11.6). Many highlights are formed on these lions' heads since they are made of brass. We can be fairly certain that Vermeer is painting indoors, and that the illumination here comes from a window – in this rare instance, on the right. We would therefore expect the highlights to take their shapes from the windows they reflect, and to approximate to rectangles. But they do not: several of them are close to perfect circles. This part of the painting seems to be rendered in very soft focus; and in a unfocused optical image, bright points of light like these, even if in reality they have other shapes, become spread out into true circles – so-called 'circles of confusion'. Their circularity derives in effect from the circular shape of the camera lens.

Hyatt Mayor came to the conclusion that Vermeer is copying in paint here an artefact of a lens – something that would not be visible looking directly at the lion's heads with the naked eye. Later, in the 1960s, Charles Seymour and Henry Beville tried to reproduce these 'circles of confusion' with an antique lion's head chair.[18] They used a nineteenth-century plate camera, suitably unfocused, to take a photograph in which some of the rectangular highlights were again turned into near circles.

These authors then – Pennell, Hyatt Mayor, Seymour – argued for Vermeer using the camera obscura from the evidence of idiosyncrasies of his painting technique, and from what seem to be results of his imitating the qualities of optical images. By the 1980s, the idea that Vermeer had indeed used the device had become quite generally accepted among art historians – with some notable exceptions – the remaining debate being about the exact nature and extent of that use. What I have discovered is that the perspective geometry of the room and of Vermeer's painted interiors can tell us much more. Let us return to the plan of the room. We saw how the projected images of six paintings at the back wall were the same sizes as the corresponding canvases. This perplexing result is very readily explained on the assumption that Vermeer is using a simple booth-type camera obscura of the design shown earlier (Fig. 11.13a). He would have set his lens at the viewpoint of the picture, and would have projected an optical image onto the back wall. The image is the same size as his canvas, because he has traced it. This is the essence of the argument of my book *Vermeer's Camera*.

The drawing of Fig. 11.14 shows how the painter might possibly have set up his camera at the back of his studio. The booth is about 90cm deep: not large, but sufficient to sit inside and work in relative comfort. If we accept the existence of a third window, glimpsed in the mirrored ball, of a similar size to the other two windows, then this camera obscura would have fitted nicely against one of its casements. Vermeer could have opened the shutter on this casement to let in the light and look at his canvas, and could have closed it again to black out the cubicle, so that he could study and trace the optical image. The positions of the viewpoints of the respective pictures differ somewhat, and it would have been necessary to move the lens slightly. Perhaps the front of the booth was curtained. The lens would have had to be large, in order to admit sufficient light to obtain a useable image: but notice how Vermeer's studio is unusually well lit, with the windows taking up a large proportion of the sidewall. The plate camera used for taking photographs in my scale model mimics this proposed camera obscura set-up: for every photograph the plate of the camera was set in the exact line of the model's back wall. (One should emphasise, nevertheless, that working manually inside a camera obscura is very different from taking a photograph.)

There are many complications that I have passed over in this brief account. It is relatively simple to make outline drawings in a booth-type camera obscura, and so obtain perspectives of extreme accuracy. But applying paint under the projected image is not so straightforward, not least because one is working in extremely low light. All the same, the fact that Vermeer seems to be imitating the qualities of optical images in paint suggests that he was not just using the camera for drawing. I have been making experiments myself with a large camera obscura with a design similar to Fig. 11.14, in order to test the feasibility of painting from the optical image, up to now just in monochrome. All this is the subject for other papers.

Let us return finally to the reflection in the mirrored sphere in *Allegory of the Faith*. Look at the precise centre of the image in the photograph taken in the model (Fig. 11.11). One can see the reflection of the plate camera that is taking the photograph. Now look in the equivalent position in Vermeer's painted version (Fig. 11.10). We have already noted the bright patch at ceiling level, which must be the upper part of the back wall of the room. Below this is a tiny black rectangular shape. Here we have at last been granted what I believe is a distant glimpse of Vermeer's camera obscura. This indistinct dark square is as much as he is prepared to reveal of his optical technique. As in *The Music Lesson*, no sign of the man himself: perhaps he is inside the booth.

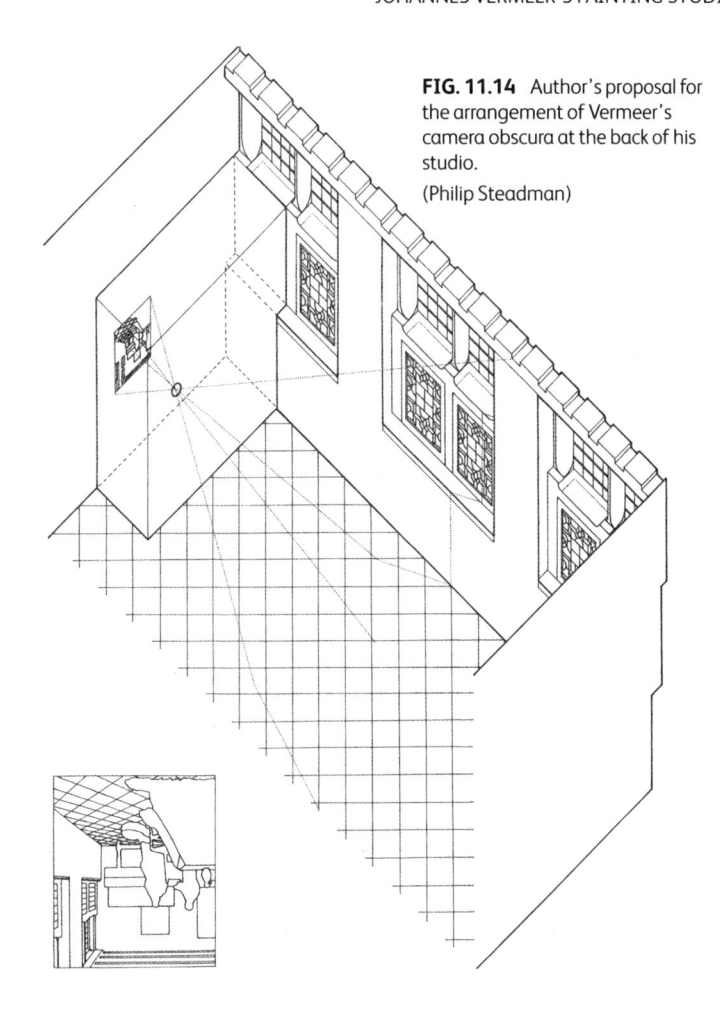

FIG. 11.14 Author's proposal for the arrangement of Vermeer's camera obscura at the back of his studio.

(Philip Steadman)

Notes

1. For Vermeer's biography, see Montias (1989).
2. Montias (1989), pp. 339–44. The trustee – perhaps significantly – was Antony van Leeuwenhoek, the pioneer microscopist.
3. Warffemius (2002), pp. 60–78.
4. See Steadman (2001), ch. 5, for a full account.
5. Montias (1989), p. 340.
6. Welu (1975), pp. 529–47.
7. Nicolson (n.d.): Vermeer's 'Lady at the Virginals' (London: Lund Humphries, Gallery Books no.12).
8. The drawings can be seen at: www.vermeerscamera.co.uk/
9. Wheelock Jr (1995), pp. 94–95.
10. Wheelock Jr (1981), p. 148. Vermeer is following the iconographical recommendations of Ripa (1644).
11. (a) *The Girl with a Wineglass*.
    (b) *The Glass of Wine*.
    (c) *Woman Writing a Letter, with her Maid*.
    (d) *Woman Standing at the Virginals*.
    (e) *The Music Lesson*.
    (f) *The Concert*.
12. For histories of the camera obscura and its uses by artists, see Waterhouse (1901), pp. 270–90; Gernsheim with Gernsheim, (1955), ch. 1; Hammond (1981).
13. Richter (ed.) (1977), pp. 115–16.
14. In particular, Worp (ed.) (1911–16).
15. Pennell (1891), pp. 294–96.
16. Pennell (1891), p. 295.
17. Hyatt Mayor (1946), pp. 15–26.
18. Seymour Jr (1964), pp. 323–31.

Bibliography

Gernsheim, H., with A. Gernsheim (1955): *The History of Photography: From the Camera Obscura to the Beginning of the Modern Era*, rev. 1969 (London: Thames and Hudson), ch. 1.

Gowing, L. (1952): *Vermeer* (London: Faber); 2nd ed. (1970).

Hammond, J. H. (1981): *The Camera Obscura: A Chronicle* (Bristol: Adam Hilger).

Hyatt Mayor, A. (1946): 'The photographic eye', in *Bulletin of the Metropolitan Museum of Art*, new series, vol. V, no. 1, pp. 15–26.

Kemp, M. (1990): *The Science of Art: Optical Themes in Western Art from Brunelleschi to Seurat* (New Haven, CT: Yale University Press).

Montias, J. M. (1989): *Vermeer and his Milieu: A Web of Social History* (Princeton, NJ: Princeton University Press).

Nicolson, B. (n.d.): *Vermeer's 'Lady at the Virginals'* (London: Lund Humphries, Gallery Books no.12).

Pennell, J.(1891): 'Photography as a hindrance and a help to art', in *British Journal of Photography*, no. 1618, vol. XXXVIII, pp. 294–46.

Richter, I. A. (ed.) (1977): *Selections from the Notebooks of Leonardo da Vinci* (Oxford: Oxford University Press).

Ripa, C. (1644): *Iconologia* (trans. D. D. Pers) (Amsterdam).

Schwarz, H. (1966): 'Vermeer and the camera obscura', in *Pantheon* XXIV, May–June 1966, pp. 170–82.

Seymour Jr, C. (1964): 'Dark chamber and light-filled room: Vermeer and the camera obscura', in *Art Bulletin* 46, pp. 323–31.

Steadman, P. (2001): *Vermeer's Camera: Uncovering the Truth Behind the Masterpiece* (Oxford: Oxford University Press).

Warffemius, A. (2002): 'Jan Vermeershuis: Een poging tot reconstructie', *Delfia Batavorum Elfde Jaarboek 2001* (Delft: Historische Vereniging Delfia Batavorum), pp. 60–78.

Waterhouse, J. (1901): 'Notes on the early history of the camera obscura', in *The Photographic Journal*, vol. XXV, 31 May 1901.

Welu, J. A. (1975): 'Vermeer: his cartographic sources', in *Art Bulletin*, vol. 57

Wheelock, A. K. (1995): *Vermeer and the Art of Painting* (New Haven, CT: Yale University Press)

Worp, J. A. (ed.) (1911–16): *De Briefwisseling van Constantijn Huygens 1608–87* (Gravenhage: Martinus Nijhoff Publishing).

MARK OSTERMAN

Mid-nineteenth-century photographic studio technique

WHY RECREATE NINETEENTH-CENTURY PHOTOGRAPHIC TECHNOLOGY?

THOSE WHO RESEARCH and recreate arcane technologies are frequently asked why. What does such activity contribute to the greater understanding of a culture? Can we, despite our studied intent to accurately recreate something from the past, shed all we know of the modern world and really put ourselves in the context of those who came before us? Can a technology truly be recreated and why does it matter?

The revival of historic photographic processes

Several nineteenth-century photographic processes have been revived in the last 30 years for three entirely different reasons: art, history and science. Beginning in the early 1970s, a group of fine art photographers in the United States began to investigate a few of the common nineteenth-century photographic printing techniques. These included hand-coated papers prepared for the cyanotype, platinum, palladium, kallitype, salt, albumen and gum printing methods.

During this era, conventional silver halide panchromatic sheet films were readily available, but were not always well matched for some of the early printing papers. Many photographers at that time relied on the higher contrast orthochromatic films used in the halftone printing industry to provide the necessary density for the process of choice. Others modified the historic printing formulas to match the modern films. The images being made with these historic processes were often a very different approach and aesthetic than photographs made in the popular era of a specific process. While modern photographers were influenced to a certain degree by historic images made with these processes, they celebrated the inevitable irregularities in their hand-coated materials, an aesthetic that was not considered attractive in the nineteenth century.

By the early 1990s, some of the more challenging primary, or so called in-camera, processes were being researched by artists and those interested in either the science or history of photography. These techniques included the daguerreotype, wet collodion negative and positive variants and, to a lesser degree, the calotype paper negative. In particular, the wet plate collodion technique has become very popular in the hands of artists in the past 15 years.

Those who use wet collodion today often list the uncertainties of the process as one of its major aesthetic attractions, and confess that serendipity plays an important role in the success of their imagery. It is important to emphasise, however, that nineteenth-century photographers would suggest that the process is only uncertain in the hands of the unskilful.

The conservation of photographic artefacts

Unlike artists, who use historic photographic processes as an old medium with a new voice, those interested in the preservation and conservation of early photography and photographic artefacts have very different reasons to revive these old techniques.

The photograph conservation community is relatively new when compared with those who conserve paper, paintings and other important artefacts. The study of photograph conservation first emerged in the early 1970s. From the very beginning, the first photo conservators observed that knowledge of how vintage photographs were made was the key to understanding the function of deterioration and potential treatment options. Now in its adolescence, the conservation community has finally come to grips with the realisation that there is no way for any one conservator to have an in-depth understanding of all the processes used to make photographs.

Since its invention in the 1820s, photography has been made on numerous physical supports, coated with a variety of binding layers, and relies on several types of image-bearing technologies. To make matters more difficult, until the early 1880s all photosensitive materials were formulated and applied by the photographers themselves, resulting in countless variations on dozens of processes.

The aesthetics of technology: techno-aesthetics[1]

Photography is the combination of art, chemistry and optics. Every decade in the technical evolution of photography contributed to the techno-aesthetics of the medium. Like all technology, what are limitations to some promise uncharted territory for others. To fully understand the approach of a particular photographer or era of photography, it is important to understand the inherent limitations and characteristics of the processes and equipment that were used.

Many curators, photo historians, collectors, dealers and archivists of vintage photography have only recently come to realise that an understanding of the process used to make a photograph contributes to the greater knowledge of both the image and the physical artefact. Even an unskilful attempt at reproducing a historic process, provided the same historic techniques and chemicals are used, can provide important visual and analytical information. Poorly processed or manipulated photographic objects have characteristic markings and stains, as do those that have been preserved in unfavourable conditions.

A depth of knowledge

Understanding what makes the difference between a reasonably good photographic print and a tour-de-force from the same negative is often very difficult to verbalise. Such critical evaluations are generally given by those who have a highly developed sense of connoisseurship, gained by years of handling original materials. However, by actually seeing historic photographic processes recreated with original techniques both skilfully and unskilfully, it is possible to observe the difference between a poorly made print in a good state of preservation compared to a well made print in poor condition.

Recreating the wet plate collodion negative process

Often the end product of an early technology does not exist as an actual artefact or even the basis of a procedure still used in modern times. This makes even the most serious attempts of process recreation speculative at best. With photography, however, nearly every process of the nineteenth century exists in private and institutional collections. It is due to the availability of these original photographic artefacts that we can confirm the validity of reconstructed historical process by analytical comparison.

The wet collodion processes is very well documented. There are numerous nineteenth-century method books published during the collodion era in French, English and German. The very first photographic journals were also introduced within only a few years of the introduction of the collodion process in 1851. If anything, the modern researcher will discover that there are so many variants of formulae, technique and artefact that it becomes difficult to identify technological trends during the collodion era. A general progression of the medium can only be understood after viewing large institutional collections of historic collodion glass negatives.

I first began research in the wet collodion negative process in 1987 as an alternative to modern silver-gelatin sheet films. My initial sources were a 1969 facsimile edition of *The Silver Sunbeam*, by John Towler (Morgan & Morgan, 1864), and an original edition of *The Negative and Ambrotype Manual* by

N. G. Burgess (1856). I did additional research at the Library Company in Philadelphia. The Library Company, founded by Benjamin Franklin, was the first successful lending library in the United States. The collection included bound issues of *The Philadelphia Photographer*, a popular American photographic journal published at the height of the collodion era.

At that time I was teaching fine art photography at George School, a private Quaker boarding school northeast of Philadelphia, PA. By 1990, when I met France Scully, I had a firm grasp of the basic collodion process and taught France the technique. In addition to collodion negatives, we were also making ambrotypes (collodion positives on glass) and ferrotypes (collodion positives on iron plates coated with asphalt). In 1991 we formed a partnership called Scully & Osterman and were married in 1993.

France Scully Osterman and I began using the collodion process for our own fine art photography. We also continued to do primary research on all aspects of wet plate collodion and the historic printing processes that had been used with collodion negatives. In 1995 we began publishing *The Collodion Journal*[2] and also started teaching group workshops in the collodion process for the general public at George Eastman House International Museum of Photography and Film, in Rochester, NY. In 1999, we moved to Rochester, where I was hired by George Eastman House to teach the technical evolution of photography to photo conservators in the Advanced Residency Program in Photograph Conservation.

My work at George Eastman House

Working at George Eastman House as the official process historian has placed me at the epicenter of historic photographic research. Much of my time is spent searching in the rare book vault or making frequent visits to the print and technology archives. This opportunity allows me to read original texts and articles about specific processes, view work made in the same era or even by the original authors, and inspect the equipment used specifically for that technique. More important, my own research darkroom at the museum is in the same building. This allows me to make experimental photographic objects based on my readings and compare these with originals in the collection under the microscope or with analysis in the conservation lab. The only thing missing to complete my research was a nineteenth-century skylight studio.

Scully & Osterman studio

Located in the Highland Park section of Rochester, NY, the Scully & Osterman Studio is a working recreation of a nineteenth-century photographic skylight studio. It was built in 1999 for research purposes and to teach historical photographic techniques to photo conservators, historians and fine arts photographers. The space incorporates common elements of vintage skylight studios built in the United States, Europe and Japan, and functions like its nineteenth-century counterparts.

In creating the skylight area, it was discovered that white walls were inappropriate for portraiture using early photographic processes. Test plates showed that too much natural light was reflecting onto the subject from the white painted walls, illuminating the shadow areas. Further reading and visits to original skylight studios revealed that interior walls in the historic studios were painted a medium-dark shade of grey blue or teal blue. The teal blue colour, when fully illuminated by daylight, photographs as a very light tone, but when shaded reflects very little actinic light.[3]

The large skylight windows required several separate adjustable shades for independently controlling the light that falls upon the background and the subject. This was evident from the period illustrations, but their effect could only be understood by experiment. It took almost two years to fully understand how to use the shades effectively. The use of reflectors and diffusers was also an important skill to be learned by experience. The studio area is fully equipped with nineteenth-century apparatus, and the darkrooms contain all the materials needed for recreating photographic processes used from 1802 to 1890.

Evolution of photographic studio portraiture

The introduction of the daguerreotype in 1839 by Louis Daguerre[4] brought the potential for an international revolution in imaging, similar to the current transition from chemical-based photography to digital imaging. Portraiture with Daguerre's process, as introduced, was difficult due to limitations in the sensitivity of silver iodide that was formed on the silver-coated copper plates. Lenses available at that time, specifically the single meniscus or achromatic lenses, were also not fast enough to make portraiture a comfortable process.

Some of the first experimental portraits made with the daguerreotype process were possible by taking advantage of a location that presented the greatest quantity of light. The earliest experimental studios were plein-air arrangements situated on balconies or rooftops. At this time, exposures were measured in minutes. A typical exposure with an iodised plate could require

15–20 minutes in full sun. Rooftop lighting was not easily controlled and while portraits from this era are exciting artefacts from a historical viewpoint, they seldom have the presence and roundness of those made only five years later under the studio skylight.

In 1840 Alexander Wolcott patented a camera that used a concave front surface mirror instead of a lens to project the reflection of the sitter upon the sensitive plate. Based on the Newtonian telescope, the mirror was much faster than glass lenses of the period and had the advantage of being free of chromatic aberration.[5] Wolcott's cameras required a very large and expensive mirror expertly ground from cast speculum metal to make a portrait large enough to compete with the typical ivory miniature or silhouette. Nevertheless, in 1840 the world's first commercial skylight studio was opened by Richard Beard using simple iodised daguerreotype plates and a Wolcott mirror camera.

The earliest skylight studios

Beard's studio was located within the Royal Polytechnic Institution in London and featured frontal lighting. This was done with direct sunlight in order to give maximum illumination to the subject and to suppress shadows. Profiles were the easiest pose for subjects to endure, but the addition of blue glass windows allowed the subjects to look towards the light source without squinting during the exposures, which were shortened to within a minute or two.

By the mid-1840s, however, the daguerreotype process and subsequent improvements in lens design were mature enough to allow subjects to be daguerreotyped within the interior of a skylight studio similar to an artist's atelier. Daguerreotype exposures by this time had been reduced to seconds rather than minutes. Within 15 minutes of a sitting, the customer left the studio with a unique image, the same plate that was exposed in the camera. It was encased in a protective housing and was relatively permanent if the actual surface was not touched or exposed to sulfurous fumes, which caused the silver plate bearing the image to tarnish.

These technical improvements allowed the daguerreotypist to use indirect light from a northern orientation rather than direct sunlight. North light illumination did not create hard shadows and was consistent from mid-morning to mid-afternoon. The first studios were lit by large side windows with many lites.[6] Skylights were also set into the existing roof, a design that almost always leaked unless special provisions were made. Many skylights were assembled with overlapping lites constructed in the same fashion as greenhouses of the period. Others had specially designed rails that channelled the water to the bottom of the skylight frame. While north was usually the preferred orientation, studios in climates where bright sun was seldom seen often had skylights with a southern exposure. Many of these were glazed with lites that were ground[7] to diffuse the light to prevent hard shadows.

The best studios in the daguerreotype era were fitted with both a sidelight and skylight. Interiors of such studios were usually painted a matte grey blue or greenish teal blue colour to subdue uncontrollable reflections.[8]

Special studio apparatus was also becoming available through suppliers to help the subjects remain motionless during the 5–20 second exposures. These indispensable props included a cast-iron head immobiliser, a posing chair and a small side table upon which the subject could lean. For the next 30 years this three-point system of support remained the standard approach for keeping the subject from moving when sitting for a portrait. Other apparatus such as white-painted cloth or metallic foil reflectors[9] and diffusers[10] were used to add light to the subject or subdue areas that were too bright.

Backgrounds in the daguerreotype era were usually a plain, medium dark colour, wool being the preferred material. The cloth was stretched over a wood frame and held by supports on either side. A very effective technique was to suspend the background and have an assistant move the frame during the exposure, creating the sense of intangible space behind the subject in the final image. By the 1850s, painted backgrounds with *trompe l'œil* landscapes were being used by some daguerreotypists. These were generally painted in monochromatic hues of warm or cool grey. Military backdrops used during the American Civil War[11] often included camp scenes. By the last quarter of the nineteenth century, these were available illustrating a variety of scenes, including seascapes and interiors.

It was the plain background, however, that was associated with a better class of studios. Thin cotton was stretched over a large wooden frame and painted in flat oil or distemper[12] paints. The frame was held by adjustable standards so that the background could be swung or tilted to allow the light from the skylight to fall upon the surface in a soft gradient. This allowed the shaded side of a subject's face to be juxtaposed against the lighter side of the background and the illuminated side of the face against the darker side. This approach created a three-dimensional effect.

The skylight required the most delicate manipulation. It was discovered that a north skylight is most effective when certain sections of the windows can be shaded. To that end, special roller blinds or adjustable festoons of cloth were fitted under the windows to allow the photographer to control the light falling on the background and the subject independently. The well-equipped skylight had a system of cords and pulleys that could be manipulated from the camera while the photographer viewed the effect upon the ground glass of the camera.

Daguerreotype studios could be found throughout the industrial world and even on the edge of many frontiers. The daguerreotype portrait became the first commercially successful photosensitive process and remained so until the mid-1850s when it was finally challenged by the classic combination of albumen prints made from collodion on glass negatives.

Photography vs daguerreotypy

In chronological parallel to the evolution of the daguerreotype, Wm. Henry Fox Talbot introduced and improved the photogenic drawing technique,[13] which eventually lead to the calotype[14] and other paper negative processes. These methods were used to make paper negatives, which in turn were contact printed onto a second sheet of sensitised paper as the final positive print. The actual term 'photography' came to be associated with processes in the nineteenth century that relied on a negative to produce a paper photographic print. Today, photography is a generic term for any process made by chemically processing a material that was made sensitive to light or other radiant energy.

While the paper negative processes were an important step on the evolutionary ladder of photography, they were not universally practised to the same degree as the daguerreotype. Portraiture in a skylight studio was nearly impossible because paper negatives required much more exposure and the market for landscape prints was not particularly strong. It is possible that the total production of paper negatives in the nineteenth century was probably less than a single year of daguerreotype-making in the United States in the early 1850s.

The industry of photography

The wet-plate collodion process was introduced in March 1851 by Frederick Scott Archer.[15] It had been prophesised by others, including Gustave le Gray[16] and Robert Bingham,[17] but Archer was the first to publish a complete formula that actually worked.

Four years before the introduction of the wet plate process, French photographer Niépce de Saint Victor had shown that albumen from chicken eggs could be used as a binder to suspend silver iodide upon sheets of glass. The process was used for landscape negatives even after the advent of collodion, but albumen plates were never used in the portrait studio due to their extremely low sensitivity to light. The use of glass as a support material, however, set a standard that lasted throughout the nineteenth and twentieth centuries.

Collodion was a liquid that was being used in the early 1850s as an adhesive for medical dressings. It was made by first treating tufted cotton in sulfuric and nitric acids. After washing, the so called 'nitrated cotton' was dissolved in equal parts ether and strong alcohol. This viscous fluid could be applied in very thin coatings and when allowed to dry was an effective adhesive. The root of

NINETEENTH-CENTURY PHOTOGRAPHIC TECHNOLOGY

FIG. 12.1

Unskillful collodion technique on a vintage carte-de-visite glass negative. These process and chemical artefacts are often the main attraction to the wet plate collodion method by modern practitioners who are looking to emphasise the hand-made elements of their imagery.

(Courtesy George Eastman House)

the word 'collodion' actually comes from a Greek word meaning 'to stick'. When dry, collodion was also flexible and perfectly clear. Le Gray, Bingham, Archer and others were correct to suggest that collodion could be used as a binder to suspend the same light-sensitive chemicals that were being used to make paper negatives.

The process as introduced by Archer

As described by Archer in 1851, the collodion was process very straightforward. A sheet of clean glass was coated with a solution of collodion containing a small percentage[18] of potassium iodide. After a few seconds to allow some of the ether and alcohol to dissipate, the coated plate was taken to a darkroom[19] and placed into a solution of silver nitrate.

In a few minutes, a compound of light-sensitive silver iodide[20] formed on and in the collodion film. The plate was withdrawn and exposed in the camera while still wet. The process was called wet plate because the collodion film will lose sensitivity if the surface is allowed to dry. After exposure, the plate was taken into the dark room and developed with an acidic solution of pyrogallic acid.[21]

After the development, the plate was washed in a tray of clean water. A sheet of thin paper was then placed in the water and in contact with the collodion side of the plate. When the plate was withdrawn from the water, the collodion would adhere to the paper. An edge of the collodion film was then peeled from the plate and folded over the paper sheet, allowing the thin collodion pellicle bearing the negative image to be detached from the glass. The paper bearing the collodion image could then be rolled upon a glass rod for subsequent processing, or it could be fixed and washed immediately. After washing, the pellicle could be stored flat, rolled or re-attached to a sheet of glass for printing onto sensitive papers.

Archer also experimented with coating the dried collodion film with a solution of gutta-percha.[22] This could be followed by a second layer of non-iodised collodion, which when dry would strengthen the film. These flexible negatives required no glass support for contact printing.

The technique of stripping collodion film was to become standard practise in photomechanical procedures well into the mid-twentieth century. The stripping step was quickly seen as an unnecessary step by most photographers, however, and the process came to be practised with a single glass plate as a permanent support for each exposure.

The formulas as published by Archer and others in the early 1850s were perfectly suited for making landscape negatives to be printed out onto hand-coated salted paper and albumen papers. Collodion plates were much more sensitive than the paper negative processes. However, the use of pyrogallic acid as a developer resulted in a plate that was much less sensitive to light than the daguerreotype. As a consequence, the wet collodion process was not generally used for commercial studio work until important improvements were made in the mid-1850s.

The improved collodion process

The substitution of an acidified solution of iron sulphate[23] as a developer made shorter exposures possible. Iron developed plates required half the exposure of plates developed with pyrogallic acid. This innovation made studio portraiture a possibility and opened the doors for photography to challenge daguerreotypy as a commercial portrait medium.

For the photographer, the studio approach and apparatus needed for illuminating and posing the sitter was nearly identical to that used by daguerreotypists. It is important to emphasise that, while the exposures with wet collodion plates were actually slightly longer than the daguerreotype, the glass negatives could be used to make innumerable albumen prints. The prints themselves were seldom as permanent as the daguerreotype plate, but they were certainly easier to view and considerably cheaper.

In the late 1850s the carte-de-visite format was introduced. No other innovation affected the market for portraiture more than this universal format. These calling card size albumen prints were mounted on thin card stock and quickly became the most popular product of the nineteenth-century portrait studio. The carte-de-visite was actively traded among friends and family and spawned the introduction of the archetypal family photograph album. Cartes were also purchased from dealers who printed images of celebrities that were collected with the same intensity as sports cards are today.

A typical sitting: making the carte-de-visite

Urban photographic studios were typically found in the top floor of a building in a commercial block. Access was generally gained through a small door on the street level identified by a sign and possibly a sample board or photographs mounted near the entrance.[24]

The customer would walk up one, two or three flights of stairs to a landing and the door to the studio. In the larger studios this would lead the sales room/lounge area decorated with displays of photographic prints and different framing options. Walking up the stairs would usually leave the subject flushed, a serious problem when using blue-sensitive photographic materials. As a result, even a slightly reddened face would photograph much darker than expected. The photographer would take special care to have the subject relax enough to allow the red tint to leave their faces before the portrait was taken.

During this time the customer would be encouraged to look at samples of other portraits made by the photographer. Posing the subject to create an impression of ease and an expression that would be the most flattering was the key to successful portraiture and repeat business. The subject was posed informally and the skylight shades manipulated to gain a general sense of the effect. A reflector to illuminate the shadow side of the face and a diffuser to reduce light from the skylight side would then be brought in if needed. The photographer would then begin to compose and focus the lens while looking at the ground glass of the camera.

When the camera was set and the subject posed, the photographer would finally move in the heavy, cast-iron head stand. This immobiliser was carefully placed so that the adjustable 'spoons' on the upper bracket were just touching behind the base of the skull. It did not actually prevent movement, so much as let the subject know when they were moving during the exposure. The subject was told to relax as best they could and the photographer walked a few feet away to the darkroom to prepare the plate.

The wet plate-era darkroom was usually illuminated by a window glazed with red or deep amber glass. An inexpensive safe light could easily be made by stretching amber-coloured calico cloth or varnished brown butcher paper over the window. Darkroom windows for negative making were usually set lower than for ambrotypes or ferrotypes[25] because the progress of development was viewed through the plate rather than on the surface plate.

Preparing the plate

Glass was cut, cleaned and polished the day before, or in the morning before the studio opened. Glass sheets were cut by the photographer to a specific format using a cutting diamond mounted in a brass ferrule, fitted to a wooden handle. The diamond was drawn over the plate along a straight edge to score the surface. Once scored, the plate was carefully fractured along the score line. After breaking the plate to size, the sharp edges were roughened with a whetstone or the edge of another sheet of glass.

The plate was then placed in a special wood vice and cleaned with a piece of linen and slurry of whiting,[26] alcohol and water. The plate was cleaned on both sides and then polished with a second piece of clean linen. Cleaned plates were occasionally coated on one side with a dilute solution of albumen to prevent peeling of the collodion film during processing. If coated with albumen, the plates were then allowed to dry on a slotted wood rack. In either case, when done, the plates were placed in a grooved storage box until needed when a customer came in for a portrait.

As the customer waited, a clean plate was removed from the box and a solution of collodion, bearing an iodide and a bromide, was poured onto the middle of the plate. The plate was then carefully tilted toward all the corners sequentially to distribute the collodion, and finally the excess was poured off from the last corner back into the original bottle. The room would smell heavily of ether.

After a few seconds, some of the alcohol and ether would evaporate, leaving a clear collodion film firmly attached to the glass. The plate was placed on a dipper and slowly lowered into a vertical glass tank of silver nitrate solution. The lid was then closed and the photographer waited for silver iodide to form on and slightly in the collodion film. During this time, the photographer would talk with the subject patiently waiting in the skylight room. After a few minutes, the photographer would return to the darkroom to check the surface of the plate. When withdrawn, the collodion film had changed from clear to what appeared as a milky white in the red-lit darkroom. If the silver solution ran off the surface in an unbroken line, the sensitive plate was removed from the tank, allowed to rest on some absorbent paper for a few seconds to drain, and placed in the light-tight plate holder.

The plate holder was taken to the camera and rested on the floor next to the camera. The

FIG. 12.2

Cutting a glass plate with a diamond. The diamond scores the surface of the glass allowing it to broken in a controlled fashion.

(Courtesy Scully & Osterman)

NINETEENTH-CENTURY PHOTOGRAPHIC TECHNOLOGY

FIG. 12.3
Pouring the collodion onto a polished glass plate. The very heart of the wet collodion process, pouring collodion onto the plate and draining the excess evenly requires a very steady hand.
(Courtesy Scully & Osterman)

FIG. 12.4
While the collodion coated plate is sensitising in the silver nitrate solution the photographer would pose the sitter, focus the camera and make adjustments to the skylight and reflector.
(Courtesy Scully & Osterman)

photographer would then re-check the focus and make minor adjustments to the pose and lighting. The wet plate had to be exposed and developed before the silver solution on the surface of the late dried, or areas of the image would not appear in the final negative. The window of opportunity depended greatly on the ambient temperature and relative humidity. A common wet collodion process-artefact was a wavy deposit of metallic silver that usually formed on the bottom corners of the plate, as it occupied the plate holder. These so-called 'oyster stains' were produced by the collodion side of the plate drawing contaminated silver solution onto the surface by capillary action – a consequence of not draining the excess silver from the plate enough before putting it into the plate holder.

With the subject ready, the focusing glass was replaced by the holder and the dark slide removed, exposing the sensitive plate to the interior of the camera. Subjects were informed that they could breathe easily and even blink their eyes from time to time during the exposure, but otherwise should remain still. The cap was removed and the photographer counted aloud the 15–30 seconds exposure required for a negative.

The cap was replaced and the photographer asked the subject to remain in the general conformation of the pose until the plate was processed. The dark slide was replaced and the plate holder taken into the darkroom. After removing the plate from the holder, it was held in one hand whilst a small cup of developer was carefully poured across the surface. Failure to cover the entire surface of the plate evenly with the developer would result in tide lines; an artefact only visible in wet-collodion plates. Within a few seconds after applying the developer, the highlights of the image would appear, followed soon after by the mid-tones.

Development was generally stopped before the deepest shadows were completely obscured. Judgments of exposure and development were based on observation and experience. A gross under- or overexposure was determined by how quickly the image appeared during processing. If the plate was technically imperfect, a new plate was coated immediately and placed in the silver solution for a second attempt while the subject was still in the studio.

If properly exposed, the development was stopped at the appropriate time by pouring fresh water over the surface of the negative. After the developer was washed from the negative, it could be inspected in normal light without ill effects. If the negative was evaluated as usable, the plate was placed in a dish of water temporarily and the photographer politely dismissed the customer. As soon as the customer left, the negative was then fixed in either sodium thiosulfate or potassium cyanide to remove the areas of unexposed and undeveloped silver iodide. After fixing, the negative was given a thorough washing in clean water and allowed to dry on a rack or dried over an alcohol lamp.

If a plate had been properly exposed but was deficient in printing density, it could be chemically improved by a variety of after-treatments. These included variant techniques of intensification and redevelopment. Intensification chemically provided additional silver density or changed the colour of the silver image to a warmer hue to create spectral density.[27] The redevelopment technique converted the silver image back to a sensitive condition. The plate was then re-exposed to light. A second development could then be used to gradually add silver to the original image. Every after-treatment left the collodion film with identifiable characteristics: the study of which could occupy an entire chapter.[28]

Once dry, the plate was heated over an alcohol lamp and a warm alcohol-based gum sandarac and oil of lavender varnish was poured over the surface using a technique similar to coating with collodion. The varnished negative was then heated to evaporate some of the

FIG. 12.5
The exposed plate is developed by slowly pouring iron sulfate across the surface. Once the visible image is fully developed, the process is stopped by washing the plate with fresh water.
(Courtesy Scully & Osterman)

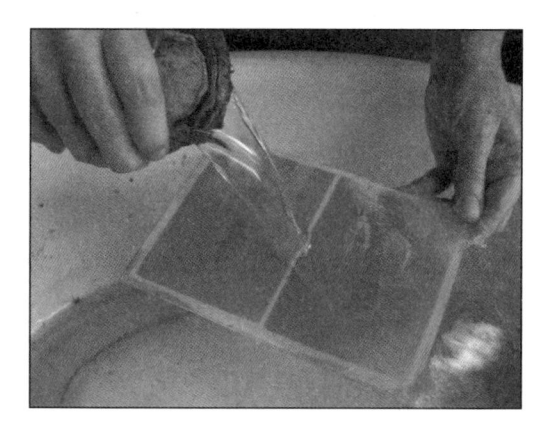

FIG. 12.6
The fixed and washed negative is dried over an alcohol lamp. Heating the plate is also necessary before pouring a protective varnish onto the surface of the plate.
(Courtesy Scully & Osterman)

alcohol and placed on a wooden slotted rack to dry completely. Printing could not be done with negatives until the varnished surface was completely dry and hard, which could take days depending on the relative humidity.

A veiled truth: retouching the negative

All of the nineteenth-century photosensitive processes had an inherent fault: the plates and papers had very limited colour sensitivity. Camera images such as daguerreotypes, calotypes and wet-plate collodion plates were based on the sensitivity of silver iodide. Printing papers generally relied on the sensitivity of silver chloride. In both cases these materials were most sensitive to blue, violet and invisible ultraviolet radiation. The warm hues of yellow, orange, brown, red and green had a marginal effect on these materials and therefore did not register on the plates and papers unless the exposure was extended for a great deal more time.

As a result, the warm tones found in human skin resulted in a false record, often leaving exposed skin looking like tanned leather. Freckles and blemishes went completely black. If the exposure and development was extended, the general effect would lighten these areas somewhat, but at the expense of subtle details in the face. While flattering when a subject had blemished, freckled or otherwise unsightly skin, the effect was not always a good argument for having one's portrait made by the camera. Overexposure was a common technique used in the daguerreotype studio, where plates could only be tinted.

In a wet-plate studio, the photographer had many more options. Correctly exposed plates with insufficient density could be redeveloped. But even when collodion plates were exposed and developed properly, inherent limitations of the process were corrected by retouching the negative. Because collodion negatives were routinely coated with a smooth, protective varnish, they required a special treatment to establish 'tooth' or texture on the surface to receive retouching. When dry and hard, this varnish coating was selectively roughened with a fine abrasive such as fine emery. This treatment was usually confined to the subject's head and occasionally the hands. The photographer would then carefully apply pencilling to the negative in the thin areas where it would print too dark otherwise. Alternatively, a sheet of thin tracing paper could be tacked onto the edges of the glass side of a negative and graphite retouching applied to the paper. This technique created a softer effect.

Samples of antique collodion negatives can also be found with names or numbers scratched into the perimeter of the image. This was done using a small stylus before the plate was varnished. The information was included as a reference to identify the subject.

Printing onto albumen paper

In 1850 the albumen printing process was introduced in France by Louis Desérié Blanquart-Evrard. Thin paper was floated upon a solution of egg white bearing a small percentage of sodium or ammonium chloride. Once coated, the albumen paper was dried in a heated room, then pressed and stored until needed. By the end of the 1850s this was a commercial trade, and photographers purchased the albumenised paper though photographic supply houses.

To make a print, the albumen paper was floated, albumen side down, upon a solution of silver nitrate in subdued light. This procedure allowed the compound of light-sensitive silver chloride to form on the albumen binder layer. The albumen would also absorb some excess silver nitrate needed for the printing-out process. Once dry, this sensitive paper was most effectively used within a few hours. It was trimmed and placed in contact with the collodion side of the glass negative in a special contact printing frame. The printing frame was designed to allow inspection of half the paper through a hinged door. The other side of the paper was held under pressure to prevent the paper and negative from moving out of the original registration.

To form the image, the printing frame was placed in the sun. In minutes a visible photograph began to print-out in dark purple photolytic silver.[29] No developer was necessary. Keeping track of the degree of printing was accomplished by removing the printing frame to an area of subdued light and opening the inspection door. If the print needed more time, the printing frame was returned to the sunlight.

When fully printed, the printing frame was opened and the print placed in a tray of tap water containing a pinch of sodium chloride. This would precipitate the excess silver nitrate in the albumen as a cloudy white residue. The print was washed in several changes of this chlorinated water until no more precipitate was visible.

The next step was to place the print into a second tray containing the gold toning solution. The toning solution was made by dissolving a few grains of gold chloride in distilled water modified to be nearly pH neutral.[30] The print remained in the toning bath until the image hue was perceptibly cooler. This usually required between 5 and 15 minutes, depending on the strength and pH of the toning bath. After toning, the print was fixed in a solution of sodium thiosulfate for several minutes and washed for no less than 20 minutes in several changes of fresh water. Failure to sufficiently fix and wash albumen prints contributed to severe fading and highlight yellowing in even a short period of time.

Finishing the print

The washed prints were taken from the water and placed between clean blotters to dry. Prints were then trimmed to size using a thick glass guide and a sharp knife or second-hand shaving razor. Frugal studios kept these trimmings for rendering the gold. The print was then dampened on the albumen side and placed face down on a clean piece of cloth. A small quantity of wheat starch paste was applied to the back of the print using a stiff brush and the print carefully positioned on a pre-cut card mount.

A blotter was placed on top of the mounted print and a brayer rolled over the surface to smooth the print and force the excess paste out the end. The blotter was removed and the excess paste wiped from the mount with a damp cloth. Similarly mounted prints were stacked and pressed together while still damp. This prevented the prints from drying out completely.

While in the mount-pressing clamp, a print burnisher was pre-heated for the final finishing. Burnishers were manufactured as two general types. The first production burnishers introduced in the early 1870s used a hand-cranked textured cylinder to pull the mounted print over a polished steel bar heated from below. Later burnishers used two cylinders; one was textured, the other was highly polished. The upper textured cylinder pulled the print over the polished cylinder which turned at a different speed and was heated from below. In both cases these burnishers relied on heat, pressure and rubbing the surface of the albumen print to produce glossy finish.[31]

To burnish a mounted print, it was first taken from the press and given a thin coating of lubricant. This was either made from Marseille soap,[32] beeswax or paraffin. The lubricant was quickly applied with a cloth and the print placed in the burnisher with the image side facing the polishing bar or cylinder. The first pass was done with very little pressure. The gate between the polished bar or cylinder and the textured cylinder was then closed slightly and the print cranked through a second time. The process was repeated several

RECONSTRUCTIONS

FIG. 12.7 (top left)
Albumen paper is floated upon a solution of silver nitrate solution to make it sensitive to light. The chloride salted albumen coating was applied using the very same technique as used for sensitising.

FIG. 12.8 (top right)
Albumen printing is a photolytic process that requires no development. The image becomes visible as light filters through the negative onto the sensitised paper. Here the image is being inspected to see if it required more exposure. The printed image is then washed in dilute salt water, toned in gold chloride, fixed in hypo and washed again.

FIG. 12.9 (bottom left)
The dry albumen print is dampened slightly and pasted upon a thin card mount using wheat starch paste.

FIG.12.10 (bottom right)
Before the mounted print is perfectly dry it is coated with a wax or soap lubricant and run through a heated burnisher. The burnisher imparts a gloss to the surface and sinks the print into the surface of the mount. This is an early style bar burnisher.

FIG. 12.11 (below)
The finished cartes-de-visite made in a skylight studio with original equipment, typical formulas and documented procedures from the mid-nineteenth century.
(Courtesy Scully & Osterman)

times, closing the gate slightly between each pass, until the desired finish was achieved.

The outer surfaces would feel dry after burnishing, but the core would still be slightly damp, requiring the print to be placed in a press or clamping block to prevent curling. Some clamping devices placed a reverse curve in the mounted print so that the tendency of the albumen print to shrink over time pulled the mount into plane.

Strangely, of all the techniques and materials used to recreate nineteenth-century photography, it is the card mount for albumen prints that represents the greatest difficulty. The vintage mounting boards often have four or even six plies occupying the same thickness as a modern sheet single-ply Bristol board. There is no manufacturer today making this type of card stock and replicating it would necessitate a separate focus of research.

Conclusion

The recreation of basic nineteenth-century photographic techniques used in commercial skylight studios is relatively easy, despite what may have been previously written in photographic histories. Most of the chemical compounds used by these photographers are still readily available and original vintage equipment can either be purchased or used as models for the production of working replicas.

Photographers in the nineteenth-century were not chemists; they were middle-class people with average educations who learned a specific skill set by necessity for their livelihood. Photographs that exist from the collodion era, however, suggest that the care given to posing the subject, negative production, printing and finishing, is inconsistent from one studio to the next and even within the same studio. Understanding the effects of daylight studio lighting on both the subject and a blue-sensitive collodion plate did not come easily in the nineteenth century, and is just as difficult a concept to master in modern times.

To achieve the level of expertise common to the best nineteenth-century photographers requires a commitment of time and dedication that most people today are unwilling to undertake. Nevertheless, even a poorly made collodion negative or albumen print can provide historians with valuable insights that are impossible to observe by any other means.

Glossary

Actinic Light: Light that has an affect on photo-sensitive materials. In the nineteenth century this included visible blue and violet light, and invisible ultra violet radiation.

Albumen: The clear yellowish protein separated from the yolks from eggs. Albumen was used as a binder for sensitive papers and plates, and as a subbing layer on collodion plates to prevent peeling.

Ambrotype: A direct positive wet collodion process on glass based on the phenomenon of the developed silver in the highlights being a light colour. When ambrotypes on clear glass are backed with dark material, the image appears as a positive. Dark glass was also used.

Daguerreotype: A direct positive process on highly polished silver-clad copper plates. This was the first commercially successful photosensitive process. It was superseded by the collodion process in the mid 1850s.

Ferrotype: A direct positive wet collodion process on iron plates previously coated with a backed asphaltum varnish. The chemicals and technique are identical to the ambrotype. Also called; Melainotype and tintype.

Head stand: A heavy cast-iron base with a rising central rod fitted with adjustable calipers. Also called an immobiliser, the calipers were placed in contact at the base of the sitters head just below the ears to steady them during long exposures.

Negative: A first generation camera image which reads with inverted tonal values. Negatives are used to produce positive images on a second photosensitive material. The negative was made obsolete by digital imaging.

Photography: A general term referring to all photosensitive process, though in the mid-nineteenth century it was specific to negative/positive processes.

Plate holder: A light tight chamber designed to ferry sensitive materials between the darkroom and the camera. Once fitted to the back of the camera, a dark slide is withdrawn to expose the sensitive side of the material to the interior of the camera. After exposure the dark slide is replaced for transport back to the darkroom.

Silver Nitrate: The basis of all silver halide photography, silver nitrate is made by dissolving pure silver in nitric acid. The washed crystals are soluble in water making it easy to combine with iodides, bromides and chlorides used in various photographic processes.

Notes

1. A term introduced by the author to describe aesthetic qualities based on the technology of a medium.
2. *The Collodion Journal* was published quarterly by Scully & Osterman from 1995 to 2002, with a total production of 24 issues.
3. Actinic light refers to ultraviolet, violet and blue light; the only light to which collodion plates are sensitive.
4. *Historique et Description des procédés du Daguerréotype et du Diorama par Daguerre*, Paris 1839.
5. Chromatic aberration is the limitation of uncorrected lenses that prevents all colours of light to focus on the same plane. Blue sensitive plates will render the camera image out of focus with an uncorrected lens.
6. Lite: An individual pane of glass in a window.
7. Made translucent by grinding the surface with an abrasive or by gluing tissue paper to the inner side.
8. The blue-grey colour is most common to be found in existing nineteenth-century studios.
9. Lead or zinc foil was glued onto thin panels of wood.
10. A round or rectangular frame of thin wood or wire upon which was stretched thin white cloth. Diffusers were mounted on poles and either held by an assistant or set into an extra iron head immobilizing stand.
11. 1861–64.
12. A water based paint made with pigments, calcium carbonate and animal glue, egg yolks or casein.
13. Talbot F.R.S, W. H. F. (1839): 'Some Account of the Art of Photogenic Drawing', in *The Athenaem*, 9 February 1839, p. 114.
14. English Patent No. 8842, Photographic Pictures.

15 *The Chemist*, March 1851, p. 257.
16 *Traite Practique de Photographie Sur Papier et Sur Verre, 1850* (Paris: Gustave le Gray), p. 43.
17 Bingham, R. (1850): *Photogenic Manipulation*, 7th edition (London: George Knight & Sons), p. 73.
18 Iodizing in that era was usually between one and two percent.
19 A room illuminated by orange, deep yellow or red light.
20 The same light sensitive compound used in the daguerreotype and calotype processes.
21 Also known as pyrogallol. Pyrogallic acid is not actually acidic and therefore required the addition of an acid as a restrainer to prevent non image reduction of silver.
22 Gutta percha is a form of natural rubber.
23 Ferrous sulfate.
24 Original skylight studios are still easily found in small towns all over the United States.
25 Ambrotypes and ferrotypes are direct positives made with the collodion process. They are not included in this paper because these processes were generally not produced by the same operators who made negatives. Photographers usually dealt with customers on a higher social level than ambrotypists and ferrotypists.
26 Calcium carbonate.
27 Because printing-out papers of the nineteenth century were only blue sensitive, a brown or orange coloured silver deposit would actually have more printing density than neutral black of the same silver content.
28 In 2007, France Scully Osterman produced twelve sets of sample collodion variant negatives with a descriptive booklet for the Didactic Negative Project funded by the Andrew W. Mellon, Collaborative Workshops in Photograph Conservation.
29 Printing-out papers rely on a compound of silver chloride with an excess of silver.
30 Typical pH modifiers include borax, sodium carbonate or bicarbonate and various acetates.
31 Before the 1870s, prints were often hot pressed between heated sheets of zinc using a conventional etching press or similar roller press.
32 A fine soap made from olive oil rather than lye and lard.

Suggested reading

The wet plate process is the most documented process of the nineteenth century. Photographic journals were introduced in the early 1850s, and there were hundreds of manuals printed in several languages. Hundreds of formula variants were used during the collodion era (1851–80), most of which worked well enough in the hands of experienced photographers. The techniques described herein and the formulas used to produce the finished image illustrated in this chapter are based on twenty years of primary research and hands-on experience with the collodion process. The formulas used are not specific to any one nineteenth-century photographic studio but rather, typical of many.

The following manuals are a small sampling of some very useful resources we have used to recreate the wet collodion process and gain an understanding of the construction and use of a skylight studio.

Bibliography

Burgess, N. G. (1865): *The Photograph Manual. A practical Treatise, containing the Carte de Visite Process and the method of taking Stereoscopic Pictures* (New York: D. Appleton & Co.).

Lea, M. C. (1868): *A Manual of Photography. Intended as a Text Book for Beginners and a Book of Reference for Advanced Photographers* (Philadelphia: Benerman & Wilson).

Pritchard F.C.S., H. B. (1882): *The Studios of Europe* (London and New York: E. & H. T. Anthony).

Robinson, H. P. (1891): *The Studio and What to do in It* (London: Piper and Carter).

Thornthwaite, W. H. (1854): *A Guide to Photography; containing simple and concise directions for obtaining Views, Portraits, & by the action of light on prepared surfaces of paper, glass and metal* (London: Horne and Thornthwaite).

Van Monkhoven, D. (1865): *Traité General de Photographie* (Paris: Librairie de Victor Masson et Fils) This text is particularly good for line art illustrations of equipment and procedures used in nineteenth-century photography.

Wilson, E. L. (1881): *Wilson's Photographics. A series of Lessons, Accompanied by Notes, on all the Processes which are Needful in the Art of Photography* (New York: Wilson & Co.).

Image credits

Listed below are the sources of images within this publication. No reproduction of material in copyright is permitted without prior contact with the publisher or owner of copyright.

MICHAEL R. BAILEY
© for figures 5.2 and 5.4

D. BAYELL
© for figure 5.5

F. BEARD
© for figure 5.7

ALLAN BROMLEY and MICHAEL T. WRIGHT
© for figure 1.5

ELIZABETH CAVICCHI
© for figures 7.1b, 7.2, 7.3, 7.4a, 4b, 7.5 (left), 7.6, 7.9, 7.10, 7.11, 7.12, 7.14, 7.15, 7.16, 7.17, 7.19 (right), 7.20, 7.21 and figures 7a, 7b and 7c in art section

E. M. CLARKE
Voltaic Battery and Pole Director', in *Annals of Electricity*, 1, 1837 (W. Annan, Lithographer), plate VIII, figure 55
For figure 7.8b (right)

JOHN CREED
© for figure 3.4

CHEV. F. M. G. DE PAMBOUR
A Practical Treatise on Locomotive Engines Upon Railways &c. (London, 1836)
For figure 5.6

DELFT HISTORICAL SOCIETY
'Jan Vermeers huis: Een poging tot reconstructie', in Warffemuis, A. (2002): *Delfia Batavorum Elfde Jaarboek 2001* (Delft: Historische Vereniging Delfia Batavorum)
© for figure 11.3

DEUTSCHES MUSEUM
Musuesinsel 1, 80538 Munich
Figures 6.4, 9.1, 9.2 and 9.3 © courtesy of Deutsches Museum, Munich

EASTMAN HOUSE COLLECTION
900 East Avenue
Rochester, NY 14607
Figure 12.1 © courtesy of George Eastman House

EUROPEAN SOUTHERN OBSERVATORY
Karl-Scharzschild-Str. 2
85748 Garching bei Munich
© for figure 10 in art section

J. A. FLEMING
The alternate current transformer in theory and practice, 2 vols. (London: 'The Electrician' printing and publishing company, 1892)
For figure 7.7

JAN FRERCKS
© for figures 8.1, 10.8 and 10.11

STAATLICHE MUSEEN ZU BERLIN
Gemäldegalerie, Matthäikirchplatz 10785, Berlin
Johannes Vermeer, *The Glass of Wine* c.1658–60, oil on canvas 65 x 77 cm
Fig 11.6 © courtesy of bpk/ Gemäldegalerie, SMB/ Jörg P. Anders

GEMEENTER ARCHIEF DELFT
(Delft Muncipal Archive)
Oude Delft 169
HB 2611 Delft
For figure 11.2

J. J. GREENENOUGH
'The first locomotive that ever made a successful trip with Galvanic Power', in *American Polytechnic Journal*, 4, 1854
For figure 7.5 (right)

PETER HEERING
© for figures 9.4, 9.5, 9.6, 9.7, 9.8 and figure 9 in art section

J. HENRY
'Contributions to Electricity and Magnetism: On Electro-Dynamic Induction No. III', in *Transactions of the American Philosophical Society*, 6, 1839, pp. 303–37
For figure 7.8b (left)

IMAGE CREDITS

HEATHER HOPKINS

© for figures 2.1 to 2.12 and figure 2 in art section

H. L. HOUTZAGER, et al

De Kaart Figuratief van Delft (Rijswijk: Elmar B V, 1997)

For figure 11.1

FRASER HUNTER

© for figures 3.3(a) and 3.3(c) and figure 3 in art section

JOHN JACKSON

© for figure 6.9

LARS KANN-RASMUSSEN

© for figure 4.4

WERNER KARRASCH

© for figures 4.1, 4.8, 4.9, 4.10, 4.11, 4.13, 4.14, 4.15, 4.16, 4.17, 4.19, 4.20 and figures 4a and 4b in art section

METTE KRYGER

© for figure 4.18

LANDESMUSEUM WÜRTTEMBERG

Schillerplatz 6
70173 Stuttgart

Figure 3.3(a) © courtesy of Württembergisches Landesmuseum Stuttgart

BILL LEUNG

© for figure 6.2

THE METROPOLITAN MUSEUM OF ART/ART RESOURCE/SCALA, FLORENCE

Vermeer Jan (1632–75): *Allegory of the Faith*, c.1670. New York, Metropolitan Museum of Art. Oil on canvas, 45 x 35in. (114.3 x 88.9cm)
The Friedsam Collection, Bequest of Michael Friedsam, 1931. Accn.: 32.100.18 © 2011. Image © The Metropolitan Museum of Art/Art Resource/Scala, Florence

For figures 11.5 and 11.10

MUSEUM OF SCIENCE & INDUSTRY IN MANCHESTER

Liverpool Road, Castlefield
Manchester M3 4FP

Figure 5 in art section © courtesy, Museum of Science & Industry in Manchester

NATIONAL MUSEUMS SCOTLAND

Chambers Street
Edinburgh, EH1 1JF

© National Museums Scotland for figures 3.1, 3.2 . 3.3(b), 3.5 and figure 3 in art section

NATURAL HISTORY MUSEUM NÎMES

13 Boulevard Amiral
Courbet, 30000 Nîmes

Figure 3.3(c) © courtesy of Nîmes Museum

CHARLES G. PAGE

'Method of increasing shocks, and experiments, with Prof. Henry's apparatus for obtaining sparks and shocks from the Calorimotor', in American Journal of Science, 31, pp. 137–41; reprinted in *Annals of Electricity*, 1837

For figure 7.8a

and

'On the use of the Dynamic Multi-plier, with a new accompanying apparatus', American Journal of Science, 32

For figure 7.18

JOE PEIDLE

© for figures 7.3f and 7.3g

ROBERT C. POST COLLECTION

© for figure 7.1a

PAUL ROJAS

© for figure 6.5

THE ROYAL COLLECTION

Johannes Vermeer, *A lady at the virginals with a gentlemen (The Music Lesson')* c.1662–65, oil on canvas, 73 x 64cm.
RC code: RCNIN 405346

The Royal Collection © 2011, Her Majesty Queen Elizabeth II for figures 11.4, 11.8 and figure 11 in art section

TONY SALE

© for figure 6.7

THE SCHOOL OF COMPUTER SCIENCE

University of Manchester
Kilburn Building, Oxford Road
Manchester, M13 9PL

© for figure 6.8

THE SCIENCE MUSEUM

Exhibition Road
South Kensington, SW7 2DD

Figure 5.3 courtesy of Science Museum, R. Stephenson & Co. collection

Figures 6.1, 6.6, 6.10, 6.12, 6.14, 6.15, 6.16, 6.19 and 6.20 © courtesy of Science Museum/SSPL

FRANCE SCULLY/MARK OSTERMAN

Scully & Osterman Studio
Rochester, NY 14620

Figures 12.2 to 12.11 and figures 12a and 12b in art section courtesy of Sully & Osterman

MARINA FRASCA SPADA

© for figures 10.9 and 10.10

KLAUS STAUBERMANN

© for figures 10.4, 10.5, 10.6 and 10.7

PHILIP STEADMAN

© for figures 11.7, 11.9, 11.12, 11.13 and 11.14

OMARI STEPHENS

© for figures 7.9 and 7.13

273

RECONSTRUCTIONS

W. STURGEON

'On the Electric Shock from a single Pair of Voltaic Plates, by Professor Henry, of Yale College, United States: Repeated, and new Experiments (28 Sep 1836)', in *Annals of Electricity*, 1, 1837

For figure 7.19 (left)

and

An experimental investigation of the influence of Electric Currents on Soft Iron … , in *Annals of Electricity*, 1, 1837

For figure 7.19 (middle)

DORON SWADE

© for figures 6.3, 6.11a, 6.11b, 6.13, 6.17, 6.18a, 6.18b and figure 6 in art section

JEFF TINSLEY

© for figure 7b in art section

UNIVERSITY OF JENA

Collaborative Research Centre 482
Carl-Zeiß-Str. 3
07743 Jena

Figures 8.2, 8.3, 8.4 and figure 8 in art section © courtesy of the Collaborative Research Centre 482, University of Jena

THE VIKING SHIP MUSEUM

Vindeboder 12
DK-4000 Roskilde

© for figures 4.1, 4.2, 4.3, 4.4, 4.8, 4.9, 4.10, 4.11, 4.13, 4.14, 4.15, 4.16, 4.17, 4.18, 4.19, 4.20, 4.21 and figures 4b and 4c in art section

ALESSANDRO VOLTA

'On the Electricity excited by the mere Contact of conducting Substances', in *Philosophical Transactions*, part 2, 1800

For figure 7.4c

and

Le Opere di Alessandro Volta (Milan: Hoepli, 1800)

For figure 7.4d

MICHAEL T. WRIGHT

© for figures 1.1, 1.2, 1.3, 1.4 and figure 1 in art section

TREVOR YORKE

© for figure 11.11 and figure 11 in art section

J. C. F. ZÖLLNER

Grundzüge einer allgemeinen Photometrie des Himmels (Berlin: Mitscher & Röstell, 1861)

For figures 10.1, 10.2 and 10.3

Every effort has been made to trace and contact the owners of the copyright of the photographs and drawings in this book. The Editors and the Publisher apologise for any inadvertent errors and will be pleased to make the necessary amendments in future editions or reprints if contacted in writing by the copyright holders. Copyright rests with those artists/authors and institutions named.

Associated keywords

Antikythera Mechanism
astronomical [instrument] 1–20
dimensions 5–7
examination [of the original] 3
gearing 1, 11, 12, 13, 14, 17
materials 7–10
radiography 3, 6
restoration 5, 7, 11, 13
technique [original and modern] 3, 6, 7, 14–17
tools [original and modern] 14–17
virtual [model, reality …] 5, 18

Roman dyeing
computer simulation 36
experimental archaeology 31, 44
Finite Element Analysis 36, 46
manufacturing 21–46
physical replica 26, 38
Pompeii 22–22, 23–26, 28, 29, 30, 31, 32–34, 36, 38, 39–40, 42, 43, 44, 45, 46
Roman, dyeing 21–46
systems 42

Deskford carnyx
boar 51, 54, 56
carnyx 50–58
copper alloy 51
Deskford 50–58
Gundestrup 53, 55
music-archaeology 51
Roman iconography 53–54
Tintignac 52, 56

Sea Stallion from Glendalough
experimental archaeology 60, 62, 71
long ship 71, 81
Nordic boatbuilding 64
reconstruction 64–70
resources 71–72
Sea Stallion from Glendalough 59–81
tool marks 65, 69
trial voyage 73–74
Viking Age 60, 61, 62, 63, 76, 81
Viking Ship Museum, Denmark 60, 61, 62, 63, 76, 81

***Planet* locomotive**
driving 93–95
fitting up 92–93
locomotive project, the 88–89
maintenance 95–97
performance 98–99
design and specification 90–92
valve setting 93

Historic computing machines
Analytical Engine 104, 108
Babbage, Charles 103, 104, 107, 108, 110–11, 112, 114–22, 125
Bombe 105
Colossus 105, 107, 109–10
Difference Engine 103, 105, 107, 110–13, 118, 121–24, 125
Fowler, Thomas 104, 107, 109, 110
Manchester 'Baby' 105, 109
Zuse, Konrad 104, 107, 108

Spiral conductor of Charles Grafton Page
acupuncture 134–35, 136, 140, 141, 147, 157, 162
contact breaker 141, 157, 159
Davis Jr, Daniel 157, 159–60, 163
electromagnetic induction 159–60, 162–63
electromagnetism 131–32, 155
electropuncture 134–35
Faraday, Michael 132, 135–39, 141, 154, 158, 162–163
frequency 143, 148, 151–154
galvanism, medical 134–35, 141
Gooding, David 132, 154
Henry, Joseph 137–39, 140, 141, 158, 161, 162–63
induction coil 127–28, 131
motor 132, 135, 140, 141, 143, 149
oscilloscope 128, 143, 144, 145–46, 149–50, 153, 154
Page, Charles Grafton 127–64
Post, Robert 157, 163, 164
resistance, electrical 134, 138, 141, 145
spiral conductor 127–160
Sturgeon, William 157–59, 160, 162–63
Volta, Alessandro 134, 162
voltage 128–29, 134, 138, 143–44, 145, 146–47, 149–152, 153, 154, 155, 163

An electrical-historiographic-didactic experiment
electrical machine 172, 179–80, 183
electricity c.1800 1172, 176, 179, 182, 185, 186, 187
Jena, University of 172, 175
Schmidt, Georg Christoph 172
textbooks 172, 176, 178–179, 182, 184, 186, 190

Materialised skills
Adams, George 199
Baker, Henry 195
Brander, Georg Friedrich 200
Crary, Jonathan 199
Dollond 197, 201–206

275

RECONSTRUCTIONS

Junker, Friedrich August 197, 201–205
Ledermüller, Martin Frobenius 195, 208
Nairne, Edward 196, 199
Priestley, Joseph 196–97
Ratcliff, Marc 197, 199, 206
Schickore, Jutta 199
Walters, Alice 198–99

Zöllner photometer
Bruhns, Carl 209, 217, 225
calcspar 215–17
Canada balsam 215, 217
Deutsches Museum 212, 214, 215, 218, 219, 220
Dove, Wilhelm 211–12
European Southern Observatory 214
Nicol prisms 211, 212, 213, 228
Nicol, William 215–16
town gas 222–23, 232
Wiedemann, Gustav 212
Zöllner, Karl Friedrich 209–31

Johannes Vermeer's painting studio
Allegory of the Faith 238, 240, 243, 248
 – mirrored ball 243
camera obscura 246, 247–48
highlights 247
Music Lesson, The 238–39, 240, 243, 245, 248
 – mirror 240, 242, 245
reverse perspective 238, 240
Thins, Maria 235
 – house in Delft 235–37, 240, 242
Vermeer, Johannes 235–48
Warffemius, Ab 236–37, 238, 243, 245

Mid-nineteenth-century photography
actinic light 255
albumen 251, 258, 260, 262, 266–69
ambrotype 254, 261
daguerreotype 252, 255–58, 260, 265
ferrotype 254, 261
head stand 261
negative 252, 253–254, 258, 259, 260, 261, 263, 264–266, 269
photography 251–69
plate holder 262–63
silver nitrate 259, 262, 266